Regulation and Control in Living Systems

Regulation and Control in Living Systems

Edited by **H. KALMUS**

Reader in Biology, the Galton Laboratory, University College,
The University of London

JOHN WILEY & SONS   London   New York   Sydney

Made and printed in Great Britain by
William Clowes and Sons, Limited, London and Beccles

Foreword

Ideas of regulation permeate the entire field of biology, as this book shows. As they arose at many times and on different occasions, it is not surprising that no uniform terminology has developed. Consequently the same words may have different meanings in separate specialist branches of biology, while analogous relations which occur at diverse levels of organization may be described in different terms. This situation has during the last few decades been aggravated by the intrusion of technological and in particular electronic terminology, which, being usually more precise or even quantitative, has captured the imagination of many biologists. Unfortunately by uncritical use it has often resulted in a state of confusion, where communication between the control engineer and the biologist has broken down.

These terminological and conceptual differences create considerable difficulties for any book on biological regulation in its general aspects. A single author may invent and apply a new unifying terminology. This—whatever its excellence—cannot serve as an introduction to further specialist reading and usually leads to more confusion. Alternatively he may use different terminologies in different chapters. Fortunately an editor of such a book can bypass this dilemma and can leave to his specialist authors the choice of most of their words. He must nevertheless try to remove a few obvious and other more insidious obstacles to mutual comprehension. His main task ought to be to restore the communication between the engineers and the biologists. This at the present juncture is difficult. If it were easy several satisfactory books of this type would have undoubtedly appeared. Nevertheless it seems to the editor that the time to attempt a synthesis has now come.

H. KALMUS

Contents

V Conclusion

I
PRINCIPLES and METHODS

1

Introduction—Control and Regulation as Interactions within Systems

H. KALMUS

Department of Human Genetics and Biometry,
University College, London

History of the Concepts

In so far as cells, organisms or societies are more than the sums of their parts, they are so by virtue of controlled interactions or regulations between these parts. These interactions can only be understood in terms of the systems within which they operate and not simply by studying the isolated parts. This Aristotelian, 'holistic' approach familiar to technologists and naturalists is in some ways easier than the atomistic, Democritean study of the parts as practised in classical physics and the biological ancillaries; a boy can 'understand' the principles of a clock in a qualitative way many years before he can grasp quantitatively the mechanical properties of the parts of the clock. To understand 'fully' the working of the clock he will later in his life require information of both the parts and the whole system. The expectation that one can understand a machine simply by understanding the properties of all its components is as sanguine as the expectation that organismic biology (Dobzhansky, 1964) will turn out to be 'nothing but molecular biology' or even 'nothing but chemistry'. The Aristotelian and the Democritean* approach will always stay in opposition to each other and will both be needed to

---

* Neo-Darwinism, which explains evolution as a result of selection acting on random changes (mutations) and random combinations of particles (the genes), is the current form of Democritean biology.

complement each other in any attempt to understand life. The anti-
thesis between holism and atomism, which at the beginning of the
century was confused by the ideological quarrel between vitalists
and mechanists, can now be resolved by the science of *Cybernetics*—
the study of systems, which appear goal directed and simultaneously
have a transparently causal structure (see also pp. 260, 287).

Control and regulation were first clearly perceived in the social
sphere, though their role in animal sociology is only just emerging
(see Wynne Edwards, Chapter 15). The noun *control* derives, according to Dr. Johnson, from French military language, where 'contre
role' meant a 'register or account kept by another [!] officer, that
each may be examined by the other'. It was also used to describe a
'check' or 'restraint' and 'power, authority or superintendence'.
These more general uses became rather similar to those of 'regulation'
and the verb 'to control' is also almost synonymous with 'to regulate'.

The Latin word *regula,* from which the English word regulation
and many similar words in other European languages presumably are
derived, meant among other things a ruler, a square, an instrument
for bone setting and more generally a norm or prescription of nature
('rule'). In Johnson's dictionary 'regulation' is defined as (i) the act
of regulating and as (ii) method: the effect of being regulated.

Most of these words have a teleological connotation: a ruler is
used to draw a line, which one wants to be straight; a bone is set to
the 'normal' pattern; any rule is designed to achieve some specific end.
As long as scientists did not doubt that the world was designed to serve
a divine purpose the word regulation was, when referring to nature,
more or less used synonymously with 'natural law'. In his famous tract
on the creation (1691) John Ray states: 'Being but stupid matter,
they [the spirits] cannot continue any regular and constant motion,
without the guidance and regulation of some intelligent being.'

Nowadays the word regulation is used in several and much more
restricted ways, mainly by the lawyer, the technologist and the biologist. In all these fields regulations may be defined as devices for the
achievement of desirable situations by either reestablishing some
disturbed equilibrium or by changing a system towards some optimal
situation. In the legal sphere this is quite obvious and we need not
deal with it here. However it should be observed that the modern
restricted use of the word regulation in biology derives from technology and sociology but not from physics. Events in the inorganic
world—to which man-made machines do not belong—though 'governed by laws' are not conspicuously regulated, claims to the contrary
like Le Chatelier's (1888) principle notwithstanding.

Many modern machines incorporate regulating devices and as these have been invented by man for specific ends we do not doubt their purpose. A centrifugal regulator ('governor') is designed to keep steam pressure constant, an automatic steering device to keep a steady course. Their efficiency and construction can be assessed with these aims in view.

In biology as in social matters the situation is often equivocal. Overall there can be two opinions of whether there is any purpose in life and many opinions of what such a purpose may be. This point has been very strongly made by Descartes (1664) who argued the '...kind of causes which is wont to be taken from the end, to have no use in Physics or natural matters; for I cannot without rashness think myself able to find out the ends of God.' (Transl. J. Ray, 1691.)

While this has been accepted as orthodoxy—at least in theory, a further assertion of Descartes will not easily be acceptable to biologists; again following Ray's translation he said:...'neither can nor ought we feign or imagine, that some of God's ends are more manifest than others; for all lie in like manner, or equally hidden in the unsearchable abyss of his wisdom.' Most of us assume that a wing is an organ of flight and that lungs serve respiration; but the roles of many organs are still controversial, e.g. the ocelli of the insects and several structures in the echinoderms. Now while it is obvious in principle how to study the functions of wings or lungs, it is rather difficult to discuss satisfactorily any regulatory device or in fact any machine, the function of which one does not know. It is however possible to describe such devices and to apply quasi-teleological provisional criteria when suspecting a regulatory function. This applies to arrangements resulting in constant internal environments, e.g. chemical or those affecting anticipatory changes (p. 155).

The logic of such approaches is sometimes rather shaky. We assume often with little evidence that the constancy (Claude Bernard, 1859; 1876) of certain measurable factors, such as body temperature (p. 87) or tissue tonicity in a living system must in some way be 'useful' or in Darwinian terms 'selectively advantageous', and survey the situation under that aspect. This, however, is all we can do: the efficiency of a device can only be assessed in reference to definite assumptions concerning its function. Thus looking at life as a biologist implies looking at it in the search for 'machinery' and not only for morphological detail or biochemical reactions.

There exist, however, considerable differences between the products of technology and the products of organic evolution. The most fundamental of these is probably the far greater complexity of the parts and interactions in living systems, which often obscure individual and specific regulatory processes (Kalmus, 1963). This complexity is a consequence of the incomplete and imperfect specialization of parts of the body, which to some extent retain their original universality and multiple functions, as seen from their faculty of regeneration (pp. 259, 313). Thus in biology we see complex reactions of complex systems to complex environmental challenges and we infer that all this serves the maintenance of life. Only exceptionally is it possible to isolate part of these processes and make this simplified system amenable to quantitative treatment. Some chapters of this book deal with such situations—others do not. Nevertheless the contents of this book are all elaborations (theorems) of the assumption or axiom (Kalmus, 1963) that organisms are in many ways regulatory machines.

Every age compares life with its contemporary machinery: Descartes' age with clocks, the late 19th century with steam engines and the last two decades with the apparatus of control and communication. The theory of these machines 'Cybernetics' (Wiener, 1948) has reached a certain maturity and both its achievements and limitations are now clearer than before. The main features of the history of machine theories of life and in particular of regulation and control appear to the author as follows.

The implements and machines available as models for organisms until the 17th century were constructed for limited practical purposes and mostly quite lacking choice control or regulation which had to be provided by the human operator. Whether or where to plough and how deep a furrow was under the control of the ploughman and not the plough. A clock too—though it must contain at least one regulatory device—was wound and set by hand. It is thus not surprising that Leibniz like Descartes thought of animals as 'automata'. The girl puppet in the 'Tales of Hoffmann' is a caricature of such a machine. Nineteenth-century technology was preoccupied with the concept of energy, and thermodynamic aspects of machinery are still involved in explanations of living systems and their evolution—often with little wisdom (Kalmus, 1966). However, from consideration of energy and efficiency certain regulatory devices were evolved like the 'governor' (p. 14) but only a few.

In modern servomechanisms and computers, on the other hand, regulatory devices have greatly multiplied enabling the latter to 'make decisions' of many kinds, and to adapt their working to fortuitous circumstances or the effects of their own activities. Quite recently a considerable number of models existing either in physical reality ('hardware') or at least 'on paper' have been designed for the explicit purpose of imitating and thus explaining certain higher performances of living systems. For instance, of the sense organs, the brain or whole animals (Grey Walter, 1961). These models are in addition to the machines which were originally invented for strictly practical technological purposes and only later used for explanations of biological functions. Thus it is no longer a question of drawing analogies between what one finds, let us say between a clock and an animal, but of asking engineers to invent machines capable of performing specific tasks, for example, the recognition of visual patterns, the avoidance of obstacles, 'feeding' or the maintenance of various activities in the face of external disturbances. This approach ('Autonomics') which in the U.S.A. has been called 'Bionomics' is very successful and a considerable part of this book is devoted to it. It provides the main justification for the inclusion of the technological chapters. Its most sophisticated development—not here described— occurred in the use of computers, which can be instructed to produce conclusions and decisions from a great variety of information, previously only dealt with by brains.

A fundamental idea, which is now common to engineers and biologists, especially physiologists is that of the closed loop or feedback. Information* concerning the effects of a machine's or an organism's own actions is perceived and 'fed back' to the machine or organism, thereby altering its subsequent performance, often in an adaptive way for instance by 'negative feedback'. In the social sphere this is of course quite commonplace. For example children are examined to see what they have learned and what they should be taught next and industrial products are inspected to see whether they are up to standard. During the last war, military scientists developed this approach and called it 'operational research'. These activities and many others result in controls and regulations of the most varied kind, and as the contents of this book show a corresponding multitude of regulating mechanisms can be found at many levels of organization.

A phenomenon observable in the evolution of a number of technologies, as well as of some biological and sociological systems is a 'transfer of control'. This usually occurs in the course of centralization and specialization, sometimes in both directions at the same

---

* The word 'information' here and in most of this book has its everyday meaning and not the special meaning of Shannon & Weaver's (1949) mathematical theory of communication. Concepts connected with this theory of discrete signals, like coding, channel, noise, redundancy etc., are not extensively discussed.

time. A craftsman, for instance a shoemaker, inspects and improves his handiwork himself, while in a shoe factory several methods of quality control may be performed in a special department. Similarly a unicellular organism can control all its functions within its structures, while many functions of a cell in a metazoon are governed from outside by nervous and hormonal influences. In the social sphere instruction has been largely removed from the oral tradition within the family and entrusted to our elaborate educational machinery. The theory of such transfers of control might merit a separate study.

The intricacy of the regulatory devices which control the functions of established organs are probably far surpassed by the complexities in the regulation of development (Chapters 10 and 11). It is therefore not surprising that the processes and devices which result from developmental processes can only be described in very general biological and even teleological terms. Additionally the logical structure of developmental events can only rarely be formulated in the form of precise flow diagrams or in quantitative terms. The situation can be roughly described by an old observation of unknown origin, but relevant to this book, namely that 'engineers deal with primitive situations in an elegant way, while biologists must tackle elegant situations with primitive means.' Nevertheless embryology is one of the classical fields of study in regulation. The experimental evidence can be summarized under two principal headings, that originally called 'regulation' (see p. 313) and 'tolerance'. The word regulation when used by embryologists refers to well-known but still astonishing facts, such as the development of a divided sea urchin's egg into two whole, though smaller, pluteus larvae and sometimes young sea urchins, or the regeneration of the tail—though a rather rudimentary one—in the lizard.

Regulations of this kind reveal a considerable tolerance in so far as identical or at least very similar organisms can be shown to arise under often drastically different conditions. If half an egg can produce a whole animal, the original whole must have been redundant both in material and in information.

The tolerance and regulatory capacity of developing systems imply that in any particular instance the individual partial processes concerned in development rarely operate under conditions which an engineer might consider optimal and that the resulting organs are never quite perfect. Nevertheless out of this general imperfection of life functional mature organisms are built. Attempts to build good machines from bad parts may still appear quite wrong to old fashioned engineers, but in modern technology self-correcting devices and redundant parts are common place. Some of the problems arising from this situation have been briefly discussed in a paper by von Neumann (1929) which deals with 'the synthesis of reliable organisms from nonreliable components'. But the complexity of life's balancing acts far surpasses these models.

Tolerance and imperfection appear as necessary prerequisites and consequences of the selection mechanism. If in the course of evolution better adapted organisms are selected in preference to less well adapted ones, these latter which have flourished at some prior time still persist. And if on the other hand stabilizing selection maintains a genetical polymorphism, this must be through the action of a balance of defects, which is another way of saying that the population carries a 'genetical load' (Muller, 1956; Haldane, 1959), and every individual has some constructional defect.

One would expect that the load of imperfection in a species would tend to be minimized and that the individual control and regulating mechanism concerned in survival would in consequence interact in specific ways, perhaps according to some general principles, e.g. in a hierarchy or another informational structure. We shall outline a few ideas in this hardly explored field at the end of this book (see Chapter 16, p. 425).

In physical or chemical experiments the choice of suitable manipulated and dependent variables is not too difficult. In biology this choice has in the past often been very clumsy—mostly owing to the prevailing 'physicism' or 'chemicism'. This applies for instance to the sense perceptions of animals including ourselves. We do not normally—at the control level—react to physical or chemical entities like radiations, odorous chemicals or sound waves, but to integrated entities like colour sensations, flower odours or biological entities like food, mates and enemies. While the 'synthesis' from the physical or chemical stimuli components to 'things' by our afferent systems is of the utmost interest, there is only limited sense in correlating the wavelength of light or the concentration of a vapour with salivation when looking for feeding regulations; the correlation of particular foods with feeding behaviour will produce a better insight into an animal's culinary habits. Such ecological–behavioral problems are more akin to the problems of economists and sociologists than to those of physicists and chemists.

A characteristic control device of the higher animals is their capacity of modifying their actions by means of stored and processed information. They have 'a memory' which is implicated in all processes of learning. We know that most of our memory resides in specific regions of the brain, but we do not know how or by what structures it is stored. So far, in general, the molecular and electronic models of memory and of the brain have been rather sterile, while the multitude of contradictory 'theories of learning' also makes one suspicious of this whole approach. For these reasons memory and learning are not extensively discussed in this book. Similarly, details of information transfer in the nervous system have also been omitted. The relations between input and output in intact biological systems are most conveniently described in terms of continuous functions. How and to what extent the discrete discharges of neurons are implicated in this relation is at present far from obvious.

Finally it might be useful to point to certain limitations of the engineering approach—or at least the approach of the contemporary engineer to the problems of control in biology. Those resulting from our ignorance of the 'purpose of it all' have already been mentioned. Almost equally obscure are the partial ends towards which organisms strive or in fact whether they strive at all. Profitable guides in this direction would probably be insights derived by psychoanalysis, introspection and meditation, but as the results of such procedures do not lend themselves to quantitative treatment only a sort of mock engineering has so far emerged. At the present time there is hardly an accepted set of axioms from which one can deduce theorems capable of dealing both with the externally observed facts ('objective facts') of biology and with the more immediate experiences of our conscious mind. There is however one characteristic, which distinguishes an organism—both as externally observed and as internally experienced—from the machines which we know, namely its open, exploratory creative and opportunist nature. Life and indeed research do not usually proceed by deductions from rigidly defined logical premises, but by means of inductive steps of discovery or guess-work. We do not know the nature of the mechanisms, which perpetually act on 'conclusions drawn from insufficient evidence' and thus cannot apply the notions of regulations and control to them. In this sense then this book is not a disguised general biology, but only deals with one of the several partial approaches to the study of life.

References

Descartes, R. (1664–1670). *Opera Philosophica*, Daniel Ezzevier, Amsterdam.

Dobzhansky, Th. (1964). 'Determinants and evolution of life.' *Proc. Nat. Acad. Sci., U.S.*, **51** (5), 907–908.

Grene, Marjorie. (1962). 'Is biology an exact science?' *The Listener*, Oct. 11th, pp. 558–560.

Grey Walter, W. (1961). *The Living Brain*, Penguin Books Ltd., Middlesex, pp. 113 ff. 241–244.

Haldane, J. B. S. (1959). 'The cost of natural selection.' *J. Genet.*, **55**, 511–524.

Hassenstein, B. (1965). *Biologische Kybernetik*, Quelle & Meyer, Heidelberg.

Kalmus, H. (1963). 'Axioms and theorems in biology.' *Nature*, **198**, 240–243.

Kalmus, H. (1966). 'Organic evolution and time.' In Frazer, J. T. (Ed.), *Voices of Time*, George Brazilier, New York.

Le Chatelier, H. (1888). *Recherches Experimentales et Theoretiques sur les Equilibres Chimiques*, Dunod, Paris.

Mittelstaedt, H. (1956). *Regelungsvogaenge in der Biologie*, Oldenbourg, Munich.

Mittelstaedt, H. (1961). *Regelungsvorgaenge in lebenden Wesen*, Oldenburg, Munich.

Muller, H. J. (1956). 'Further studies bearing on the load of mutations in man.' *Acta Genet.*, **6**, 157–168.

Ray, J. (1691). *The Wisdom of God, Manifested in the Works of the Creator*, Samuel Smith, London.

Shannon, C. E. and Weaver, W. (1949). *The Mathematical Theory of Communication*, University of Illinois Press, Urbana.

von Neumann, J. (1929). 'Probabilistic logics and the synthesis of reliable organisms from unreliable components.' In Shannon, C. E. and McCarthy, N. (Eds.), *Automata Studies*, Princeton University Press, New Jersey.

Wiener, N. (1948). *Cybernetics: or Control and Communication in the Animal and the Machine*, M.I.T. Press, Cambridge, Massachusetts, 2nd ed., 1961.

See also Further Reading, p. 436.

2

Regulation and Control in Engineering

B. R. WILKINS

*Department of Anatomy, University College, London**

Introduction

Before considering regulatory mechanisms in biological systems, it is useful to discuss regulatory systems employed in engineering. This provides a frame of reference by defining and showing the context of the engineering terminology that has been freely drawn upon in the description of biological systems, and also enables us to consider how engineering concepts may usefully be employed in the analysis of such systems.

Engineering systems, at least in their simpler forms, can be analysed mathematically and their behaviour described as a function of changes in various parameters. As a result techniques have been developed that enable a system to be described mathematically on the basis of measurements on the whole system, even when we do not know details of the individual components, a situation frequently encountered in biology (see Chapter 3).

Also arising out of this mathematical analysis, it is possible to predict, from the form of the equations, limitations in control systems, and in particular that some combinations of circumstances could lead to failure. One form of failure is instability, and this not only prevents the mechanism from achieving its purpose but can also cause the whole mechanism to be destroyed by a process analogous to an explosion: a process that, once having started, cannot be stopped. It is not always obvious from a qualitative examination of a system that

* Present address: Department of Electronics, University of Southampton.

it could become unstable, as some early constructors of servo-mechanisms (and even of bridges) discovered to their cost, so that the conditions that give rise to instability, as well as the techniques available for combating it, constitute important parts of control circuit theory.

A further point that emerges from the mathematical treatment is that some features of a system may be inevitable, not just because it is an engineering system (with need to consider, for example, tolerances in manufacture, which are perhaps equivalent to biological variation) but because it is a physical system; that is, because it possesses properties such as mass and inertia which must exist in any system, including a biological one. Limitations introduced by factors such as these, therefore, must also apply to biological systems, and this can suggest fruitful lines for investigation and experiment (see p. 124).

Finally, from a generalized consideration of engineering systems, universal concepts have arisen concerning the whole basic nature of control processes and these again can be applied to biological systems, using the methods of information and communication theories.

The concept of the interrelation of engineering, mathematics and biology is by no means new, but it has received a particular impetus since the Second World War with the development of electronic computers and control systems, coupled with the observation that animal control systems have important similarities to servomechanisms (Rosenblueth and coworkers, 1943). It has subsequently shown itself to be of great value both to biology and to engineering, and has been regarded as a new discipline under the title of *Cybernetics*, which is defined (Wiener, 1949) as the study of communication and control in animals and machines.

Regulation and Control

Having, so to speak, established the credentials of the subject in the present context, we can now move to a consideration of the essential features of regulating and control systems.

A basic general system is shown in Figure 1. The controller, whose behaviour is determined by instructions, controls the flow of energy from the power supply to the controlled system. A distinction is often made between a *regulator*, which is required to keep a controlled quantity constant and independent of changes of power supply or disturbing factors in the environment, and a *controller*, which is required to make the controlled quantity vary in accordance with changes of instruction (see p. 4 ff.).

An example of a regulator, and also one of the earliest examples
of a control system in engineering, is the steam engine governor
made by James Watt. He wanted his engine to run at a constant
speed irrespective of the load it was driving, but was faced with the
fact that with a constant steam supply the speed of the engine falls
as the loading increases. The governor makes the steam supply increase if the speed starts to fall below the desired value, and decrease
if the speed starts to rise above it, thus turning the engine into an
almost constant speed machine.

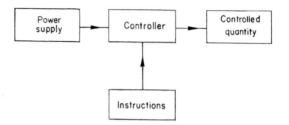

Figure 1. Block diagram of a basic general control system. The controlled
quantity is varied by the power supply under the direction of the controller,
which in turn is governed by the instructions.

A typical example of a controller is the 'lock-on' radar system, in
which the aerial abandons its normal regular search pattern when a
target is found, and instead follows the target wherever it may move.

Both these examples have biological parallels. A regulator keeps
the mammalian blood temperature very nearly constant despite wide
variations in the temperature of the environment, while the system
controlling the eyes enables them to 'lock-on' to 'targets' which
may represent advantage or danger.

In a regulator, the quantity to be controlled has a natural tendency
to vary and must be kept constant, while in a controller the quantity
has a natural tendency to remain constant or to vary randomly and
must be made to follow specified variations. The distinction between
a regulator and a controller, however, is not fundamental, since each
type of system can be regarded as a special case of the scheme represented by Figure 1, differing only in the form of the instructions.

Information Processing

The idea of a mechanism receiving and carrying out instructions
may call for some explanation. These terms have been deliberately

used to emphasize the importance of the concept of information processing.

If we consider a thermostatic water heater we find that power is supplied to the heater only if the water temperature is below a certain level. If this desired level is changed by manually altering the setting of the thermostat, we are in fact giving an instruction to the system that the water temperature is henceforth to be maintained at the new level. This instruction will be carried out, but it is important to notice that the range of instructions that can be obeyed is strictly limited by the physical construction of the system, and furthermore, that the instruction has to be couched in the right language. The only instruction that can be obeyed by the thermostat is one telling it to maintain the water temperature at a specified level, and the only language that can be understood by the system is the one in which the required temperature is indicated by the position of a control knob. In a similar way, we can regard all control systems as essentially concerned with the processing of information, and they are all also subject to similar limitations.

Looking again at Figure 1, which can be taken to represent either a controller or a regulator, we can say that essentially the controlled system is driven by an energy source, and that the ultimate function of the controller is to control the supply of energy to the controlled system, in accordance with the instructions. These may be inbuilt and fixed, or derived from outside and variable. It should be noted that the energy supplied to the controlled system is in general much greater than the energy required to insert the instructions into the controller. In other words, a controller generally contains an amplifier. In fact, an amplifier can itself be regarded as a basic control system (as shown in Figure 2) since energy is supplied to the load from the power supply unit, but the rate of supply of energy, and the variations in this rate of supply of energy, are dictated by the input voltage, which is an instruction signal requiring very little power for its transmission.

It is instructive to consider in a little more detail some characteristics of an amplifier treated as a control system. If in the circuit of Figure 2(a) we have a constant supply voltage, V, constant resistors R_L and R_O, and constant characteristics of the valve, then there will be a one-to-one relationship between the input voltage V_{in} and the output voltage V_{out}. In other words, under these conditions, for any required value of V_{out} there is a particular value of V_{in} necessary, so that if we had a knob which controlled the value of V_{in}, it could be calibrated in terms of V_{out}. Unfortunately this simple arrangement, although it might be good enough immediately after calibration, is incapable of maintaining precise control for any prolonged period because a variation in any of the circuit elements or of the power supply or in the characteristics of the valve would cause a change in

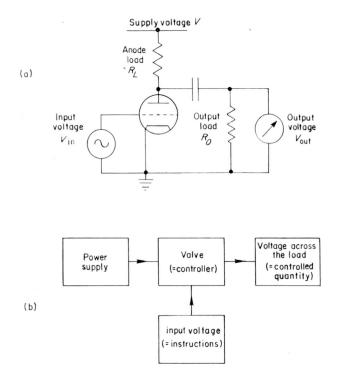

Figure 2. Simple valve amplifier. (a) Circuit diagram. (b) Block diagram show-ing how an amplifier contains all the elements of a control system.

the function relating the output to the input. Such variations will inevitably occur as functions of temperature and of ageing of the components.

Closed Loop Systems

Consider now a system for controlling the rate of flow of fluid along a pipe. The first obvious requirement is a stop valve, operation of which will vary the rate of flow. The simplest approach to the prob-lem is to provide the stop valve with a pointer and a dial, and then to calibrate this by measuring the rate of flow resulting from various dial settings. To obtain any given rate of flow we have then first to consult the calibration chart to obtain the appropriate dial setting (i.e. translate the instruction into the right language for the machine) and then to apply this setting to the valve. This arrangement is illustrated in Figure 3 and for many practical purposes would be quite satisfactory as it stands, particularly bearing in mind its great merits of cheapness and simplicity. Its performance, however, is vitally dependent upon the calibration curve: alterations in the calibration function will cause the output to vary unpredictably from that which is intended. Factors that will inevitably affect the calibration function include silting of the pipe, wear in the valve, changes in the fluid supply pressure and many other unpredictable variations.

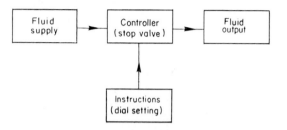

Figure 3. Elementary system for controlling the flow of fluid along a pipe: this system is critically dependent on the unpredictably varying relationship between dial setting and control action.

One way in which we can achieve accurate control of this system is to fit a flow meter to the output and then to employ a human operator at the stop valve with instructions to keep the flow at the specified level. This modified system is shown in Figure 4, and it is vital to recognize the following essential changes that have been introduced.

1. The output (i.e. the flow rate), which is the controlled quantity, is continuously measured.

2. The difference between this output and the intended output (the instruction) is measured. In this case it may consist of perhaps no more than 'too much', 'correct' or 'too little'—an imprecise measurement, but nevertheless a definite one.

3. If the output is not correct, the controller is actuated in such a way as to reduce the error.

The whole cycle is then repeated until the output has reached the desired level.

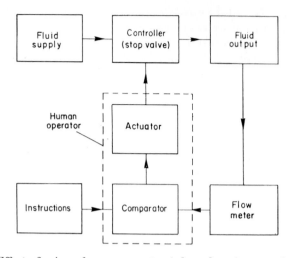

Figure 4. Effect of using a human operator (whose functions are shown within the broken lines) to improve the control system of Figure 3. The need to maintain calibration between dial setting and flow rate is removed.

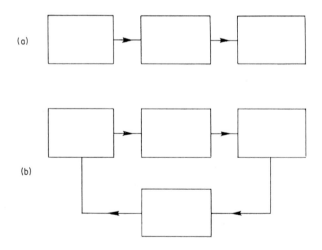

Figure 5. Fundamental patterns of connection in control systems. (a) Open loop system. (b) Closed loop system.

An important point to notice here is that with this modification to the system it no longer matters that the calibration function between pointer setting and flow changes with time. Any change, whatever the cause, produces a difference between the intended output and the actual output, and when this is detected by the operator, he acts so as to reduce it to zero.

The essential difference between the system of Figure 4 and that of Figure 3 is brought about by the existence of a closed loop of dependency from pointer to fluid flow and meter, and back again through the human operator and stop valve to the pointer. In other words, information about the output of the system is used to influence the input: this feature is known as *feedback*, and is a fundamental technique in control systems. A system containing a feedback loop of this kind is known as a *closed loop system*, as opposed to the system of Figure 3 which is known as an *open loop system*. This terminology is not altogether a happy choice, and could perhaps be confusing: a loop is after all not a loop unless it is closed. Nevertheless, since this is the standard terminology we must adopt it, emphasizing again that in an open loop system, as shown in Figure 5(a), there is in fact no loop at all, but only a set of one or more elements in series; while a closed loop system, as shown in Figure 5(b), contains a group of elements arranged in such a way that the input to the system is partly dependent on the output.

Automatic Control Systems

In order to see how this process can be made automatic, and so arrive at a convenient general description of a feedback control system, a number of features of the system should be noted. First of all, however complex the control system, there is one specific quantity or property that is being controlled, and in order to control it, it must be capable of measurement, either directly or indirectly. Secondly, the controlled system is driven by a controller which responds to signals at its input, and this controller may be electrical, mechanical, hydraulic or one of several other types. The controlled quantity may also be of any of these types, and need not necessarily be of the same type as the controller. For example, in a position control system, the controller may be an electric motor, while the controlled quantity may be the angular deflection of a rod—a situation analogous to that of a sun navigating animal (see p. 109). In such a case, the measurement of output angle must be converted by a *transducer* to an electrical signal, since these signals constitute the only language that this controller understands. In a similar way the instruction to the system telling it the required output must also be put into the same language. This transformed input signal is called the *command signal*. Finally, the output signal and the command signal must be combined so as to measure the difference between the two and transmit this difference, the *error signal*, to the controller. This point of comparison is known as the *summation point*, or *summing junction*.

A diagram of a general feedback control system embodying these features is shown in Figure 6. We are now in a position to summarize the essential difference between open loop and closed loop control. In an open loop system, control is exercised assuming a constant relationship between the command signal and the controlled quantity,

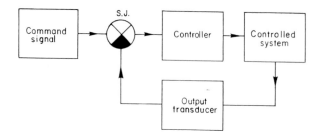

Figure 6. Essential components of an automatic closed loop control system. The black quadrant of the summing junction (S.J.) indicates a negative input.

whereas in a closed loop system the controller is actuated by a signal depending on the error between the actual output and the required output, no matter how that error might have arisen.

Time Lags in Control Systems

On the simple view of control systems that has so far been presented, it would appear that any measurable quantity can be controlled as precisely as we like by adding a feedback loop, the only obvious limitation being the resolution available in the measuring device. Unfortunately, however, we have left out of account a feature that is inevitably present in any physical system and that not only places severe limitations on performance but can even under certain circumstances cause the behaviour of the system to alter drastically and destructively.

In any physical system, between the application of an input signal and the appearance of the corresponding output signal there is a finite time delay. Even the finite velocity of light introduces a delay—a problem that is having to be considered now in connection with control of satellites and space craft (Smith, 1962). Thus if we wish to control from the earth the activities of a mechanism on the moon, nearly a quarter of a million miles away, then any command signal originating on earth would take about $1\frac{1}{4}$ seconds to reach the moon, and when the system had responded, the information about its response would take another $1\frac{1}{4}$ seconds to get back. Thus even without taking into account the delays in the system itself, there is a minimum of $2\frac{1}{2}$ seconds delay round the loop, and this will obviously place limitations on the range of control that can be exercised. In more ordinary circumstances control systems are often required to move some mechanical device possessing inertia, and a fundamental feature of a body with a finite inertia is that its velocity cannot be changed instantaneously.

In order to appreciate the significance of these time lags in a control system we might consider the operation of a room heating system

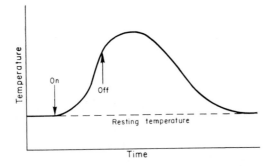

Figure 7. Changes in room temperature resulting from adjustments to a radiator: an example of a control process with inherent time lags.

using water filled radiators. If a radiator is turned on from cold, it takes an appreciable time to become hot and to raise the room temperature. Furthermore, having got hot it gives up its heat gradually and continues to heat the room for some time after being turned off. This is illustrated in Figure 7.

Consider now the situation depicted in Figure 8 in which the room temperature with the radiator off is T_1 and the required room temperature is T_2. If we assume that the radiator is capable of heating the room to a temperature much higher than T_2, then the variation of temperature with time may be as shown. The radiator is turned on, and the temperature begins to rise. When the temperature reaches T_2 the radiator is turned off, but, as noted above, it continues to heat the room, raising its temperature above T_2 and impelling the inhabitants to open the windows. This has the desired effect of reducing the temperature, and when it has fallen to T_2 the windows are closed.

However, the temperature continues to fall towards its resting level T_1, and the radiator has to be put on again. This cycle repeats indefinitely, the actual temperature always chasing the target value but never maintaining it: a process known as *hunting*. Similar time lag difficulties must exist in biological temperature regulation systems (see p. 87), owing to the thermal capacity of the body. Notice that this is a closed loop system (closed once more through the human operator) and that if we open the loop (by refraining from adjusting the radiator or windows) the temperature will settle to an equilibrium value at which the rate of supply of heat from the radiator is equal

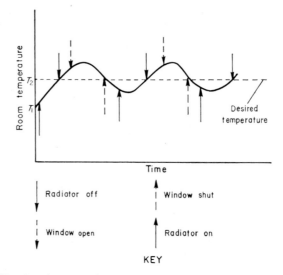

Figure 8. Hunting in control systems. By manipulating the radiator and window, the room temperature is kept close to the desired value without maintaining it.

to the rate of loss of heat to the outside atmosphere. This equilibrium value depends on a very large number of factors, so that even although different rates of heat supply, and so different settling temperatures, may be obtained with different settings of the tap, any calibration of the tap in terms of temperature would be hopelessly unreliable.

Instability of Control Systems

Suppose now that we have a position control system, the details of which need not at the moment concern us, but which conforms to the block diagram of Figure 6, and consider the effect of time lags in this system.

A convenient way of finding out about the lags in the circuit is to open the loop, and to measure the feedback signal (which is proportional to the position that is being controlled), comparing this with the command signal. If then we apply a command signal that varies with time, we can obtain an output signal that also varies with time, but is out of step with the input, as shown in Figure 9.

If now we progressively increase the frequency of the input signal, we may reach the stage shown in Figure 10(b) at which the output, instead of following the input is exactly out of step with it. If the loop were now closed, then the command signal would be reinforced by the feedback signal instead of opposed by it. This would result in the

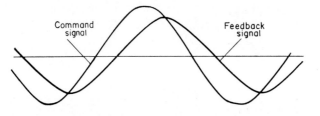

Figure 9. Input and output signals for a linear open loop control system. If the input signal is sinusoidal, then the output signal is also sinusoidal, differing from the input only in amplitude and phase by amounts depending on the frequency of the signal.

controller giving a larger drive to the controlled system, thus moving it even further from the right position, so producing a larger feedback signal and so on. The result is depicted in Figure 10(c), where the output is making bigger and bigger oscillations; these, if not prevented, can sometimes lead to the system's self-destruction. Such a system is said to be *unstable*. The howl in a badly adjusted public address system is a well-known example of the phenomenon of instability.

Thus, although a closed loop system can provide precise control, it can also suffer from hunting or instability which, even if not destructive, would make the system useless for control purposes. The avoidance of such behaviour is a major concern of control engineers: it can be done, but we should notice that avoidance of instability requires additional equipment, design effort and cost, so that an open loop system, which cannot suffer from this defect, can find a use when the control does not need to be very precise, or when the operating conditions of the system are bound to remain very nearly constant.

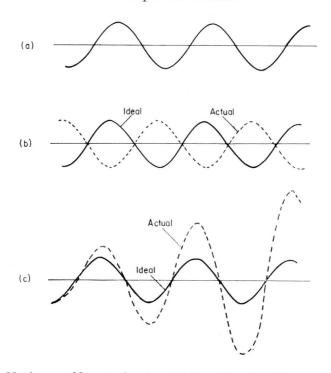

Figure 10. An unstable control system. (a) Input signal. (b) Output signal with the feedback loop open. At the frequency shown the feedback signal is 180° out of phase with the input. (c) When the loop is closed the feedback signal reinforces the input signal instead of opposing it, and the system goes into uncontrollably increasing oscillations.

The volume control of a radio receiver is one common example of this situation. There also exists a need for open loop control when the response has to be so fast that feedback is not possible. Examples of this kind do not generally arise in engineering, but in biology they are not uncommon, as in prey capture mechanisms (see p. 111), or in some activities requiring rapid coordination, such as playing tennis.

Compensation for Time Delays

When we come to consider the mathematics of control systems, the development of satisfactory criteria for stability will be a major aim, but it is useful to derive first in qualitative terms one important device for improving system performance.

If we have the position control shown in Figure 11, and apply to it a sudden change of command signal, as shown in Figure 12(a), the response may be as shown in Figure 12(b). The physical explanation of this form of response is similar to that for the temperature system response shown in Figure 8: the signal applied to the motor cannot move the motor and load instantaneously, but causes a gradual build

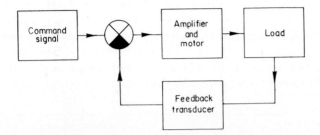

Figure 11. A closed loop position control system.

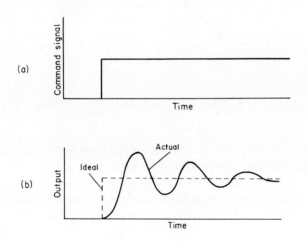

Figure 12. Response of a position control system to a sudden change of command signal. (a) Input: a step function. (b) Output, showing a damped oscillation about the required value instead of the ideal step.

up of speed. When the load has reached the required position, the error signal is reduced to zero, but the momentum of the moving load carries it on past this position—a phenomenon known as *overshoot*. When the load passes the required position, the error signal changes sign, and tries to move the motor in the opposite direction. This again cannot be achieved instantaneously, but the load is eventually

2+R.C.L.S.

brought to rest, and moved back towards the desired position. However, as before, when it reaches the desired position, it is again travelling with some finite velocity, and its momentum carries it on to an overshoot in the opposite direction. The overshoots become progressively smaller because of frictional effects, and eventually the system settles down to its proper position after executing a number of these *damped oscillations*. This system is not actually unstable, but the form of response is far from ideal.* It is not difficult to see where the fault lies: the drive to the motor is not being reversed until the error passes through zero, and this is too late. The fact that the velocity takes some time to build up after the application of a signal must be catered for by reversing the drive to the motor *before* the load

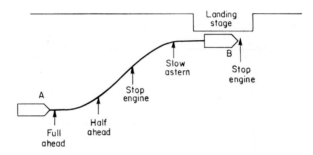

Figure 13. Method of handling a motor boat to bring it alongside a landing stage.

gets to the required position, so that by the time it does get there, the velocity has already been reduced to zero.

A familiar example of this is the operation of bringing a boat alongside a landing stage. This is illustrated in Figure 13, and a typical pattern of engine manipulation is shown. Notice that the engine is stopped some time before the landing stage is reached, and the velocity is finally reduced to zero by a judicious application of reverse engine. In this example, we have another position control system, in which the boat is the load and the motor and human operator constitute the controller. To this control system has been applied a command signal instructing it to move the load from A to B. Three forms of response pattern are illustrated in Figure 14. The response marked 'actual' is the one obtained from the position control system of Figure 11. The response marked 'ideal' is the perfect response, where

---

* Although oscillations are a nuisance in a control system, they can be put to good use as a timing mechanism (see Chapter 7, p. 155).

the load follows the command signal exactly. This, we know, is unobtainable because an instantaneous change of position of a mass would require infinite power, but we can hope to achieve the response marked 'desired', which is the one obtained from the handling of the boat shown in Figure 13. In order to achieve this response, as we have seen, the system must not wait for the load to reach its goal, but

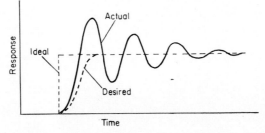

Figure 14. Three forms of stable response to a step input. The ideal would be a step output, but this is physically unrealizable. The best compromise between this ideal and the actual oscillatory response is the critically damped response.

must take action in advance: in other words some kind of anticipatory mechanism must be called into play (see pp. 83, 155). How this can be achieved in practice will emerge from the mathematical treatment.

Improvement by Learning

One final point may be noted from this boat example. The handling of the boat in the way described is no easy matter, especially when there are tides and winds to take into account as well as the individual idiosyncrasies of the particular boat. A man getting into a boat for the first time, even if aware of the problems and of the general way of dealing with them, is unlikely to perform ideally. The same man a few years later, after he has had some experience, may be an expert. This change is brought about by learning, and although we cannot go into the subject here, it is perhaps worth mentioning that the construction of adaptive control systems, and of learning systems in general, is occupying the attention of many control system engineers at the moment. Such studies derive much from the study of learning in animals, and besides having valuable engineering implications, hold out the hope that they may also ultimately repay the debt by helping to cast some light on the problems of learning and memory mechanisms in man.

References

Rosenblueth, A., Wiener, N. and Bigelow, J. (1943). 'Behaviour purpose and teleology.' *Phil. Sci.*, **10**, 18.

Smith, K. U. (1962). *Delayed Sensory Feedback and Behaviour*, Saunders, London.

Wiener, N. (1949). *Cybernetics*, M.I.T. Press, Cambridge, Massachusetts, and Wiley, New York.

3

Basic Mathematics of Control

B. R. WILKINS

*Department of Anatomy, University College, London**

The basic mathematics used in the analysis of control systems, introduced in this chapter, can be regarded as the formal justification for many of the statements made in Chapter 2. Although detailed knowledge of this material is not essential for the understanding of the rest of this book, some of the concepts and technical terms introduced here will be referred to in later chapters.

One way of conducting an analysis of a control system is to use the known physical properties of the component elements to set up a differential equation relating the output to the input. We can then obtain the time response of the system to any input signal by solving the differential equation for that particular input function. This method of analysis is of very widespread use, but it does have limitations. Perhaps the most important of these limitations is that the method is applicable in general only to linear systems. A linear system may be defined as one for which a change in input produces a proportional change in output: in other words, if an input x produces an output y, then an input $2x$ produces an output $2y$. A linear system also obeys the superposition law: if an input x produces an output y and an input w produces an output z, then inputs x and w acting together will produce an output of $y + z$.

In practice, no physical system is truly linear, although in very many cases linearity can be taken as an approximation over the working range. Nevertheless, the consideration of linear analysis is worthwhile for two reasons: first, that non-linear analysis is mathematically always difficult and sometimes impossible; and second, that

* Present address: Department of Electronics, University of Southampton.

linear analysis will often give a good indication of the behaviour of the system, even when the assumption of linearity is not entirely justified.

Speed Control System

The linear analysis procedure may be demonstrated by considering the simple speed control system shown in Figure 1. The controlled variable is the angular speed of a load, which might be a machine tool or a wheel. The load is turned by a motor whose output torque is proportional to the current fed into it by the amplifier. Since the amplifier responds only to voltages, the input and output quantities, which are speeds, have to be converted to voltages by means of suitable transducers. To measure the output (the speed of the load) a tachometer generator is coupled to the output shaft. In practice, servomotors are often made for this purpose with a tachometer

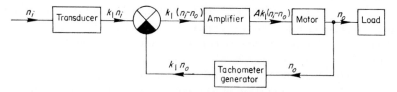

Figure 1. Schematic diagram of a simple speed control system.

generator built into the motor housing. This generator will produce a voltage proportional to n_o, the speed of the motor, the constant of proportionality being denoted by k_1. The input, n_i, which is the required speed, can be transduced by a potentiometer that is calibrated so that the voltage delivered is proportional to n_i with the same constant of proportionality k_1. These two voltages are subtracted, and the difference, which is proportional to the difference between the required and actual speeds, is fed to the amplifier. This speed difference is the error, and for this reason a system such as this is referred to as an error actuated system.

We can now set up the equations for this system. The voltage applied to the amplifier is given by

$$v = k_1(n_i - n_o)$$

Also, if the amplifier is linear, the current supplied to the motor is

$$i = \frac{Av}{R} = \frac{Ak_1}{R}(n_i - n_o)$$

where R is the resistance of the field winding and A is the ratio of output voltage to input voltage for the amplifier. This ratio is called the *gain* of the amplifier.

The torque developed by the motor is

$$T = k_2 i = \frac{A k_1 k_2}{R} (n_i - n_o) \qquad (1)$$

This torque is absorbed in accelerating the load, in accordance with the relationship

$$T = J \frac{dn_o}{dt} \qquad (2)$$

where J is the total inertia of the motor and load.

Combining equations 1 and 2 gives

$$\frac{dn_o}{dt} = \frac{A k_1 k_2}{J R}(n_i - n_o)$$

or

$$\tau_1 \frac{dn_o}{dt} + n_o = n_i \qquad (3)$$

where
$$\tau_1 = \frac{J R}{A k_1 k_2} = \text{a constant.}$$

This equation represents the operation of the system, and by substituting the appropriate function for n_i and solving we can obtain the response of the system to any given input.

One input function that is often used as a test of a system's performance is the step function: a sudden change in the demanded speed. If a step function of magnitude N is applied to the system at time $t = 0$, then the response is found by solving

$$\tau_1 \frac{dn_o}{dt} + n_o = N$$

with
$$n_o = 0 \quad \text{at} \quad t = 0$$

The solution to this equation is

$$n_o = N(1 - \varepsilon^{-t/\tau_1}) \qquad (4)$$

where ε is the exponential constant.

The quantity τ_1, which has the dimensions of time, is called the *time constant* of the system.

Figure 2 shows the step function change in n_i and the resulting change in n_o (the response of the system) given by equation 4. This system is said to be a first order system, because its performance is described by a first order differential equation. The two electrical networks shown in Figure 3 are also first order systems, for their

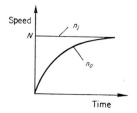

Figure 2. Exponential response of a first order control system to a step input.

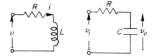

Figure 3. Two electrical examples of a first order system.

operating equations are

$$\tau_2 \frac{di}{dt} + i = \frac{v}{R} \tag{5}$$

where

$$\tau_2 = \frac{L}{R}$$

and

$$\tau_3 \frac{dv_o}{dt} + v_o = v_i \tag{6}$$

where

$$\tau_3 = CR$$

both of which are first order differential equations.

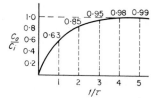

Figure 4. Generalized step response applying to any first order system.

Since equations 3, 5 and 6 are all of the same form, the step response of each system is also of the same form, and may be expressed as

$$C_o = C_i(1 - \varepsilon^{-t/\tau})$$

where C_o is the output variable, C_i is the magnitude of the input step, and τ is the time constant.

By plotting C_o/C_i against t/τ we obtain the 'normalized' step response shown in Figure 4, and this will then apply to any first order system.

Position Control System

Figure 5 shows a position control system, whose equation can be obtained by applying the same methods as with the speed control system. The two systems are in fact very similar, the only essential difference between the two being in the output transducer, for we have now to control the position of the load instead of its speed; this requirement is easily satisfied by replacing the tachometer generator by a potentiometer.

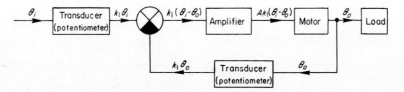

Figure 5. Schematic diagram of a simple position control system.

In analysing this system, we will include the effects of friction by supposing that the torque necessary to overcome friction is given by

$$F = fn_o = f\frac{d\theta_o}{dt} \qquad (7)$$

where θ_o is the angular position of the load and f is a constant.

The equation of the system can now be obtained, noting that the input to the amplifier is proportional to the error in position. Thus, with the same notation as before, the torque developed by the motor is given by

$$T = \frac{Ak_1k_2}{R}(\theta_i - \theta_o) \qquad (8)$$

where θ_i is the demanded position.

This torque has to overcome the frictional forces as well as accelerating the load, so that load equation becomes

$$T = J\frac{d^2\theta_o}{dt^2} + f\frac{d\theta_o}{dt} \qquad (9)$$

2*

The equation of the system is obtained by combining equations 8 and 9, giving

$$J \frac{d^2\theta_o}{dt^2} + f \frac{d\theta_o}{dt} = \frac{A k_1 k_2}{R} (\theta_i - \theta_o)$$

which may be written

$$\frac{d^2\theta_o}{dt^2} + 2\zeta\omega_n \frac{d\theta_o}{dt} + \omega_n^2\theta_o = \omega_n^2\theta_i \tag{10}$$

where

$$\omega_n^2 = \frac{A k_1 k_2}{JR}$$

$$\zeta = \frac{f}{2J\omega_n}$$

This is a second order linear differential equation, and the system is therefore described as a second order system. Other examples of second order systems are shown in Figure 6: in each case the input and output of the system are related by equation 10, with values of ω_n and ζ shown in the diagram.

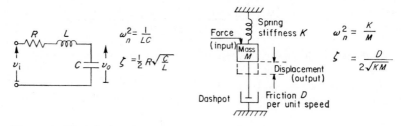

Figure 6. Electrical and mechanical examples of a second order system.

The step response of a second order system is obtained by solving the equation

$$\frac{d^2\theta_o}{dt^2} + 2\zeta\omega_n^2 \frac{d\theta_o}{dt} + \omega_n^2\theta_o = \omega_n^2\Theta \tag{11}$$

where Θ is the magnitude of the input step.

The solution of this equation may be expressed as

$$\theta_o = \Theta + A\varepsilon^{p_1 t} + B\varepsilon^{p_2 t} \tag{12}$$

where A and B are constants and p_1 and p_2 are the roots of the equation

$$p^2 + 2\zeta\omega_n p + \omega_n^2 = 0$$

which is called the *characteristic equation*.

Thus p_1 and p_2 are given by

$$p_1, p_2 = \omega_n[-\zeta \pm \sqrt{\zeta^2 - 1}] \tag{13}$$

There are several different forms this solution can take, depending on the values of these roots, which in turn depend on the value of ζ.

If $\zeta = 0$, corresponding to a system with no friction, the roots are purely imaginary, given by

$$p_1, p_2 = \pm j\omega_n$$

where
$$j = \sqrt{-1}$$

and the complete solution of equation 11 is

$$\theta_o = \Theta[1 - \cos \omega_n t]$$

This response, shown in Figure 7, exhibits continuous oscillations about the target value, of peak to peak amplitude 2Θ and angular

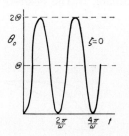

Figure 7. Step response of a second order system with no damping ($\zeta = 0$): a continuous oscillation of constant amplitude.

frequency ω_n, which is for this reason called the *natural angular frequency* of the system.

For a system with friction, that is $\zeta > 0$, there are three possible cases to consider. If ζ lies between 0 and 1 the roots given by equation 13 are complex, and after substituting in equation 12 and simplifying we can write the solution as

$$\theta_o = \Theta \left\{ 1 - \frac{1}{\alpha} \varepsilon^{-\zeta\omega_n t} \sin [\alpha\omega_n t + \phi] \right\} \tag{14}$$

where
$$\phi = \tan^{-1} \frac{\alpha}{\zeta}$$

and
$$\alpha = \sqrt{1 - \zeta^2}$$

This expression represents an oscillation about the desired position, the oscillations becoming progressively smaller because of the exponential term, as illustrated in Figure 8. These are known as damped oscillations, and the degree of damping depends on the exponential term. This in turn depends on the value of ζ, which for this reason is called the *damping factor*. As the damping factor becomes larger, the number and the amplitude of the oscillations decrease as shown in Figure 8.

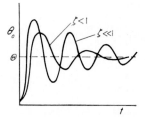

Figure 8. Step response of a second order system which is underdamped $(0 < \zeta < 1)$.

If $\zeta > 1$, then we see from equation 13 that the two roots are real so that the solution given by equation 12 is simply an exponential approach to the desired value, shown in Figure 9, the approach be-

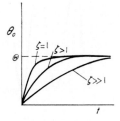

Figure 9. Step response of a second order system which is overdamped $(\zeta > 1)$ or critically damped $(\zeta = 1)$.

coming progressively slower as the damping factor is increased. The borderline case between these two forms of response is when $\zeta = 1$, and the roots given by equation 13 are real and equal. In this case equation 12 no longer represents the most general form of the solution of equation 11, and in fact the complete solution is

$$\theta_o = \Theta[1 - (1 + \omega_n t)\varepsilon^{-\omega_n t}]$$

This is also an exponential approach to the desired value, and as shown in Figure 9, it represents the fastest response that does not exhibit any oscillation. This condition is known as *critical damping*.

Higher Order Systems

If the methods that have been employed with these two simple control systems are applied to more complicated systems, it can, and often does, turn out that the system equation is of third or higher order. A third order equation may be written

$$\frac{d^3\theta_o}{dt^3} + a\frac{d^2\theta_o}{dt^2} + b\frac{d\theta_o}{dt} + c\theta_o = c\theta_i$$

where a, b and c are constants of the system.

As before, the step response of the system can in general be written as

$$\theta_o = \Theta + A\varepsilon^{p_1 t} + B\varepsilon^{p_2 t} + C\varepsilon^{p_3 t} \tag{15}$$

where Θ is the magnitude of the input step, A, B and C are constants, and p_1, p_2 and p_3 are the roots of the characteristic equation

$$p^3 + ap^2 + bp + c = 0$$

The possible forms of response represented by equation 15 depend, just as with the second order system, on the different possible combinations of roots of the characteristic equation. Since this is a cubic, one of the roots must be real; the other two can both be real, or both imaginary, or both complex. Thus possible forms of step response are

$$\theta_o = \Theta(1 + A\varepsilon^{\alpha t} + B\varepsilon^{\beta t} + C\varepsilon^{\gamma t}) \tag{16}$$
if all the roots are real;

$$\theta_o = \Theta[1 + A\varepsilon^{\alpha t} + B\sin(\omega t + C)] \tag{17}$$
if one root is real and two are imaginary;

and $$\theta_o = \Theta[1 + A\varepsilon^{\alpha t} + B\varepsilon^{\beta t}\sin(\omega t + C)] \tag{18}$$
if one root is real and two are complex,

where in each case A, B and C are constants depending on the constants of the system, and α, β, γ and ω are related to the roots of the characteristic equation.

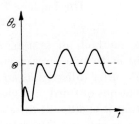

Figure 10. Step response of a third order system whose characteristic equation has two purely imaginary roots. A continuous oscillation is superimposed on the exponential approach to the target value.

Assuming for the moment that α, β and γ are all negative, then equation 16 represents an exponential approach to the demanded value, similar in form to the responses shown in Figure 9. Equation 17 contains an exponentially decaying term and also a constant amplitude oscillation, as shown in Figure 10. Equation 18 contains a damped oscillation superimposed on the exponential decay term: this response is shown in Figure 11.

If, however, α, β or γ are positive then a new situation arises. If, for example, in equation 18, β is positive, then the term

$$B\varepsilon^{\beta t} \sin (\omega t + C)$$

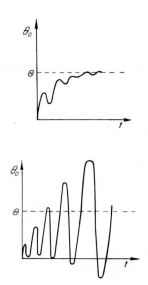

Figure 11. Damped oscillatory form of step response of a third order system.

Figure 12. Unstable form of step response of a third order system.

instead of being a decaying oscillation becomes a growing oscillation. The response in this case is depicted in Figure 12, and is an example of complete instability.

The Laplace Transform as a Means of Solving Differential Equations

The methods of system analysis that have been used so far are subject to two limitations with practical systems. The first is that high order differential equations (and many practical systems have equations of third or higher orders) are laborious to solve; the second is that it is difficult to relate the solution to the physical situation so as to decide how the system should be modified to avoid instability or to improve the response. In an effort to overcome these limitations, the methods of the Laplace transform are used.

The Laplace transform is a mathematical device for changing the variable in the system equations from time, t, to a new variable, s, which is simply an algebraic number (although not generally real).

The particular cunning of the device is that the operations of differentiation and integration, which make manipulation of the original equations so awkward, become transformed into algebraic functions.

The Laplace transform of a function of time, $f(t)$, is given by

$$\mathscr{L} f(t) = \int_0^\infty f(t).\varepsilon^{-st}\, dt = F(s) \qquad (19)$$

Using this definition, the list of transforms of common functions shown in Table 1 can be compiled. This list is of course far from exhaustive, and is included here mainly for the purposes of illustration. For practical purposes, comprehensive tables of transforms have been prepared and published.

Referring to the transforms given in Table 1, it is interesting to

Table 1. Short list of Laplace transforms

	Function $f(t)$	Laplace transform $F(s)$
1	H (constant)	$\dfrac{H}{s}$
2	ε^{-pt}	$\dfrac{1}{s+p}$
3	$\sin \omega t$	$\dfrac{\omega}{s^2+\omega^2}$
4	$\cos \omega t$	$\dfrac{s}{s^2+\omega^2}$
5	$\varepsilon^{-pt}.f(t)$	$F(s+p)$
6(a)	$\varepsilon^{-at}.\sin \omega t$	$\dfrac{\omega}{(s+a)^2+\omega^2}$
6(b)	$\varepsilon^{-at}.\cos \omega t$	$\dfrac{s+a}{(s+a)^2+\omega^2}$
7	$\dfrac{d}{dt}\Big[f(t)\Big]$	$sF(s) - f(0)$
8	$\dfrac{d^2}{dt^2}\Big[f(t)\Big]$	$s^2F(s) - sf(0) - f'(0)$
9	$\displaystyle\int_0^t f(t)\, dt$	$\dfrac{1}{s}F(s)$
10	$f(t-\tau)$	$\varepsilon^{-s\tau}.F(s)$

note that transforms 3 and 4 can both be derived from transform 2, for since

$$\varepsilon^{-j\omega t} = \cos \omega t - j \sin \omega t,$$

we have

$$\mathscr{L}[\cos \omega t - j \sin \omega t] = \mathscr{L}(\varepsilon^{-j\omega t}) = \frac{1}{s + j\omega} \quad \text{by transform 2.}$$

$$\therefore \ \mathscr{L}[\cos \omega t - j \sin \omega t] = \frac{s - j\omega}{s^2 + \omega^2} \tag{20}$$

Separating the real and imaginary parts of equation 20 then gives the transforms already shown as numbers 3 and 4 in the table.

Transform 5 is a useful general result, two particular cases of which are listed as transforms 6(a) and 6(b). In transforms 7 and 8, $f(0)$ and $f'(0)$ are the values of $f(t)$ and $df(t)/dt$ respectively at time $t = 0$, and are known as the *initial values* of the function and its derivative. In many practical cases, these initial values are zero, and it is important to notice that in this case every successive differentiation of a function becomes transformed into a multiplication by s. Conversely, as indicated by transform 9, integration becomes transformed into division by s. Finally, transform 10 represents a delay time, or reaction time, which is particularly important in biological systems.

The usefulness of the Laplace transform method lies in the fact that no function has more than one transform and no transform corresponds to more than one function. Furthermore, Laplace transforms obey the normal rules of algebra, so that equations can be transformed and manipulated algebraically in such a way that on being transformed back to a time function the solution of the differential equation emerges.

As an example of the method, consider again the second order system, whose equation was

$$\frac{d^2\theta_o(t)}{dt^2} + 2\zeta\omega_n \frac{d\theta_o(t)}{dt} + \omega_n^2\theta_o(t) = \omega_n^2\theta_i(t)$$

where $\theta_o(t)$ and $\theta_i(t)$ are functions of time.

Taking Laplace transforms, assuming that the initial values of $\theta_o(t)$ and $d\theta_o(t)/dt$ are zero, we get

$$(s^2 + 2\zeta\omega_n s + \omega_n^2)\theta_o(s) = \omega_n^2\theta_i(s)$$

where $\theta_o(s)$ and $\theta_i(s)$ (functions of s) are the transforms of $\theta_o(t)$ and $\theta_i(t)$.

Thus,

$$\theta_o(s) = \frac{\omega_n^2 \theta_i(s)}{s^2 + 2\zeta\omega_n s + \omega_n^2} \qquad (21)$$

If now we consider the response of the system to a step disturbance of magnitude Θ then from transform 1 of the table

$$\theta_i(s) = \frac{\Theta}{s}$$

$$\therefore \theta_o(s) = \frac{\omega_n^2 \Theta}{s(s^2 + 2\zeta\omega_n s + \omega_n^2)} \qquad (22)$$

In order to convert this back to a time function, we split the right-hand side into partial fractions. The way in which this is done depends on whether or not the quadratic factor has real factors. If we suppose first that there are real factors, then equation 22 can be expressed as

$$\theta_o(s) = \frac{A}{s} + \frac{B}{s - p_1} + \frac{C}{s - p_2} \qquad (23)$$

Where A, B and C are constants whose values can be determined by equating coefficients, and where $(s - p_1)$ and $(s - p_2)$ are the factors of $s^2 + 2\zeta\omega_n s + \omega_n^2$. Thus

$$p_1, p_2 = \omega_n[-\zeta \pm \sqrt{\zeta^2 - 1}]$$

Equation 23 can now be transformed, using transforms 1 and 2, giving the solution

$$\theta_o(t) = A + B\varepsilon^{p_1 t} + C\varepsilon^{p_2 t} \qquad (24)$$

If on the other hand the quadratic term has no real factors, equation 22 can be expressed as

$$\theta_o(s) = \frac{A}{s} + \frac{B(s + \zeta\omega_n)}{(s + \zeta\omega_n)^2 + (1 - \zeta^2)\omega_n^2} + \frac{C}{(s + \zeta\omega_n)^2 + (1 - \zeta^2)\omega_n^2}$$

where $(s + \zeta\omega_n)^2 + (1 - \zeta^2)\omega_n^2$ is a rearrangement of the quadratic factor $s^2 + 2\zeta\omega_n s + \omega_n^2$ to make the form of the equation fit the list of transforms. Transforming now, using transforms 1, 6(a) and 6(b), we get

$$\theta_o(t) = A + B\varepsilon^{-\zeta\omega_n t} \cos \alpha\omega_n t + C\varepsilon^{-\zeta\omega_n t} \sin \alpha\omega_n t \qquad (25)$$

where

$$\alpha = \sqrt{1 - \zeta^2}$$

The solutions obtained by the Laplace transform method, equations

24 and 25, may be compared with equations 12 and 14, which are the solutions obtained by the more classical methods. The convenience of the Laplace method is particularly evident when we consider that a full table of Laplace transforms would contain an entry enabling direct conversion from equation 22 to the solution (including the values of the constants A, B and C) without even the labour of working out partial fractions.

Action of a System Expressed in Operational Form

Perhaps the most important feature of the Laplace transform is its usefulness in representing the action of a system in a form that is at once general and readily usable.

If we consider a system subjected to an input disturbing function $\phi_i(t)$ and producing an output function $\phi_o(t)$, then in general there will be a differential equation relating the output to the input, of the form

$$a_n \frac{d^n \phi_o(t)}{dt^n} + a_{n-1} \frac{d^{n-1} \phi_o(t)}{dt^{n-1}} + \cdots + a_1 \frac{d\phi_o(t)}{dt} + a_0 \phi_o(t)$$

$$= b_m \frac{d^m \phi_i(t)}{dt^m} + b_{m-1} \frac{d^{m-1} \phi_i(t)}{dt^{m-1}} + \cdots + b_1 \frac{d\phi_i(t)}{dt} + b_0 \phi_i(t)$$

If we take Laplace transforms of both sides, we get

$$[a_n s^n + a_{n-1} s^{n-1} + \cdots + a_1 s + a_0] \phi_o(s)$$

$$= [b_m s^m + b_{m-1} s^{m-1} + \cdots + b_1 s + b_0] \phi_i(s)$$

assuming that all the initial values are zero.

$$\therefore \ \phi_o(s) = \frac{b_m s^m + b_{m-1} s^{m-1} + \cdots + b_1 s + b_0}{a_n s^n + a_{n-1} s^{n-1} + \cdots + a_1 s + a_0} \cdot \phi_i(s) = \frac{B(s)}{A(s)} \cdot \phi_i(s)$$

$$(26)$$

where $A(s)$ and $B(s)$ are polynomial functions of s.

It is important to notice that in obtaining equation 26 we have made no assumption about the form of the input disturbance, $\phi_i(t)$. If we rewrite equation 26 in the form

$$\phi_o(s) = G(s) \cdot \phi_i(s)$$

where
$$G(s) = \frac{B(s)}{A(s)}$$

then we see that $G(s)$ represents the way in which the system operates on the input of the system, $\phi_i(s)$, in order to produce the output, $\phi_o(s)$. This quantity is called the *transfer function* of the system, and is in general a ratio of two polynomials in s, although in particular cases it can assume simpler forms. An integrating device, for example, would have a transfer function $G(s) = 1/s$, while an ideal amplifier (or a practical amplifier within specified limits of operation) would have a transfer function $G(s) = K$, where K is simply a constant. At the other extreme a complete description of even quite a simple system often involves third or fourth order polynomials, while an elaborate system might not be capable of description at all except with the aid of a computer.

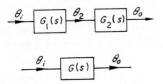

Figure 13. Combination of transfer functions for two interconnected systems: $G(s) = G_1(s) \cdot G_2(s)$.

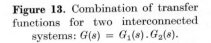

If we consider now two systems connected in series as shown in Figure 13, the transfer functions of the systems being $G_1(s)$ and $G_2(s)$, then with the notation of Figure 13 we have

$$\frac{\theta_2(s)}{\theta_i(s)} = G_1(s)$$

and

$$\frac{\theta_o(s)}{\theta_2(s)} = G_2(s)$$

Multiplying these equations together (an operation that can be shown to be valid) gives

$$\frac{\theta_o(s)}{\theta_i(s)} = G(s)$$

where

$$G(s) = G_1(s) \cdot G_2(s)$$

$G(s)$ is the overall transfer function of the system, and the fact that it is made up simply of the product of the transfer functions of the components is a further reason for the usefulness of the technique. Consider, for example, the system shown in Figure 14, which represents a position control system. The input v_i is fed to an amplifier to produce a motor field current I_f. This in turn gives rise to a motor torque T, which causes the load to take up a position θ. The differential equations relating input and output in each block are known, and in the classical method of solution these are written down as a set of simultaneous equations, and the unwanted variables I_f and T are eliminated, leaving a single equation relating θ to v_i. This is the equation of the system, and its solution for a particular function v_i is the response. With the Laplace transform technique, however, the transfer functions $G_1(s)$, $G_2(s)$ and $G_3(s)$ are written down for the original equations, and multiplied together to give the system transfer

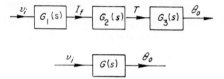

Figure 14. Block diagram of a position control system, illustrating formation of complete system equation from the transfer functions of the component parts.

function $G(s)$. The response to a given disturbance can then be extracted by reference to a table of transforms: this response has been obtained without ever having to formulate the equation explicitly at all. In a complicated system this can represent a very considerable saving of mathematical labour.

General Consideration of Stability

If a system has a transfer function given by

$$\frac{\theta_o(s)}{\theta_i(s)} = G(s) = \frac{A(s)}{B(s)}$$

where $\qquad B(s) = b_n s^n + b_{n-1} s^{n-1} + \cdots + b_1 s + b_0,$

then the step response of the system will be obtained by taking the inverse transform of the expression

$$\theta_o(s) = \frac{\Theta \cdot A(s)}{s \cdot B(s)} \tag{27}$$

where Θ is the magnitude of the input step.

To obtain this response as a time function, the expression must be separated into partial fractions, having first factorized the denominator. Any polynomial with real coefficients can be factorized into linear and quadratic terms with real coefficients, so we may write

$B(s) = (s + p_1)(s + p_2)\cdots$

$$(s^2 + 2\zeta_1 \omega_1 s + \omega_1{}^2)(s^2 + 2\zeta_2 \omega_2 s + \omega_2{}^2)\cdots \tag{28}$$

where $p_1, p_2, \cdots, \zeta_1, \zeta_2, \cdots, \omega_1, \omega_2, \cdots$ are all real. Thus the partial fraction expansion of equation 27 will take the form

$$\theta_o(s) = \frac{a\,\Theta}{s} + \frac{b}{s + p_1} + \frac{c}{s + p_2} + \cdots$$
$$+ \frac{ks + l}{s^2 + 2\zeta_1\omega_1 s + \omega_1^2} + \frac{ms + n}{s^2 + 2\zeta_2\omega_2 s + \omega_2^2} + \cdots \tag{29}$$

where $a, b, c, \cdots, k, l, m, n, \cdots$ are constants.

Each of the terms of equation 29 can now be transformed directly back to the corresponding time functions, and it is useful to consider the effects arising from each type of term.

The inverse transform of the first term, $a\,\Theta/s$, is $a\,\Theta$, which is not time dependent. This represents the *steady state response*, provided the system is stable, and is the desired output of the system. All the remaining terms are time dependent: this part of the solution, which for a stable system eventually becomes zero, is called the *transient response*.

The transient response is made up of terms of two types, each of which has already been discussed in some detail. Linear terms correspond to exponential time functions, and quadratic terms correspond to time functions that may be exponential or oscillatory, and if oscillatory may have an amplitude that decays or grows or remains constant with time. In order that the system should be stable, we require that all time dependent terms should fall to zero with increasing time, and this means that all exponential terms of the form ε^{pt} must have the exponent, p, negative; and all oscillatory terms must be decaying oscillations and not growing ones. The limiting case, where the oscillation remains constant (known as *limiting stability*), while not actually unstable in that it remains within bounds, is obviously useless as a control system and so must also be avoided.

If we compare the terms in equation 29 with equations 4, 12 and 14, which give the solutions for first and second order systems, we see that the conditions necessary to ensure stability are that $p_1, p_2, \cdots$ should all be negative and that $\zeta_1, \zeta_2, \cdots$ should all be positive. This condition may be expressed more concisely and usefully if we notice that the quadratic terms can be factorized as

$$s^2 + 2\zeta\omega s + \omega^2 = (s - p_1)(s - p_2)$$

where p_1 and p_2 are the roots of the equation

$$s^2 + 2\zeta\omega s + \omega^2 = 0 \tag{30}$$

If p_1 and p_2 are real, the quadratic becomes a pair of linear terms, and the stability requirement is that both p_1 and p_2 should be negative. If they are not real, they must be a pair of conjugate complex quantities, leading to the solution given by equation 14. It will be observed from this equation that $\varepsilon^{-\zeta\omega t}$, which is the term responsible for the amplitude of the oscillation, is derived from the real part of the solution of equation 30, and that if the oscillation is to decay, then this real part must be negative. Thus the criterion for stability of a system with transfer function $G(s) = A(s)/B(s)$ is that every real root of the equation $B(s) = 0$ should be negative, and that every complex root should have a negative real part. There are several techniques available for testing polynomials to see if they satisfy this criterion.

Measurements on Practical Systems

The theoretical treatment that has been presented so far is valuable, but in some practical situations it cannot be applied. Any component, for example, whose transfer function is not known will block the calculation. In such a case we need to be able to take measurements on the component and to deduce from these measurements sufficient information to enable response and stability assessments to be made. A technique commonly employed for this purpose is to measure the frequency response of the component.

The frequency response technique is based on the fact that if a sinusoidal signal is applied to any linear component, then the output after the transients have died away will also be sinusoidal, differing from the input only in magnitude and phase. Without attempting a rigorous proof of this statement, we can see the truth of it in general terms by remembering that

$$\frac{d}{dt}(\sin \omega t) = \omega \cos \omega t = \omega \sin \left(\omega t + \frac{\pi}{2}\right)$$

Similarly, all differentiations of sinusoidal functions yield functions that differ from the original in size and phase but not in shape. Frequency response measurement consists of applying a sinusoidal input and measuring the change of amplitude (the gain) and the change of phase, repeating the measurements at a number of different frequencies. This enables graphs to be drawn showing the variations of gain and phase shift with frequency. From these graphs the essential properties of the component under test can be inferred.

The obvious extension of this technique is to apply it to a complete system rather than to a single component, and so obtain the characteristics of the complete system in one operation. There is, however, one serious objection to this procedure: if the system is unstable, then it will go out of control as soon as it is switched on; at best no measurements are possible, and at worst the system might destroy itself. To see how these difficulties can be overcome, consider the generalized system shown in Figure 15. The frequency responses of

Figure 15. A generalized feedback system, illustrating the relationship between the various frequency responses that can be measured.

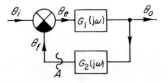

the two parts of the system are denoted by $G_1(j\omega)$ and $G_2(j\omega)$: they are written in this way because it can be shown that the frequency response can be derived from the transfer function by substituting $j\omega$ for s. There are now three distinct frequency response functions that can be defined.

$$\text{Direct F.R.} \qquad = \frac{\theta_o}{\theta_e} = G_1(j\omega)$$

$$\text{Open Loop F.R.} \qquad = \frac{\theta_f}{\theta_e} = \frac{\theta_f}{\theta_o} \times \frac{\theta_o}{\theta_e} = G_2(j\omega) \times G_1(j\omega)$$

$$\left.\begin{array}{l}\text{System F.R.} \\ \text{(Closed Loop F.R.)}\end{array}\right\} = \frac{\theta_o}{\theta_i}$$

$$\text{Since } \theta_e = \theta_i - \theta_f,$$

$$\therefore \frac{\theta_e}{\theta_o} = \frac{\theta_i}{\theta_o} - \frac{\theta_f}{\theta_o}$$

$$\therefore \frac{1}{G_1(j\omega)} = \frac{\theta_i}{\theta_o} - G_2(j\omega)$$

$$\therefore \frac{\theta_o}{\theta_i} = \text{System F.R.} = \frac{G_1(j\omega)}{1 + G_1(j\omega).G_2(j\omega)} \qquad (31)$$

$$= \frac{\text{Direct F.R.}}{1 + \text{Open Loop F.R.}}$$

If, as is often the case, the output is fed directly back to the input, then $G_2(j\omega) = 1$ and equation 31 becomes

$$\frac{\theta_o}{\theta_i} = \frac{G_1(j\omega)}{1 + G_1(j\omega)} \qquad (32)$$

Thus, the system frequency response can be deduced from open loop measurements; these measurements can be made without fear of instability, because a system without feedback cannot be unstable.

From equation 32 we can see that if at any frequency

$$1 + G_1(j\omega) = 0$$

then θ_o/θ_i will increase indefinitely, or to put it another way, there will be a finite output with no input. This is a condition of instability,

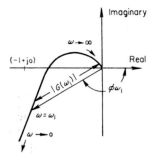

Figure 16. Typical Nyquist diagram: open circuit frequency response plotted on the complex plane.

which is included in the general statement known as *Nyquist's criterion* of stability. If we measure the open loop frequency response over a wide range of frequencies, the results may be plotted in the complex plane as shown in Figure 16, to give what is known as the *Nyquist diagram*. At any angular frequency ω_1 we draw a vector with length equal to the magnitude of the gain, $|G_1(j\omega_1)|$, and making an angle ϕ_{ω_1} with the real axis. The locus of the tip of the vector as

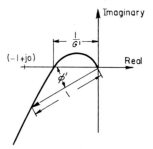

Figure 17. Gain and phase margins, G^1 and ϕ^1, obtained from the open loop frequency response.

the frequency is varied gives the open loop frequency response, $G_1(j\omega)$, and the Nyquist criterion is that for stability this curve must not enclose the point $(-1 + j0)$.

Equation 32 shows the mathematical significance of this point, and the physical interpretation can be seen in Figure 16. If at some frequency the curve passed through the point $(-1 + j0)$ then at this frequency the feedback signal would be equal to the input signal and exactly out of phase with it. This would mean that the feedback signal would be reinforcing the input signal instead of opposing it: this was the situation indicated in qualitative terms in Chapter 2 as the one necessary to produce instability.

The extent to which the open loop frequency response of a stable system fails to enclose the point $(-1 + j0)$ is a measure of the damping in the system, and two quantities that are sometimes quoted are those indicated in Figure 17. The *phase margin* is the angle by which the phase shift would have to be increased to reach 180° at the frequency at which the gain is unity. The *gain margin* is the factor by which the gain would have to be increased to reach unity at the frequency at which the phase shift is 180°.

Modification of Unsatisfactory Systems

If a system proves to be unsatisfactory when its open loop response is measured, it must be modified. One way of doing this is to insert, somewhere in the loop between the error signal and the feedback signal, an additional component (generally an electrical network) whose effect is to modify the open loop frequency response curve to some more suitable shape; that is, to produce satisfactory gain and phase margins. For this purpose, it turns out that the Nyquist diagram is not the most satisfactory form of presentation of the frequency response data. Instead we may use a pair of graphs, showing separately the variations of gain and of phase shift with frequency. It is convenient to use logarithmic scales for gain and for frequency; a pair of graphs plotted in this form is often referred to as a *Bode diagram*. The advantages of the logarithmic scales are many: they allow a range up to 10 c.p.s., which is the main frequency range of operation of a servo system, to occupy an appreciable fraction of a graph which may have to extend up to 100 or even 1000 c.p.s.; they convert the operation of combining the effects of two series components from multiplication, which is laborious, to addition, which is simple; and they change power law curves into straight lines. Thus, if we have a network whose frequency response is given by

$$\frac{\theta_o}{\theta_i} = \frac{1}{1 + j\omega T}$$

the gain of the network is

$$G = \left|\frac{\theta_o}{\theta_i}\right| = \frac{1}{\sqrt{1 + (\omega T)^2}} \tag{33}$$

For very low frequencies we may say that if $\omega \ll 1/T$ so that $\omega T \ll 1$ then we may neglect $(\omega T)^2$ compared with 1, and in this case

$$G \doteqdot 1 \quad \text{or} \quad \log G \doteqdot 0 \tag{34}$$

For very high frequencies on the other hand, if $\omega \gg 1/T$,

$$G \doteqdot \frac{1}{\sqrt{(\omega T)^2}} = \frac{1}{\omega T} \tag{35}$$

$$\therefore \ \log G = -\log(\omega T) \tag{36}$$

Thus equation 33 may be represented approximately by two straight line segments given by equations 34 and 36. This is shown in

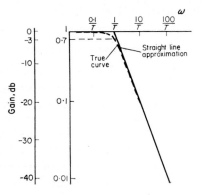

Figure 18. Gain curve for network with frequency response $(1 + j\omega T)^{-1}$, showing approximate curve consisting of two straight lines intersecting at the corner frequency, $\omega = 1/T$.

Figure 18 (note the logarithmic scales on the axes). At the 'corner frequency' when $\omega = 1/T$ then the true value of G from equation 33 is

$$G = \frac{1}{\sqrt{1 + 1}} \doteqdot 0 \cdot 7$$

Gain is commonly measured in electrical engineering in *decibels*, using the relationship

$$\text{Gain (in db)} = 20 \log_{10} G.$$

Thus the slope of the high frequency part of Figure 18 may be expressed as $20 \log_{10}(\tfrac{1}{2}) = -6$ db/octave, since, from equation 35, doubling the frequency (increasing by one octave) halves the gain. Also, the gain at the corner frequency may be expressed as

$$\text{Gain} = 20 \log_{10} \frac{1}{\sqrt{2}} = -20 \times \tfrac{1}{2} \log 2 = -3 \text{ db}$$

The dotted curve in Figure 18 shows the true shape of the curve given by equation 33, for comparison with the straight line approximation.

The phase change in this network may also be approximated by straight lines in the same way, giving the result shown in Figure 19;

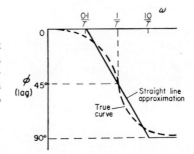

Figure 19. Phase curve for network with frequency response $(1 + j\omega T)^{-1}$, showing straight line approximations. The pair of curves shown in Figures 18 and 19 are known as a Bode diagram.

the dotted curve again shows the true function for comparison. The phase shift when $\omega = 1/T$ may be seen to be

$$\phi = \tan^{-1}\left(\frac{\omega T}{1}\right) = 45°$$

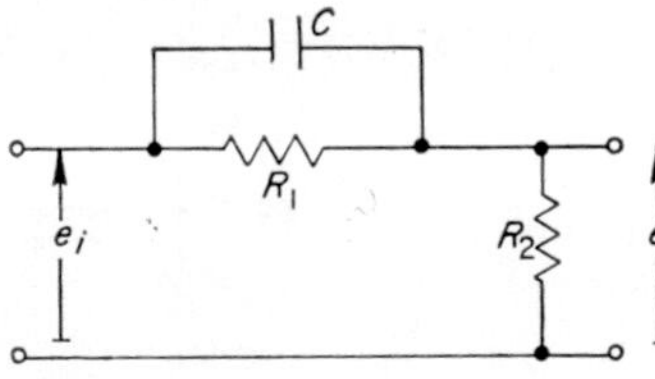

Figure 20. Phase advance network, having frequency response

$$\frac{e_o}{e_i} = \frac{\alpha(1 + j\omega T)}{1 + j\alpha\omega T}$$

where $T = R_1 C$ and

$$\alpha = \frac{R_2}{R_1 + R_2}.$$

One network that can be used to improve stability is shown in Figure 20. The frequency response function is given by

$$\frac{e_o}{e_i} = \frac{\alpha(1 + j\omega T)}{1 + j\alpha\omega T}$$

where $T = R_1 C$ and $\alpha = \frac{R_2}{R_1 + R_2}$

The straight line approximations for this network, which is known
as a phase advance network, are shown in Figure 21, taking the value
of α as 0·1 (a typical value). Now if we have a system whose open loop
frequency response is as shown by the solid lines in Figure 22, we may

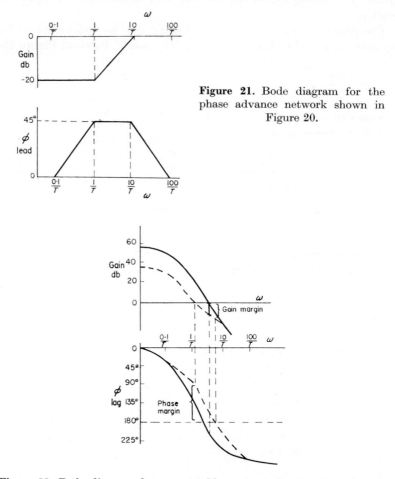

Figure 21. Bode diagram for the
phase advance network shown in
Figure 20.

Figure 22. Bode diagram for an unstable system, showing how the phase
advance network of Figure 20 can make the system stable.

notice that the gain is greater than 0 db at the frequency at which the
phase lag is 180°; this indicates that the system is unstable. Inserting
the network of Figure 20 in the open loop modifies the frequency
response as shown in the dotted curves in Figure 22: this modified
system is now stable, with gain and phase margins as indicated.

There are many other possible networks that can be employed to improve system performance: their effects can be estimated quickly and easily by these graphical techniques.

Non-Linearities in Practical Systems

So far we have discussed only linear systems, and although there is no room to discuss here non-linear systems in any detail, it is worthwhile to point out some commonly occurring forms of non-linearity in practical systems.

One inevitable form of non-linearity is saturation. An amplifier cannot give an output that is K times its input for all inputs: there

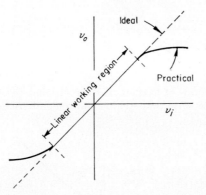

Figure 23. Saturation characteristic. Saturation is necessarily a feature of any physical system, but there may be a working region over which an approximately linear relationship exists.

must be a limitation to the power output available, and this will result in the characteristic shown in Figure 23. This kind of non-linearity must inevitably be present in any form of amplifying device, whether electronic or any other physical form, although there may nevertheless be a working region in which the operation is linear.

A more difficult form of non-linearity is Coulomb friction. This is a frictional force opposing motion, and with a constant magnitude. This frictional force should be distinguished from the force described by equation 7, which is viscous friction, which also opposes motion but in which the force is proportional to velocity. These two forms of friction are shown in Figure 24: viscous friction is a linear effect whereas coulomb friction is not. The effect of coulomb friction in a system is that an error cannot be corrected unless it is large enough

to cause enough torque to overcome the coulomb friction. This
results in a *dead space*: an area of indeterminacy within which the
system will settle in a chance position.

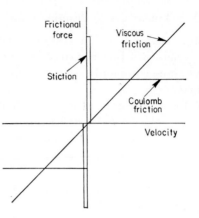

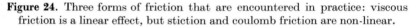

Figure 24. Three forms of friction that are encountered in practice: viscous
friction is a linear effect, but stiction and coulomb friction are non-linear.

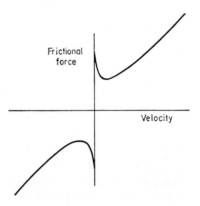

Figure 25. Complete friction characteristic, consisting of the sum of the three
components shown in Figure 24.

Yet another form of frictional force is stiction, which is effective
only when the system is at rest; that is to say, it opposes the initiation of motion, but has no effect on a system that is already moving:
this effect also is shown in Figure 24. If we add all three forms of
friction in Figure 24, we obtain the common form of friction curve
shown in Figure 25.

Finally we should mention two effects which though physically different are mathematically similar: hysteresis and backlash. *Hysteresis* was a term originally used to describe a property of magnetic materials, that the relationship between magnetic flux and magnetizing force is not constant, but depends on the magnitude and direction of previous applications of magnetizing force. A typical magnetic hysteresis characteristic is shown in Figure 26. A similar

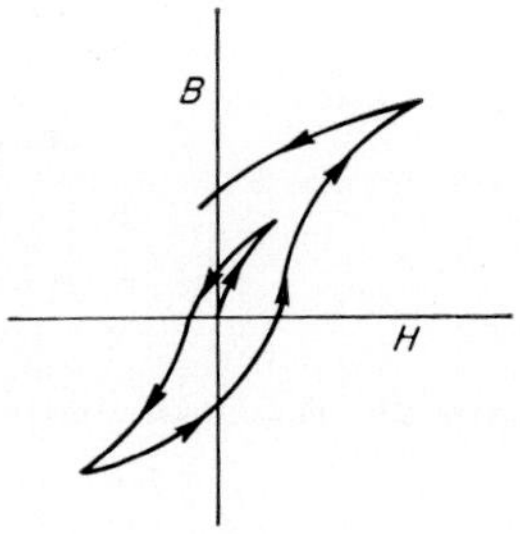

Figure 26. Hysteresis in magnetic materials. The relationship between B (magnetic flux density) and H (magnetic field strength) is not only non-linear but also depends on the past history of magnetization.

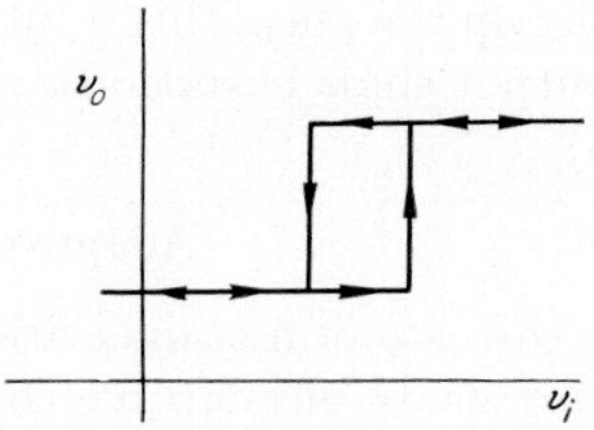

Figure 27. Characteristic of relay or electronic switching circuit. This effect is also sometimes described as hysteresis by analogy with the magnetic property illustrated in Figure 26.

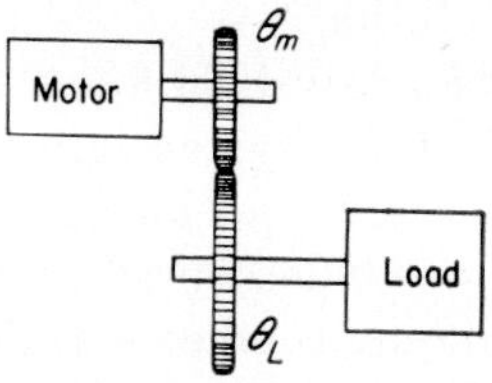

Figure 28. Gearing inserted between motor and load in a position control system, so that the position of the motor (θ_M) is no longer identical with the position of the load (θ_L).

form of curve to this is the one shown in Figure 27, which shows the relationship between input and output for a relay. A basic feature of a relay is that the voltage necessary to operate it is substantially greater than that needed to hold it closed. Electronic trigger circuits that perform the same function also have this feature, and because of the general similarity with Figure 26 this property is also often spoken of as hysteresis.

A similar effect is also produced by *backlash*, which is a property inherent in gear trains. If as in Figure 28 we have a motor shaft, whose position is θ_M, coupled through gearing to a load, whose position is θ_L, then the relationship between θ_M and θ_L which ideally should be a straight line through the origin, is actually as shown in Figure 29.

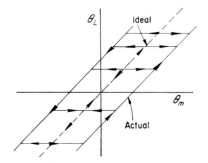

Figure 29. Backlash characteristic associated with the arrangement of Figure 28.

Whenever the motor reverses its direction of travel, the gears have to take up the play; this results in a transitional period during which the output shaft is stationary although the motor shaft is moving.

Adaptive Control Systems

As in the case of non-linearities, there is not room here to make any detailed study of adaptive control systems, but it may be useful to give a general statement of the problem as it exists and of approaches to its solution.

In the analysis of linear systems, we have assumed that the 'constants' of the system are truly constant under all conditions. This may not be true: an extreme example is the control system of a rocket or guided missile, which is required to operate in completely different environments at different stages of its flight. Even the auto-pilot of an aircraft, which may operate at all heights between 0 and 80,000 ft, experiences gross differences between operating conditions at different times in a flight. This means that a control system that is satisfactory under one set of conditions may be quite unsatisfactory under other conditions, even to the point of being unstable.

Another variation in operating conditions that can be significant is in the type of input signal experienced. A system that is adjusted to give an optimum response for a sinusoidal input may require quite different adjustments for optimum operation when it is subjected to step or to random inputs.

In all these cases there is clearly an advantage to be obtained if the adjustment of one or more critical parameters can be made automatically with reference to prevailing conditions instead of being preset on the work bench; this is the basis of self-optimizing, or adaptive, control systems.

A block diagram of one such system is shown in Figure 30. The lower part of the diagram shows a normal control system. The adaptive controller operates an adjustment, which might be, for example, a potentiometer that varies the damping ratio. While the system is operating, a perturbation generator makes continuous small changes

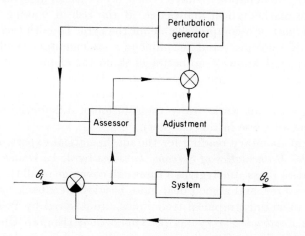

Figure 30. A typical form of adaptive control system. By making continual small changes in setting of some parameter and correlating these changes with the resulting changes in performance, an optimum setting of the parameter can be automatically arrived at.

in the setting of the adjustment, and these changes result in continuous small changes in the error of the system. The box marked 'assessor' has the function of correlating the error changes with the perturbation, and deciding in which direction to alter the adjustment so as to improve performance, on the basis of some predetermined criterion. A system of this kind is known as a hill-climber: it will find the peak of performance if there is one provided that certain conditions are met by the problem.

This is a control system whose operation can be likened to a process of learning in animals: the perturbations represent exploratory trials; the assessor represents a measurement of success or failure; the correcting device represents reinforcement of favourable behaviour

3+R.C.L.S.

and elimination of unfavourable behaviour. In a sense, therefore, we may regard this system as being a model of a learning process. This is one of the ways in which the apparently widely different fields of engineering and biology have come together during the last two decades; it is not perhaps too much to hope that as engineering learning systems are gradually improved, we may derive from them some ideas about the principles of organization of biological ones.

Conclusion

In writing this chapter, I have aimed my remarks at a reader with moderate mathematical knowledge, at the risk of making it trivial for the expert or incomprehensible for the tyro. To both these groups I can only apologize, and recommend supplementary reading. For the beginner, I know of no better book on the elements of Calculus than *Calculus Made Easy* by Sylvanus P. Thompson (Macmillan, 1927). This assumes only fourth-form knowledge of mathematics to begin with, but works up to the solution of differential equations (including all those in this chapter).

As supplementary reading for the mathematical expert, *Introduction to the Mathematics of Servomechanisms* by J. L. Douce (English Universities Press, 1963) gives a thorough coverage of all the material in this chapter, and a good deal more besides. Feedback systems in general, including graphical techniques, are covered by P. H. Hammond in *Feedback Theory and its Applications* (English Universities Press, 1958). Thereafter there is an abundance of books dealing with advanced aspects.

II
CELLS and INDIVIDUALS

4

Molecular Regulation in the Cell

D. LEWIS, F.R.S.

Department of Botany, University College, London

The Problem

Life hangs upon a thread. This is literally true for unicellular organisms including human egg and sperm cells. Body cells have a second chance as they contain two threads. The thread is of course the whole set of chromosomes which can usefully be pictured as a long molecule of deoxyribonucleic acid, DNA. The memory and permanent information of the cell, which has frequently to be called upon for synthesizing the right enzymes in the right amount, is carried in the sequence of the four different bases of which DNA is built up. Genetical studies have established that the life of the cell depends upon the intactness of this molecule, and if the cell is to reproduce, this molecule must be able to replicate exactly. Because there is only one, or at most two, copies of DNA in the cell the control of its replication is crucial for orderly growth and cell division.

Watson and Crick (1953), when they postulated the double helical structure of DNA and the pairing rules that the base adenine pairs with thymine and cytosine with guanine, also suggested a mechanism of replication of the molecule. It was suggested that at replication the two strands of the double helix open out and that free bases in the cell pair up with the two opened-out strands of the double helix. In this way two new molecules are formed, both of which have one strand of the original molecule and one newly synthesized strand. The two strands in DNA are different and complementary to one another so the newly synthesized strand is different in the two new molecules. This semi-conservative replication has since been proved by several entirely different types of experiment (Meselsen and Stahl, 1958; Taylor, 1957; Cairns, 1963).

In the test tube the chemistry of DNA replication has been elucidated mainly by Kornberg and his colleagues (Kornberg, 1960).
Kornberg showed that the synthesis of DNA *in vitro* required an
enzyme DNA polymerase which he obtained from the bacteria
Escherichia coli, the four deoxyribonucleotide triphosphates,
adenosine triphosphate, A.T.P., guanosine T.P., cytosine T.P. and
thymidine T.P., and a primer molecule of DNA to act as a template.
The primer DNA, before it was effective as a template, had to be
gently heated to open up the strands of the double helix.

In the living cell, Maaløe (1961) has shown that the synthesis of
protein is necessary for the initiation but not for the continuation of
DNA synthesis and it may be presumed from this that the DNA polymerase and other unknown enzymes must be synthesized afresh
before each new cycle of DNA replication. The regulation of DNA
synthesis could be by the regulation of the synthesis of these enzymes
by mechanisms which are described later, p. 64.

The synthesis of DNA in Kornberg's *in vitro* system and in the
living cell have many striking similarities. But there appears to be
one difference which, even if found later to be due to insensitive
methods of measurement of DNA synthesis *in vitro*, dramatically
emphasizes the crucial importance of an accurate system of regulation in the cell. *In vitro* DNA synthesis stops as soon as any one of the
four nucleotide triphosphates becomes exhausted, which indicates
that the absence of any one of the four building blocks brings the
process to a halt. *In vivo* a comparable experiment has been done with
Escherichia coli by using a mutant which cannot synthesize thymine,
an essential base of DNA. If such mutants are supplied with all the
raw materials for growth but deprived of thymine most of the cells
die. By carefully timed experiments it has been found that there is a
short period of about four minutes out of the twenty minutes of the
division cycle of *E. coli* when DNA synthesis is not progressing. Depriving the cell of thymine during this period of non-replication does
not kill the cell. During the replication period depriving a cell for as
little as 2 minutes kills the cell. This is similar to the exhaustion of
one of the four nucleotide triphosphates in Kornberg's system.

One explanation of 'thymineless death' is that DNA synthesis
comes to a halt as in the Kornberg system and other synthetic processes continue. Later when the thymine is supplied, the cell is so
unbalanced that it cannot recover. Alternatively once DNA synthesis
has started it cannot stop until it has completed the whole molecule.
In continuing its replication it either misses out the thymine bases,
or breaks the pairing rules and substitutes one of the other bases.

After a few incorrect or missing bases the information carried in the DNA would be so garbled as to kill the cell.

The brevity of the effective period of thymine starvation during DNA synthesis makes the second alternative more credible. Since the whole molecule of DNA in *E. coli* has 10^7 base pairs and replicates in 16 minutes from one end (Cairns, 1963), the number of base pairs synthesized in one minute would be 6×10^5. Of these approximately one quarter would require thymine so that there could be about $1 \cdot 5 \times 10^5$ mistakes in this time which gives ample opportunity for a lethal mistake to occur. On the alternative hypothesis suspension of DNA synthesis for one minute while other synthetic processes continue would seem unlikely to cause an irreparable change in the cell. This view, that once DNA synthesis has started in *E. coli* then it must go on uninterrupted until completion, has been advocated by Maaløe (1961) and if it is true then the regulation of DNA synthesis must be exclusively in the initiation of synthesis. The picture in higher organisms is complicated by the fact that DNA synthesis starts at several different points in the thread and the start of synthesis at the different points is not completely synchronized (Taylor, 1957; Lima-de-Faria, 1961).

The function of DNA is twofold: it must be a permanent record handed on to each new cell, and it must when required pass on a message to the cell. This function of producing a message is also regulated. In higher organisms certain basic types of proteins, histones, form complexes with DNA at certain stages of the division cycle. Bonner (1964) has shown that DNA complexed to its specific histone is unable to make a message. Its activity therefore can be regulated and probably by the presence or absence of specific histones.

Most or all metabolic and synthetic processes in the cell are mediated by enzymes, which are also macromolecules. The regulation of enzyme synthesis and activity is not so immediately vital as the regulation of DNA because instead of one copy of DNA there may be as many as thousands of copies of an enzyme. Nevertheless, as Davis (1961) has so ably discussed, the regulation of enzymes is extremely important in the economy of the cell and, in the long run, can be equally vital. Furthermore their regulation might eventually be found to be of prime importance in the differentiation of cells.

The problem of regulation in the cell with our present state of knowledge therefore revolves around those aspects of regulation which are so dramatically apparent in DNA. They can be stated as the regulation of the synthesis and activity of macromolecules.

Regulation of Enzymes

All the important chemical reactions in the cell, synthesis, catabolism, polymerization and depolymerization are catalysed by specific enzymes, and it is not surprising therefore that our present knowledge of regulation in the cell is almost entirely derived from experiments on regulation of the activity and the synthesis of enzymes. The experiments so far have been confined to biosynthetic and catabolic enzymes, but the broad principles of regulation which are emerging will probably eventually include polymerizing, depolymerizing and activating enzymes. Because enzyme reactions are time dependent, enzyme regulation may also be the basis of biological clocks.

The presently accepted model for enzyme regulation has been derived from the work of Jacob and Monod (1961) on the genes controlling the synthesis of β-galactosidase in *Escherichia coli*. β-Galactosidase is an inducible enzyme which is only synthesized in large amounts after the cell has been induced by the presence and absorption of the substrate or of a similar molecule that acts like the substrate. It is a catabolic enzyme which normally hydrolyzes the β-galactoside lactose, into galactose and glucose. Two ancillary enzymes are involved in the utilization of β-galactosides, these are a permease which is necessary for the entry of the β-galactosides into the cell and β-galactosideacetylase the function of which in the system is not quite clear. The genes which encode the amino acid sequences of the three enzymes are closely linked in a group.

Several gene mutations convert this inducible system into a constitutive one, so that the enzymes are synthesized at full rate even when the inducer is absent. One set of these mutations is in a gene called a regulator gene, i^+. Cells which carry one or two copies of the mutated gene i^-, and one normal, i^+, gene are inducible. In genetic terms the constitutive i^- gene is recessive to the dominant, i^+. Other constitutive mutants are of another gene, the operator, O. These mutant alleles are dominant, or to be more precise, both the mutant and the normal wild type allele are themselves both expressed when present together in the same cell. Furthermore they show a *cis–trans* position effect, so that the operator gene regulates the synthesis of the enzymes encoded only by those genes on the same chromosome and contiguous to the operator. The general scheme is illustrated in Figure 1 and the *cis–trans* effect in Figure 2.

The whole cluster of genes for the three enzymes in the β-galactosidase system is switched on or off by the action of the single regulator and operator gene. Such a cluster of genes is called an operon.

From the genetical evidence Jacob and Monod postulate that the regulator gene produces a repressor substance, which has two active sites; one is specific for binding to the inducer and the other site specific for binding to the operator. The two sites are not entirely independent in their binding properties because the repressor cannot

Model I
Regulator gene
R
Operator gene
Structural genes
O
A
B
Genes
Repressor
Messengers
Proteins
Repression or induction
Metabolite

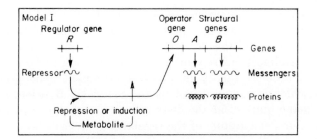

Figure 1. The regulator-operon model of regulation based upon the hypothesis that the repressor acts on the gene and prevents the transcription from gene to messenger. (From Jacob and Monod, 1961.)

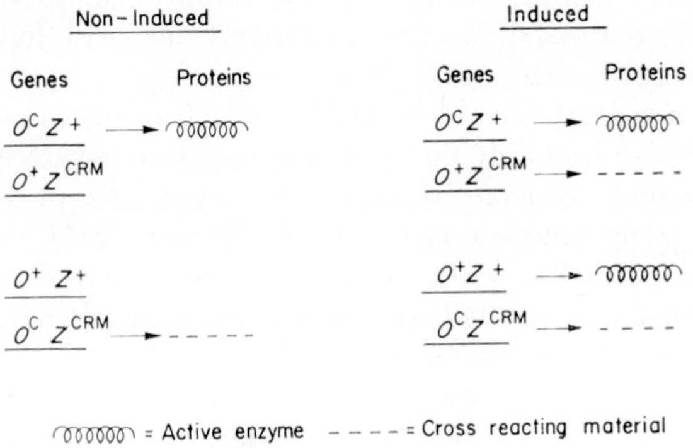

Non-Induced
Induced
Genes
Proteins
Genes
Proteins
O^C Z+
O^+ Z^CRM
O^+ Z^+
O^C Z^CRM
O^C Z+
O^+ Z^CRM
O^+ Z +
O^C Z^CRM
= Active enzyme
= Cross reacting material

Figure 2. The action of the operator gene O on the β-galactosidase structural gene Z shows a *cis–trans* position effect. The O gene controls only the Z gene which is contiguous to it on the same chromosome. O^C = constitutive O^+ = inducible.

be bound to inducer and operator simultaneously. The repressor–inducer complex does not bind to the operator, thus leaving the operon free to be transcribed, the free repressor binds to the operator preventing transcription. Although the repressor has not been isolated or its precise mode of action known, the general repressor hypothesis has been applied to several different systems with great success, and it will have a profound effect on our thoughts about

regulation for a long time. At present the hypothesis of the regulator and operon system rests on genetical evidence alone. This is the mutation of the regulator and operator genes, the dominance relationship of the mutants, the *cis–trans* position effect and deletion mutations which show that deletion of the regulator gene causes the system to be constitutive.

If this hypothesis is proved to be correct then it will be a remarkable example of brilliant interpretation based upon an old and largely abandoned genetical theory. For the postulate of the repressor substance rests almost entirely on a modified version of Bateson's presence and absence theory in which a dominant gene produces a product which is not produced by the recessive. Bateson, however, did not make quite clear the distinction between the gene and gene product. Further mutants of the regulator gene have been described which support the repressor hypothesis by Willson and colleagues (1964). A dominant mutation of the regulator gene causes loss of ability to be induced by lactose. This is interpreted as a mutation affecting the repressor so that it cannot bind with β-galactosides. The important theories arising from interpretation of dominance show that an up to date modification of a general theory of dominance can have important heuristic and predictive value.

There is a second type of regulation which operates on β-galactosidase, and also probably on most enzymes that produce glucose as an end-product. Glucose represses the synthesis of the enzyme whether β-galactoside is present or not (Monod, 1947). A more complex source of carbon such as lactose is not utilized until all the simpler forms are metabolized. Only then, and when the complex carbon source is present, is the enzyme synthesized to degrade it. The glucose repression is not altered by constitutive mutations of the regulator and operator genes of the induction system (see Mandelstam, 1962; Loomis and Magasanik, 1964). It is probable that there is a separate genetic system for glucose repression, and that the repression and induction system are the same in principle as a comparison with the regulation of biosynthetic systems will show. A possible example of one and the same operator and regulator gene for repression and induction is to be found in the D-serine deaminase system in *Escherichia coli*. One step mutants show a pleotrophic effect on induction and repression.

There are other examples of enzymes which are induced by their substrate and repressed by their end-product. Nitrate reductase is a good example, it is induced by nitrate and repressed by ammonia (Morris and Syrett, 1963).

The two systems of regulation of catabolic enzymes by induction and repression provide together a fine and economical regulator of cellular metabolism.

Biosynthetic Systems

Microorganisms must synthesize some twenty amino acids, about six purines and pyrimidines and perhaps twenty vitamins and co-factors. The number of enzymes specifically required for the synthesis of a single amino acid such as histidine is at least nine. If we assume conservatively that on the average six enzymes are required for each synthetic product and if we add the twenty amino acid activating enzymes required for protein synthesis the total is three hundred. If other enzymes are included such as polymerases the number may be nearer five hundred. The synthesis of these enzymes must be a great drain on the resources of the cell, and an economical control of their synthesis and use would appear to be an absolute necessity for survival. The synthesis of enzymes is such a drain that mammals, for whom a supply of some synthetic metabolites is always available in the food, have lost the capacity to synthesize the metabolite. Davis (1961) has pointed out that the amino acids which are not synthesized in mammals are only those with long pathways demanding the most enzymes. Apart from these reasons of economy of enzymes in the cell there is an equally important problem of the regulated production of the end product. Not only could over production be wasteful but it could be disastrous if it led to a serious imbalance in the cell. Two systems of regulation working together meet these exacting requirements of enzyme and end-product control. But it must be remembered that the regulation system also makes demands on the resources of the cell and if it is too elaborate and costly will defeat its own purpose. A comparison of the two main regulation systems, enzyme repression and end-product (feedback) inhibition will show how the regulation system has been fashioned with strictest economy.

Repression of enzyme synthesis

Unlike the catabolic enzymes, where all the known facts of an inducible catabolic enzyme can be illustrated in one system, β-galactosidase, it is not yet possible to illustrate all the features of biosynthetic enzyme regulation in any one system. This is in part due, as we shall see later, to differences between biosynthetic pathways and the genetic system controlling them, but it is mainly due to the incompleteness of the experimental data.

The operon concept is best illustrated in the synthetic pathway ending in histidine. Histidine is synthesized from phosphoribosyladenosine triphosphate in ten steps with an enzyme mediating each step (cf. Ames, Garry and Herzenberg, 1960). In *Salmonella typhimurium* there are eight genes which code these enzymes and all eight genes are tightly linked in a cluster (Hartman and coworkers, 1960a, b).

The level of the histidine synthesizing enzymes in *Salmonella* is low in wild-type cells growing on minimal medium or on medium enriched with histidine. A mutant with a block in the histidine synthesis system, however, when grown on suboptimal amounts of histidine has a much higher level of the histidine enzymes. This means that the endogenous level of histidine in wild-type cells is enough to give maximum repression of enzyme synthesis and only when the synthesis of histidine is reduced will full enzyme synthesis proceed. Only the end product and not the intermediates causes this repression. Furthermore all the five enzymes tested in detail are repressed by histidine and to similar and varied degrees according to the concentration of histidine supplied. The maximum release from repression is seventy-five times fully repressed level. In Ames's terminology the enzymes are coordinately repressed (Ames and Garry, 1959).

A deletion mutation at the extreme end of the gene at one end of the cluster prevents all the other seven undamaged genes from functioning so that none of the enzymes are synthesized (Ames and colleagues, 1963). This end gene is the operator gene and it is presumed that its partial deletion has rendered it incapable of initiating the transcription of the operon. Secondary mutations were found which restored the function of the seven undamaged genes so that the seven enzymes were synthesized and they were not subject to histidine repression. On analysis it was found that one of these secondary mutants was an extended deletion which has removed not only the whole of the operator but also genetic material outside the operon. The other was a duplication of the remainder of the operon, and this duplication was carried on an extra chromosomal fragment. In both secondary mutations the reasonable interpretation is that the seven genes of the operon are able to function and are not subject to histidine repression because they have been removed from the influence of the operator. The whole cluster can be interpreted as an operon as in the β-galactosidase system so that the whole of the ten enzymes are controlled by the state of the operator which is switched on or off by the action of its end-product, histidine.

If the operon is a unit of transcription of the genetic material then the state of the repressing system would have the same effect on all the genes in the operon, hence the coordinate repression, which has been found. The exact nature and position of the operator is not known (cf. Beckwith, 1964). It may be part of the nucleotide sequence of the first structural gene or it may be a gene in its own right. Jacob and coworkers (1964) have found that the operator is not part of the nucleotide sequence of the first structural gene. A regulator gene similar to the one found in the β-galactosidase system has not yet been identified in the histidine system.

The regulator–operon concept has to be modified if it is to be of general application because tight clusters of genes for a synthetic pathway such as the histidine system are the exception in bacteria and are not known in fungi, plants and animals. For most synthetic systems the genes are scattered throughout the chromosomes, with perhaps two genes closely linked (Pontecorvo, 1959).

The arginine synthetic pathway in *Escherichia coli*, studied by Vogel (1961), Maas (1961), Maas and coworkers (1964a, b), is a diffuse system with the genes scattered throughout the chromosome. As in the histidine system all the seven enzymes which catalyse the synthesis from glutamate to arginine are repressed only by the end-product arginine. But unlike the histidine series the enzymes are not coordinately repressed. Low levels of arginine produce a seven-hundred fold increase in one enzyme (5) and only a thirty fold increase in another of the enzymes (Maas, 1961); although Gorini and coworkers (1961) think that if enzyme 5 is excluded it may be found that under better experimental conditions the other enzymes in the system will show coordinate repression.

A regulator gene, similar in action to the regulator gene in the β-galactosidase system has been found and the allele conferring repressibility is dominant over the allele for non-repressibility (Maas and colleagues, 1964). It is argued as in the β-galactosidase system that the regulator gene because of its dominance and lack of position effect, produces a cytoplasmic repressor substance. In the arginine system it is assumed that the repressor substance is only active as a repressor when it is coupled to arginine, whereas in the β-galactosidase system the repressor is only active when it is not coupled to β-galactoside.

An unusual mutant in *E. coli* has been found not to be repressed by its own synthesized arginine but to be repressed by arginine supplied in the medium. Furthermore only the synthesized arginine was incorporated into protein (Sercarz and Gorini, 1964). This indicates that there is some separation of arginine in the cell so that there are two pools of arginine one for repression and one for protein synthesis. The normal cell must have a controlled interconnection between the pools so that endogenously formed arginine can overflow into the repressor pool. The mutant may have a block in the interconnection. A similar two pool system has been postulated for carbamyl phosphate mutants in *Neurospora crassa* (Charles, 1964).

Nothing is known about the operator in the arginine system, but if each structural gene had the same operator then the same repressor could act on all the genes, as in fact it does. In this case if the repressor acts at the gene level we would expect coordinate repression because the repressor would have the same affinity for all the operators.

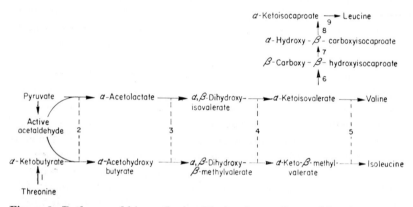

Figure 3. Pathway of biosynthesis of isoleucine, valine and leucine showing two parallel pathways mediated by the same enzymes, 2, 3, 4 and 5, and a branch pathway mediated by enzymes 6, 7, 8 and 9. (From Freundlich and coworkers, 1962.)

If complete lack of coordinate repression is proved for the arginine system it will throw doubt on the explanation based on the same operator for all the genes.

The economy of the control system itself is shown in all these systems, a common operator for the histidine system of 9 enzymes and a common regulator for the arginine system of 7 enzymes. The extreme importance of regulation of enzymes to the cell can be seen from branched and common synthetic pathways. The synthetic pathways of the amino acids isoleucine, valine and leucine are all connected and have both a common pathway for valine and isoleucine and a branched pathway for leucine (see Figure 3).

For repression of enzymes 2, 3 and 4 which are shared by the common pathway to all three amino acids the presence of all three acids is necessary. Clearly if one of the amino acids repressed these enzymes there would be no synthesis of the other two amino acids which would then be in short supply. The enzymes in the branch pathway 6, 7, 8 and 9 to leucine are repressed by leucine alone. For maximum repression of the common enzymes threonine is necessary as well as leucine, isoleucine and valine (Armstrong and Wagner, 1963). This system of regulation by three end-products of all the genes in a complex synthetic system is but another example of fineness and economy of the control system.

End-product inhibition of enzyme activity

The finest control of synthesis of the end-product is by the end-product inhibiting the action, not the synthesis of one of the enzymes in the pathway (Umbarger, 1961). The main facts are:

1. Only the end-product, or a close structural analogue of the end-product, and not an intermediate inhibits.
2. The inhibition is on the first enzyme of the pathway or if there is a branch in the pathway on the first enzyme after the branch.

Reference to Figure 3 illustrates these points; isoleucine inhibits enzyme 1, valine inhibits enzyme 2 and leucine inhibits enzyme 6. None of the intermediates have inhibitory effects.

The system as a regulator of synthesis of the end-product and intermediates is a much finer control than repression which is primarily to conserve the synthesis of unwanted enzymes, and the sensitivity of the two systems is different. It has been shown that the enzymes that synthesize arginine are inhibited by arginine supplied at levels between 1 and 10 μg/ml but enzyme repression did not occur until the level of arginine had exceeded 10 μg/ml. End-product inhibition has a strict economy, which is a feature of many control systems, so that only one enzyme has to be inhibited and only the first in the system will stop the synthesis of all the intermediates and end-product (see p. 76). There is no evidence for any regulator or operator gene in end-product inhibition, and there is no reason to think that any such genes are necessary. End-product inhibition should be regarded more as an example of the well-known affinity of a small molecule for an enzyme which acts as an enzyme inhibitor. In all the known examples there is no evidence that a special gene product such as a repressor is required.

End-product inhibition illustrates the very important concept of allosteric proteins which will be discussed more fully later. Briefly, it is assumed that many enzymes have two or more active sites, one binding the substrate and the other binding the inhibitor, for instance the end-product. In this way, two entirely different small molecules can bind to the same molecule.

Molecular Regulation in Plants and Animals

Although no regulatory systems in higher organisms, fungi, plants or animals have been analysed fully both genetically and biochemically, there is much evidence in favour of systems similar to those found in bacteria. On the genetical side, the dissociator and activator genes described in maize plants by McClintock (1956) control the expression of other genes. Their action has certain parallels with the regulator and operator in bacteria. The dissociator gene has an effect only on the gene immediately contiguous with it and is therefore like the bacterial operator; the activator gene, on the other hand, may be located anywhere in the genome and still operate on its dissociator gene.

The maize system differs in that the dissociator gene has been found to be somewhat mobile, but this difference may only be fortuitous as its mobility was the very means of detecting it, and it may be the exception for dissociator genes to be mobile in maize. On the biochemical side in plants, Bonner and his colleagues (1964) have revealed a system of repression of protein synthesis in the pea seed. DNA extracted from cotyledons which synthesize globulin is competent to act as a template for messenger RNA which in turn is the template for globulin synthesis. DNA from pea tissue which does not synthesize globulin is not a competent template, but if the histone is removed from the DNA then its competence is restored. This is the first experimental evidence for an old theory that histones are the gene regulators at least in higher organisms.

There are innumerable examples of genes in plants and animals which limit the expression of other genes in certain tissues. The specific S protein which controls self- and cross-fertilization in plants is one which shows features of a regulation system. A recessive mutation closely linked to the structural gene for the S protein affects the S protein in the pollen but has no effect on the same protein in the style (Lewis, 1960). This, however, is not an example of protein repression but of inhibition of activity.

In animals the best example of regulatory systems is found in haemoglobin. Not only has the molecule been found to change its shape by the binding of oxygen which is an example of the allosteric effect of Monod and Jacob, to be explained later, but the three different forms of the haemoglobin molecule, adult, foetal and embryonic (cf. Huehns and Shooter, 1964) are all the result of different requirements for the binding and release of oxygen in the different conditions of life from embryo to the adult state. The high foetal haemoglobin trait (Neel, 1961) in which adults have a high proportion of foetal haemoglobin, and β-thalassaemia in which haemoglobin synthesis is at a low rate are both genetically determined. Both appear to be due to mutations affecting the regulatory system, high foetal haemoglobin is probably due to an impairment of the repressor system and thalassaemia to a block in the release from repression. The high foetal haemoglobin gene is linked to the structural genes for two of the polypeptides of haemoglobin and it appears to have a *cis–trans* position effect similar to that of an operator gene (Neel, 1961). Zuckerkandl (1964) has elaborated a detailed scheme for haemoglobin regulation by a system of two operons of the bacterial type, one regulating the α- and β-chain synthesis and the other regulating the γ-chain. Much evidence is still required to substantiate this scheme in detail.

Several reports that hormones may be effector substances which regulate enzyme synthesis and activity in animals have recently appeared. Tomkins and coworkers (1963) have demonstrated that steroid hormones influence the state of polymerization of glutamic dehydrogenase obtained from beef liver. The state of polymerization in turn alters the activity of the enzyme.

The release of genes from repression has been observed cytologically in the giant chromosomes in insects. Certain bands of the chromosomes puff up and synthesize RNA at a high rate. The pattern of band puffing is characteristic of different tissues and stages of development of the insect. More recently Becker and Clever have shown that the hormone ecdysone causes the puffing of certain bands. (For review, see Beermann and Clever, 1964.)

One of the main differences between bacteria and higher plants and animals is in the organization of the genetic material. In bacteria there is one circular chromosome of DNA which is uncomplexed with histone and is not surrounded by a nuclear membrane. In higher organisms there are many chromosomes of complex structure consisting of DNA, RNA and histones, and surrounded by a nuclear membrane. It is therefore highly probable that, although the regulatory systems in bacteria and higher organisms are similar, important differences will be found. Fungi have the complex nucleus of higher organisms and in these organisms systems of regulation have been found, for example tyrosinase in *Neurospora crassa* (Horowitz and colleagues 1961) and alkaline phosphatase in other fungi (cf. Pontecorvo, 1963). The fungal systems have not been fully analysed and when they are it may be found that they bridge the gap between bacteria and higher organisms.

The Nature and Action of the Repressor

No repressor has yet been isolated so that its possible nature can only be conjectured from indirect observations. Pardee and Prestige (1959) showed that cells of *Escherichia coli* can become repressed, that is they can synthesize the repressor substance, in the presence of 5-methyl tryptophan. This analogue of tryptophan inhibits protein synthesis. From this Jacob and Monod (1961) concluded that the repressor is not a protein and suggested that it may be a polyribonucleotide. However, the same authors have more recently (1963) favoured the view that the repressor is a protein, and this has received some support from experiments on the effect of specific inhibitors of protein and ribonucleic acid synthesis on the repression of the enzyme, alkaline phosphatase (Gallant and Stapleton, 1964). The inhibitors of protein synthesis, chloramphenicol and canavanine, prevent the formation of the repressor, but 5-fluorouracil, which inhibits RNA synthesis, allows the formation of the repressor. Although the evidence is conflicting Monod and coworkers (1963) favour a protein as the specific repressor, and have developed an extremely useful hypothesis of allosteric proteins. Such proteins, and it is assumed that most enzymes and globulins are of this type, can combine with not only the substrate but other small molecules which are entirely different from the substrate. These small molecules may be the effectors such as the end-product which combines with the repressor. The concept is particularly applicable to end-product inhibition in which the first enzyme in a synthetic pathway recognizes not only its own substrate, but the end-product, which has no similarity with the substrate. Kinetic studies and the differential effect of heat and mercurials on inhibition and on activity indicate that there are different sites on the protein for the two different molecules; the interaction of the two molecules on the protein is believed to be due to a change in shape of the protein by the binding of the end-product.

The site of action of the repressor is also not completely settled. The most favoured theory is that the repressor binds to the operator gene and thereby prevents the transcription of the genes in the operon. The operon is considered to be a unit of transcription. Again the evidence for the site of action is meagre and inconclusive. It has been claimed that non-induced cells of *Escherichia coli* lack a specific messenger RNA which binds to a specific DNA whereas lactose-induced cells contain the messenger. However, Beckwith (1964) has evidence that the operator mutations allow messenger RNA synthesis but block the translation of the messenger RNA to protein. There is little doubt that the most economical way for the cell to control enzyme synthesis is by repressing messenger RNA synthesis, otherwise the RNA would be wasted.

Although it is easy to envisage the concept of allosteric proteins which bind with two simple molecules such as an amino acid and an intermediate, it is more difficult to see how the protein can bind to the DNA of the operator. The bonding between nucleic acids appear to be by hydrogen bonds which obey the Watson–Crick pairing rules. The inescapable nature of the pairing rules for nucleic acids is seen in the translation process where twenty different amino acids have to line up in a specific order dictated by the messenger RNA. This is done by twenty specific transfer RNA molecules which attach to the amino acids. The actual pairing and alignment between the amino acids and the messenger is by hydrogen bonding of the transfer RNA to the messenger. The difficulty of the repressor binding to the operator would not arise if sites on the repressor protein became attached to similar transfer RNA molecules; the pattern of transfer RNA molecules would then bind specifically to the specific DNA of the operator. If we assume twenty different transfer RNAs, one for each amino acid, it would be necessary to have only about four transfer RNAs for each specific protein repressor. This would give enough specificity and variation to meet the requirements of the regulation in the cell. If the transfer RNAs had only one triplet code, four transfer RNAs would give only twelve paired bases with the DNA; this would provide an extremely small energy of binding, and undoubtedly more base pairs would be required to hold the protein firmly to the DNA. This could be met by more than three coding bases on each transfer RNA or more transfer RNA molecules per repressor molecule. These hypothetical transfer RNAs cannot be the same as the transfer RNAs for the free amino acids used in protein synthesis, because the transfer RNA is attached to the amino acid by the carboxyl group and it is this group which condenses with the amino group of the next amino acid in the chain. Before condensation can occur the last transfer RNA must detach, leaving only one transfer RNA on each growing polypeptide chain. But transfer RNAs which attach to the R group of the amino acids may fulfil the requirements.

The Economy of Regulation

One of the striking features of repression and induction of enzymes is that there is always some synthesis of enzyme even when the system is maximally repressed or non-induced. It may be that the gene must be derepressed to replicate and at this time a messenger RNA is synthesized which would give the small amount of enzyme that is always found. The main point is that no genes appear to be completely switched off but only reduced to a low rate of activity. A basic assumption of the regulator–operon system of control is that the regulator gene needs to produce only a very low number of repressor molecules. The number, theoretically, can be as low as would ensure at least one molecule per set of chromosomes to be distributed to each daughter cell at cell division. Fifty or a hundred would be enough to ensure this. If the repressor did not work at the gene transcription level, but at messenger RNA translation, then many more repressor molecules would be required. This is another reason for preferring the DNA transcription theory.

This small production of repressor substance, which would suffice for the system, could be met by the regulator gene working at a rate equivalent to the low rate obtained with maximum repression in enzyme producing genes. There is therefore no need for a regulator system of the repressor. This removes one of the main conceptual difficulties of a regulation system in which a regulator gene R regulates another gene A, for the question arises—what about gene R? It will also need regulating by a gene X. All that is required is that the regulator should always be maximally repressed, but how this is brought about is another problem about which we have no information. In general, the controlling system can be envisaged as a regulator gene producing about fifty protein molecules per cell, and these regulate, through the operator, the synthesis of some four to nine enzymes. These enzymes can be produced at rates which vary a thousand fold in β-galactosidase to only twenty-five fold in other enzymes.

Another aspect of the limited number of molecules of repressor is that different levels of repression can easily be obtained by the differences in concentrations of the effectors, i.e. end-product in a synthetic system or substrate in a catabolic system. Clearly the level of repression will bear a direct relationship to the percentage of repressor molecules in the cell which are bound to the effector. It is also not difficult to see how cellular differentiation may be effected by this system. If, for example, the synthesis of an effector is stopped in a multicellular organism, before few cell divisions have occurred no effector will be present and repression or induction will occur.

Another variable of great importance in regulation is the life of the messenger RNA. At one extreme there is the β-galactosidase messenger in *E. coli* which at most produces only about ten enzyme molecules per molecule of messenger, and at the other extreme is the messenger for haemoglobin which lasts for at least 24 hours after the reticulocyte has lost its nucleus. Different systems will require messengers with different lifetimes; if regulation of protein synthesis is always at the gene messenger level then the systems with a long-lived messenger will have a crude regulatory system which has a long lag between the switch off event and the cessation of protein synthesis. The cells producing haemoglobin are specialized for this purpose and there is no need to switch off haemoglobin production, hence a short lived messenger would be a wasteful system of haemoglobin production.

Finally there is the balance sheet of number of genes in an organism and the known and inferred requirements. If we assume an estimate of five-hundred enzymes in *Escherichia coli* and on the average one regulator for five enzymes we have five-hundred genes for the coding of the enzymes and one-hundred for the regulators. This is well within the capacity of the DNA present in the cell. The discrepancy here, particularly in higher organisms, is that the amount of DNA is many orders of magnitude greater than can be envisaged for structural and regulatory processes. It is probable that the regulation and organization in a higher organism is much more demanding than any we can infer from bacteria. Certainly the whole process of learning and memory with its multitude of variation and interconnections requires specific information with a high order of specificity which we now only associate with macromolecules of the nucleic acid and protein type. It is possible that much of the genetical material in higher animals is in reserve for learning and memory.

The next ten years will see the whole field of regulation differentiation advance so rapidly that many of the problems which loom large now will be solved.

References

Ames, B. N. and Garry B. (1959). 'Coordinate repression of the synthesis of four histidine biosynthetic enzymes by histidine.' *Proc. Nat. Acad. Sci. U.S.*, **45**, 1453.

Ames, B. N., Garry, B. and Herzenberg, L. A. (1960). 'The genetic control of the enzymes of histidine biosynthesis in *Salmonella typhimurium*.' *J. Gen. Microbiol.*, **22**, 369–378.

Ames, B. N., Hartman, P. E. and Jacob, B. F. (1963). 'Chromosomal alterations affecting the regulation of histidine biosynthesis in *Salmonella.' J. Mol. Biol.*, **7**, 23–42.

Armstrong, F. B. and Wagner, R. P. (1963). 'Repression of the valine–isoleucine pathway in *Salmonella.' Proc. Nat. Acad. Sci. U.S.*, **49**, 628–633.

Beckwith, J. R. (1964). 'A deletion analysis of the *Lac* operator region in *Escherichia coli.' J. Mol. Biol.*, **8**, 427.

Beermann, W. and Clever, U. (1964). 'Chromosome puffs.' *Scientific American*, **210**, No. 4, 50–58.

Bonner, J. (1965). *The Molecular Biology of Development*, Oxford University Press, London.

Cairns, J. (1963). 'The bacterial chromosome and its manner of replication as seen by autoradiography.' *J. Mol. Biol.*, **6**, 208–213.

Charles, H. P. (1964). 'Relationships between certain pyrimidines and arginine mutants of *Neurospora*, as revealed by their response to carbon dioxide.' *J. Gen. Microbiol.*, **34**, 131–142.

Davis, B. D. (1961). 'The teleonomic significance of biosynthetic control mechanisms.' *Cold Spring Harb. Symp. Quant. Biol.*, **26**, 1–10.

Freundlich, M., Burns, R. O. and Umbarger, H. E. (1962). 'Control of isoleucine, valine, and leucine biosynthesis, 1. Multivalent repression.' *Proc. Nat. Acad. Sci. U.S.*, **48**, 1804–1808.

Gallant, J. and Stapleton, R. (1964). 'Physiological evidence on the nature of the repressor of alkaline phosphatase synthesis in *Escherichia coli.' J. Mol. Biol.*, **8**, 431–441.

Gallant, J. and Stapleton, R. (1964). 'Derepression of alkaline phosphatase synthesis by chloramphenicol and canavanine inhibition.' *J. Mol. Biol.*, **8**, 442.

Gorini, L., Gunderson, W. and Burger, M. (1961). 'Genetics of regulation of enzyme synthesis in the arginine biosynthetic pathway of *Escherichia coli.' Cold Spring Harb. Symp. Quant. Biol.*, **26**, 173–182.

Hartman, P. E., Loper, J. C. and Sermon, D. (1960a). 'Fine structure mapping by complete transduction between histidine-requiring *Salmonella* mutants.' *J. Gen. Microbiol.*, **22**, 323–353.

Hartman, P. E., Loper, J. C. and Sermon, D. (1960b). 'Complementation mapping by abortive transduction of histidine-requiring *Salmonella* mutants.' *J. Gen. Microbiol.*, **22**, 354–368.

Horowitz, N. H., Fling, M., Macleod, H. and Watanabe, Y. (1961). 'Structural and regulative genes controlling tyrosinase synthesis in *Neurospora.' Cold Spring Harb. Symp. Quant. Biol.*, **26**, 173–182.

Huehns, E. R. and Shooter, E. M. (1964). 'Haemoglobin.' *Science Progress*, **207**, 353–374.

Jacob, F. and Monod, J. (1961). 'Genetic regulatory mechanism in the synthesis of proteins.' *J. Mol. Biol.*, **3**, 318–356.

Jacob, F., Ullman, J. and Monod, J. (1964), *Compt. Rend.*, **258**, 3125.

Kornberg, A. (1960). 'Biologic synthesis of deoxyribonucleic acid.' *Science*, **131**, 1503.

Lewis, D. (1960). 'Genetic control of specificity and activity of the *S'* antigen in plants.' *Proc. Roy. Soc. (London), Ser. B*, **151**, 468–477.

Lima-de-Faria, A. (1961). 'Initiation of DNA synthesis at specific segments in the meiotic chromosomes of *Melanoplus*.' *Heredity*, **47**, 674–694.

Loomis, W. F. J. and Magasanik, B. (1964). 'The relation of catabolic repression to the induction system for β-galactosidase in *Escherichia coli*.' *J. Mol. Biol.*, **8**, 417–426.

Maaløe, O. (1961). 'The control of normal DNA replication in bacteria.' *Cold Spring Harb. Symp. Quant. Biol.*, **26**, 45–52.

Maas, W. K. (1961). 'Studies on repression of arginine biosynthesis in *Escherichia coli*.' *Cold Spring Harb. Symp. Quant. Biol.*, **26**, 183–191.

Maas, W. K., Maas, R., Wiame, J. M. and Glansdorff, N. (1964a). 'Studies on the mechanism of repression of arginine biosynthesis in *Escherichia coli*. 1. Dominance of repressibility in zygotes.' *J. Mol. Biol.*, **8**, 359–364.

Maas, W. K. and Clark, A. J. (1964b). 'Studies on the mechanism of repression of arginine biosynthesis in *Escherichia coli*. II. Dominance of repression in diploids.' *J. Mol. Biol.*, **8**, 365–370.

Mandelstam, J. (1962). 'The repression of constitutive β-galactosidase in *Escherichia coli* by glucose and other carbon sources.' *Biochem. J.*, **82**, 489–493.

McClintock, B. (1956). 'Controlling elements and the gene.' *Cold Spring Harb. Symp. Quant. Biol.*, **21**, 197–216.

Meselsen, M. and Stahl, F. W. (1958). 'The replication of DNA in *Escherichia coli*.' *Proc. Nat. Acad. Sci. U.S.*, **44**, 671.

Monod, J. (1947). 'The phenomenon of enzymatic adaptation.' *Growth*, **11**, 223–289.

Monod, J., Changaux, J. P. and Jacob, F. (1963). 'Allosteric proteins and cellular control systems.' *J. Mol. Biol.*, **6**, 306–329.

Morris, I. and Syrett, P. J. (1963). 'The development of nitrate reductase in *Chlorella* and its repression by ammonia.' *Arch. Mikrobiol.*, **47**, 32–41.

Neel, J. V. (1961). 'D-haemoglobin genes: a remarkable example of the clustering of related genetic functions on a single mammalian chromosome.' *Blood*, **18**, 1769.

Pardee, A. B. and Prestige, L. S. (1959). 'On the nature of the repressor of β-galactosidase synthesis in *E. coli*.' *Biochim. Biophys. Acta*, **36**, 545–547.

Pontecorvo, G. (1959). *Trends in Genetic Analysis*, Columbia University Press, New York.

Pontecorvo, G. (1963). 'The Leeuwenhoek Lecture. Microbial genetics: retrospect and prospect.' *Proc. Roy. Soc. (London), Ser. B*, **158**, 1–23.

Sercarz, E. E. and Gorini, L. (1964). 'Different contribution of exogenous and endogenous arginine to repressor formation.' *J. Mol. Biol.*, **8**, 254–262.

Taylor, J. H. (1957). 'The time and mode of duplication of chromosomes.' *Am. Naturalist*, **91**, 210–221.

Tomkins, G. M., Yielding, K. L., Talal, N. and Curran, J. F. (1963). 'Protein structure and biological regulation.' *Cold Spring Harb. Symp. Quant. Biol.*, **28**, 461–471.

Umbarger, H. E. (1961). 'Feedback control by end-product inhibition.' *Cold Spring Harb. Symp. Quant. Biol.*, **26**, 301–311.

Vogel, H. J. (1961). 'Aspects of repression in the regulation of enzyme synthesis: pathway-wide control and enzyme-specific response.' *Cold Spring Harb. Symp. Quant. Biol.*, **26**, 163–171.

Watson, J. D. and Crick, F. H. C. (1953). 'Genetic duplications of the structure of deoxyribonucleic acid.' *Nature*, **171**, 964.

Willson, C., Perrin, D., Cohn, M., Jacob, F. and Monod, J. (1964). 'Non-inducible mutants of the regulator gene in the "Lactose" system of *Escherichia coli*.' *J. Mol. Biol.*, **8**, 582–592.

Zuckerkandl, E. (1964). 'Controller-gene diseases: The operon model as applied to β-thalassaemia etc.' *J. Mol. Biol.*, **8**, 128–147.

5

Physiological Control Systems

H. KALMUS

Department of Human Genetics and Biometry
University College, London

and

B. R. WILKINS

*Department of Anatomy, University College, London**

General

It would be presumptuous as well as futile to attempt to cover in detail in a single chapter all the control systems that are to be found within the province of Physiology. We can hope to achieve no more than a rather cursory examination of some of the systems that have been more extensively studied. In addition, by referring these systems to the general principles that were outlined in Chapters 2 and 3 we may be able to further an appreciation of some of the difficulties and pitfalls that lie in wait for the unwary investigator.

The idea of applying mathematical methods, and in particular the analytical techniques of servoengineering, to biological systems is generally credited to Wiener (1949), but it is clear that, as so often happens in the history of Science, the same idea was taking shape almost simultaneously in various other parts of the world. Inter-disciplinary cooperation was approached from both sides. In the first place engineers were faced with cases of semi-automation in which a human operator was included as part of the control system. Typical problems of this kind arose during the war, with the develop-ment of gun-laying systems in which a human operator had the task of tracking the target; that is to say, the task of maintaining the target in the centre of the gunsight. In an early plea for the use of

* Present address: Department of Electronics, University of Southampton.

servotheory in the description of human motor behaviour, Ellson (1949) pointed out that the transmission characteristic of the human operator when included in the control loop in a tracking task is of vital importance to the overall stability of the system. He cited a particular case of a gunsight computer which magnified the tracking errors of the operator. The need to regard the human operator himself as a control system had earlier been stressed by Craik (1948) who pointed out specific analogies between parts of the human organism and components of a control system: sensory devices, computing and amplifying systems and mechanical output devices.

An attempt to calculate the transfer function of a human operator in a tracking task by measuring his step response was made by Searle and Taylor (1948). They presented subjects with a target line drawn on a sheet of paper that was moved behind an opaque panel containing a thin viewing slit. The subjects had to follow this line with a pencil, and step function stimuli were presented when the position of the line was abruptly changed.

The major difficulty of applying control theory to this situation, a difficulty of which Searle and Taylor were well aware, is that biological systems are in general far from linear. One interesting non-linearity that emerged from this study was the range effect. When presented with a step command signal, the subject produced no response until a reaction time had passed. If this reaction time had been constant under all conditions, it could be mathematically described in the transfer function (see Chapter 3; a constant delay time τ is represented by the transfer function $\varepsilon^{-s\tau}$), but Searle and Taylor found that if different amplitude steps were presented in a random order, then there was a significantly shorter delay time for medium amplitude steps than for large or small steps. The accuracy of the response was also affected: Taylor (1949) reported that under the same conditions small steps were overcompensated while large steps were undercompensated. Further studies by Taylor and Birmingham (1948) used a different form of presentation: the target was a spot on a cathode ray tube, and the subject was required to maintain it in a central position by manipulating a control lever (one dimensional movement only). They measured the position, speed, acceleration and rate of change of acceleration of the control lever, and concluded that the time relations of the corrective motions were such that once started they were controlled without visual or kinaesthetic guidance. They suggested on the basis of this finding that the control process in the target tracking task is intermittent rather than continuous. In this, they agreed with the earlier suggestions of

Craik (1947), who also advanced the idea that the apparent continuity obtained under some conditions is due to central processes that extrapolate from past data. These anticipatory mechanisms (p. 157) were also noted by Mayne (1951), who pointed out that the reaction time observed in the response of a subject attempting to follow a step could be removed by presenting a repetitive square wave stimulus, the response then appearing like a straightforward servoresponse without delay. With a sine wave stimulus, the delay time is also absent, and this response too looks like a normal servoresponse. The sine wave response is also subject to a further modification, for if the sinusoidal stimulus persists for a short time then even the phase lag disappears from the response. Thus, as Greene (1954) observed, the transfer function of the human operator in this tracking situation is not only highly non-linear, but also changeable with time and experience.

All this might seem very discouraging for one who had hoped to be able to systematize the study of biological control systems by the application of analytical engineering techniques. However, there are two points to be considered before giving way to despair. In the first place, the tracking situation, although perhaps a simple task from our own subjective viewpoint, is far from being a simple mechanism in fact. Most biological systems turn out to be more or less complicated, but it is possible to select much more tractable ones than the hand–eye tracking system. Secondly, the benefits accruing from an engineering approach are not to be reckoned only in terms of the neat mathematical expressions we can derive to describe 'well-behaved' systems, but more significantly by the method of attack on the problem. This method encourages the formulation of theories that are sufficiently explicit to be rigorously tested. This point will appear several times during the course of the chapter.

The crossing of the interdisciplinary boundaries, as suggested above, was by no means a unilateral affair. Simultaneously, engineers were beginning to look at biology and biologists at engineering. Thus, Pringle and Wilson (1952) examined the operation of a tactile receptor in the cockroach leg by measuring its response to sinusoidal and step changes of stimulus. From their results they calculated the transfer function of the receptor, and were able to identify a phase advance component, which would have important implications for the stability of reflex activity. A similar method of study was also applied by Chapman and Smith (1963) to the same receptor, using more refined techniques and obtaining slightly different results but still using the transfer function as the means of describing the action of

the receptor. The transfer function technique has indeed been urged
by Pringle (1962) as the manner in which receptors generally can be
most usefully described.

A particularly fruitful line of enquiry also launched by biologists
at about the same time that Cybernetics was being christened, was
the re-afference principle due to von Holst and Mittelstaedt (1950),
recapitulated (in an English version) by von Holst (1954). They
found themselves in difficulty when attempting to explain a number
of apparently straightforward effects by means of the reflex theory,
which was at that time generally accepted. This was the theory
that behaviour could be explained in terms of reflex components
each of which was the predictable response to a particular stimulus.
They claimed that in many situations this theory was not merely
inadequate but in fact almost invariably gave the wrong answer.
An alternative is the theory of plasticity, which suggests that everything is connected to everything else in a network in the central
nervous system, and this they also found to be unsatisfactory except
in so far as it refuted the reflex theory. The first system that they
examined typifies the problem as they observed it. If a fly is placed
inside a striped cylinder and the cylinder is rotated, then the fly responds by turning in the same direction as the rotation. This is an
optomotor reaction (about which we shall have more to say later)
and was taken to be a classical example of a reflex response to visual
stimulation. If however the cylinder is held stationary, and the fly is
induced by non-visual means to turn, then it is observed that it has
no difficulty in doing so, although the changes of visual stimulation
produced by his movement are such that the optomotor response, if
it appeared, would be antagonistic to the original movement. One
possible explanation of the observed unhampered movement could
be that optomotor responses are inhibited during voluntary locomotion, and this was tested in a simple but ingenious experiment by
Mittelstaedt (1949) who rotated the head of a dung fly (*Eristalis
tenax*) through 180° about its long axis and secured it in position so
that the right and left eyes were effectively interchanged (see
Figure 1 (*a*), (*b*)). When this head-inverted fly was tested in the
rotating cylinder, its optomotor response was reversed as expected
—it now turned in the direction opposite to that of the rotation.
However, if the 'reflex-inhibition' theory was correct, then this head-
inverted fly should have performed normal locomotion in a static
cylinder because a change in sense of the optomotor response could
not have any effect if the response were inhibited. The experimental
finding, however, was that once the fly started to move in either

direction it speeded up and continued to move rapidly in small circles until exhausted. That this effect was indeed due to the opto-motor response was demonstrated by placing a head-inverted fly in an optically homogeneous cylinder instead of a striped one. Deprived of visual information, the fly's locomotor behaviour was then normal.

The analysis of this problem proposed by von Holst and Mittel-staedt was first to divide the total pattern of nervous activity into two parts: the activity produced by the receptors they named *afference* and the activity directed towards the effector organs they named *efference*. They then subdivided the afference into *ex-afference*, which

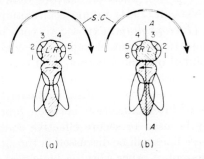

(a)   (b)

Figure 1. Method of refuting the 'reflex-inhibition' theory of optomotor re-sponses. By rotating the head of the fly through 180° about the axis *A–A*, visual information about the direction of movement of the striped cylinder (*SC*) is inverted. (From Mittelstaedt, 1949.)

is that part of the afference produced by changes in the external world, and *re-afference*, which is that part of the afference caused by the muscular activity of the organism itself. They then suggested an organization for optomotor control consisting of a command signal originating in the higher nervous centres, and producing the efference in the motor centre. This efference, besides activating the effector organs, also, it is postulated, leaves in the central nervous system an *efference copy*, with which the re-afference is compared. This compari-son then finally determines what further efference should be pro-duced. Thus the efference copy produced by a normal fly moving in a stationary cylinder is exactly matched by the re-afference due to the movement of visual excitation across the eyes. With the head inverted, however, the re-afference does not match the efference copy because of the change of sign, and so additional efference is generated in the original directional sense. This must obviously build up until it reaches the limit imposed by the mechanical capabilities of the animal's muscular system.

The same kind of organization was also applied by von Holst and Mittelstaedt to a number of other 'reflex' systems including the right-ing reflex, fixation behaviour, optokinetic nystagmus, orientation reflexes in fish and mammals, muscle control and human perceptual properties of size constancy and stability of the visual world. In these early papers, the re-afference principle was expressed only in general qualitative terms, but we can now recognize that re-afference is a form of feedback and that the operation of comparison between efference copy and re-afference in order to determine a course of action is basically very similar to the operation of an error actuated control system (see p. 30). These similarities were not lost upon the authors, and Mittelstaedt (1954), taking as an example the orientation system of a fish (*Pterophyllum*) exposed to both light and gravity, discussed the application of control engineering concepts to biological problems and pointed out the advantages of this approach. He has also applied the method himself to a number of different systems in insects, fish and humans, concentrating particularly on the different possible patterns of interconnection between functional units that could conceivably be used to secure effective control. Examples of his work along these lines will be discussed in the later sections of this chapter. A fair amount of other work on biological systems has been done using analytical techniques derived from mechanical systems, but much of it can receive no more than a brief mention here.

Stark and Herman (1961) have applied a sinusoidal light stimulus to the tail ganglion of a crayfish, and from the nerve impulse response have deduced the transfer function of the receptor. On the output side, a sinusoidal analysis has been applied by Machin and Pringle (1960) to the operation of the flight muscle of a beetle. The conceptual importance of engineering systems in this particular investigation has been stressed by Pringle (1964).

The muscular control of limbs in humans has been investigated by several workers: qualitative descriptions of the system, indicating the functions of the various component parts, have been given by Merton (1953), Hammond, Merton and Sutton (1956) and Taylor (1957); a qualitative description using the methods of servoanalysis has also been attempted by Hammond (1960).

Maldonado (1964), by means of a mathematical analysis of the patterns of velocity and acceleration used by an octopus when mak-ing an attack, has been able to show that the final pattern of the attack movement is unaffected by changes in visual information once the attack is initiated, indicating that the control mechanism in question is essentially open loop as far as vision is concerned.

The control of writing and drawing has been investigated by Smith and Smith (1962) using a closed loop television system. As he performed his tasks, the subject viewed his hand on a television screen instead of having a direct view; this enabled the experimenters to introduce delay electronically and to study the effects on his performance of varying this delay. Similar studies, differing mainly in the methods used to introduce the delay, had previously been made by Bergeijk and David (1959), and by Kalmus, Fry and Denes (1960). A more detailed and systematic approach to the subject of hand–eye coordination has been conducted by Stark, Iida and Willis (1961) and Stark, Payne and Okabe (1962), using a digital computer to compose complex stimuli and to analyse the ensuing responses.

Two general reviews which have appeared fairly recently have much the same viewpoint as the present book: the mathematical techniques of control theory as they may be applied to biological systems have been summarized by Machin (1964), and the contributions of Cybernetics to Biology have been reviewed by Hassenstein (1960).

The remainder of this chapter will be devoted to a rather more detailed examination of some specific systems, placing particular emphasis both on the use of various servoanalytical techniques and on the practical restrictions on the use of such techniques in biological systems.

Temperature Regulation

The mechanism by which warm blooded animals, man in particular, are able to maintain an almost constant internal temperature despite wide variations in the temperature of the environment has attracted attention from physiologists for many years. No attempt will be made here to present a comprehensive review of the vast literature of the subject. Such reviews have been published by Bazett (1949), Selle (1952) and Hardy (1961), among others. It is, however, of particular interest in this volume to consider briefly some recent work on the subject by Benzinger and his collaborators, not only because of its value in clarifying the problem, but also because it has provided elegant demonstrations of some of the hazards inherent in the analysis of biological control systems.

Involved in the overall temperature regulation system there can be distinguished three distinct subsections, each of which can be treated separately. In a warm environment, normal temperature is maintained by acceleration of heat loss through the physical channels of heat transfer: this may be called physical heat regulation. In a cold environment, normal temperature is maintained by acceleration of metabolic heat production processes: this is called chemical heat regulation. The third subsystem is more vague than these two but is none the less important. It may be called behavioural heat regulation and is the mechanism by which more or less conscious sensations of hot and cold are reacted to by the removal or addition of clothing, by exercise or by change of environment. Although it is highly improbable that this last system could ever be analysed quantitatively, a qualitative examination of some of its elements will be touched on later (see pp. 92, 432).

In examining physical heat regulation, Benzinger's (1959) approach was to identify three components of the control system and to attempt

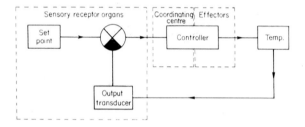

Figure 2. The elements of the human temperature regulating system as distinguished by Benzinger interpreted as an engineering control system.

to find the quantitative relationships that connect them. The components he distinguished were sensory receptor organs, effector organs and a coordinating centre in the central nervous system connecting the receptors to the effectors. This division may be compared with the typical make-up of engineering control systems discussed in Chapter 2. The comparison is illustrated in Figure 2, and it is interesting to observe that the command signal (which in a regulator is simply the constant reference level that is to be maintained) is incorporated with the output transducer and the summation point in the sensory receptors. These receptors thus have the property that the output (nerve impulse frequency) is related not to the absolute temperature, but to the amount by which the temperature deviates from the set point.

There seems little doubt that the two effector components of physical heat regulation are vasomotor activity, which adjusts the blood flow between the interior of the body and the surface, and sudomotor activity, which enables excessive heat to be dissipated by the evaporation of sweat. Since the last century a centre in the anterior hypothalamus has been known to be intimately concerned with the production of thermoregulatory responses, and this may well represent the main coordinating centre of the system.

The most interesting feature of the problem is that two quite distinct receptor systems have been identified. The skin all over the body is richly endowed with receptors sensitive to cold and warmth, and these could provide the information necessary for the coordination centre to compute an average skin temperature. On the other hand, the existence of central temperature-sensitive sites has been demonstrated in various animals by a number of different workers who have elicited thermoregulatory responses by heating parts of the brain, either generally by heating the head arteries, or more locally by applying warm and cold probes to the hypothalamus.

The classical view was that both cutaneous and central thermoreceptors played some part in the initiation of thermoregulatory responses, and it was Benzinger's first aim to assess the relative importance of these two components by making simultaneous measurements of vasomotor and sudomotor activity, and internal and skin temperatures. The measurements of the effector mechanisms were made by the use of a gradient layer calorimeter, described by Benzinger and Kitzinger (1949, 1963). Skin temperature was taken as the average of the temperatures at ten different points on the body measured with thermocouples. In the first series of tests, internal temperature was taken as the rectal temperature, but the results obtained, as described by Benzinger (1961, 1964), were neither intelligible nor reproducible: no correlations could be discovered between the regulatory responses and either skin or rectal temperature. This could only mean that the measured responses were dependent on some third variable, and that dependence on the measured temperature was either absent or else obscured by the relationship with the third variable. It is worth noting here that the statement by Benzinger (1959, 1961) and Benzinger and coworkers (1963) that the disordered results could be due simply to a joint relationship between the response and both the measured temperatures is not strictly correct. Any ordered relationship that existed would emerge from a graphical plot, provided both stimulus variables were represented. Although a bivariate function requires a three-dimensional graph for its full expression, it can be represented in a two-dimensional graph by a contour plot—a fact used by Benzinger himself in studies of other systems, in which a bivariate dependence does in fact occur.

After this first set of experiments had failed to reveal any stimulus–response relationships, the value of the rectal temperature as an indicator of the controlled internal temperature of the system was questioned. Experimental evidence from Magoun and colleagues (1938) had previously demonstrated in cats that local heating of the brain elicits heat loss responses and their results suggested that the temperature sensitive area was localized in the anterior hypothalamus. It had been assumed that although the rectal temperature would not

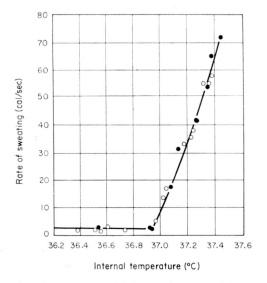

Figure 3. Variation of sweat rate with internal temperature, measured at the tympanic membrane. Open circles: subject at rest. Closed circles: subject exercising. The relationship is both consistent and reproducible. (From Benzinger, 1961b.)

be identical with the controlled internal temperature, it would nevertheless be correlated with it. This assumption had in fact been made by nearly all the early investigators of temperature regulation, but it was found by Benzinger and his group (fully reported by Benzinger and Taylor, 1963) that variations in rectal temperature completely failed to reflect variations in intracranial temperature measured at the sphenoid sinus (about an inch from the presumed position of the hypothalamic temperature sensor). The sphenoid temperature showed rapid changes when the subject entered a hot bath or swallowed a quantity of iced water, whereas the rectal temperature responded only after a long delay, and with the rapid changes smoothed out.

The sphenoid sinus is not a very convenient place to take measurements, but it was found that a thermocouple inserted into the external auditory canal and held against the tympanic membrane registered the same rapid changes and so this was taken as the measure of internal temperature for the subsequent experiments.

With this new and more accurate measurement of internal

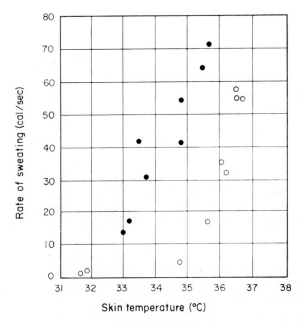

Figure 4. Variation of sweat rate with skin temperature, measured as the average of ten temperatures from all parts of the body. Open circles: subject at rest. Closed circles: subject exercising. No consistent relationship emerges. (From Benzinger, 1961b.)

temperature, Benzinger (1959) found the kind of consistent and reproducible stimulus–response relationships for which he had been searching. Figure 3 shows the rate of sweating plotted against the internal temperature, and it displays a relationship that is independent of the subject's state of rest or work. By plotting the rate of sweating against the skin temperature the results shown in Figure 4 were obtained, showing that within the range covered the response was completely unrelated to the skin temperature.

With the vasomotor response, the results were qualitatively similar, although less clear cut; this may be due to local effects of skin temperature upon cutaneous vasodilatation without central control.

Nevertheless, the results suggested a consistent relationship between internal temperature and blood flow, with skin temperature having no more than a modifying effect.

Benzinger concluded from his results that physical heat regulation mechanisms are activated by the intracranial temperature, detected by the hypothalamic temperature sensor, and that the warm receptors in the skin play no part in the responses. These conclusions appeared to be at variance with the observations of various workers such as Randall and Hertzman (1952), Brebuer and Kerslake (1960) and Hertig and Belding (1961) who had reported changes of sweat rate in response to changes of skin temperature, and this conflict led to a fuller investigation, covering a wider range of combinations of internal and skin temperatures. The results, reported by Benzinger (1961), confirmed the original conclusion that peripheral warm receptors played no part in eliciting the response, but demonstrated that sweat production due to an elevated intracranial temperature could be inhibited by simultaneous stimulation of the cutaneous cold receptors. This conclusion has also been supported by experiments conducted on rabbits by von Euler (1964).

The same calorimetric and thermometric techniques have also been applied by Benzinger, Pratt and Kitzinger (1961) to the chemical temperature regulation system—the increase of metabolic heat production in response to cold. The results show that this response is a function of both internal and skin temperatures. The combined operation of the chemical and physical systems is illustrated in Figure 5 and we may summarize the interpretation of this diagram in the following way. Increased metabolic heat production is elicited by cold stimulation of the skin, unless it is inhibited by an elevated intracranial temperature, while sudomotor activity is elicited by an elevated intracranial temperature, unless inhibited by stimulation of the cold receptors of the skin. Cutaneous vasodilation is also elicited by elevated intracranial temperatures, and in this case central inhibition of the response has not been reported. It seems probable that the direct vasoconstrictory effect of low temperature stimulation of the blood vessels themselves is sufficient to provide the necessary inhibition without the need for central pathways.

The third regulatory system, the behavioural system, depends on sensations of warmth and cold reaching the cortex. An investigation of the origins of such sensations has led Benzinger (1963) to the conclusion that cold sensations arise from peripheral stimulation, while warm sensations are produced by a rise in central temperature. This behavioural system, although not readily quantifiable, is important as a coarse temperature control; taken together with the extraordinarily sensitive autonomic control, it enables an enormous total range of environmental conditions to be tolerated without fatal results.

The work of Benzinger and his colleagues, of which this has been necessarily a very condensed account, brings out a number of points

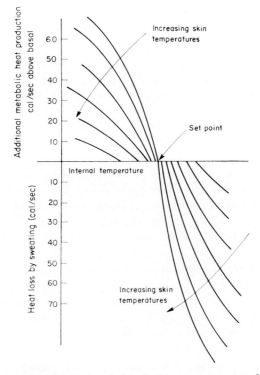

Figure 5. Operation of physical and chemical temperature regulation systems. The rate of heat production or loss depends primarily on the internal temperature, but is reduced if the skin temperature is high while the internal temperature is low or if the skin temperature is low while the internal temperature is high. (Redrawn from data in Benzinger, 1961a and Benzinger and colleagues, 1961.)

that are of interest when considering any biological control system. First of all, when dealing with a regulatory system in an intact animal, the detection of changes in the controlled variable is inherently difficult because, by definition, the system specifically prevents, or at least minimizes, such changes. Thus a change in heat production or heat loss equivalent to 10% of the basal metabolic rate is brought about by a change in internal temperature of between 0·01°c and 0·02°c, and all the changes diagrammed in Figure 5 occur within a range of internal temperatures of only about 1°c. The resolution of the temperature measuring equipment of an experimenter has to be very fine if the operation of the system is not to be missed altogether.

The second important feature of this work concerns the identification of the driving stimulus for the responses. The situation is confused by the fact that internal and skin temperatures are not independent. A change in one is accompanied, under normal conditions, by a change in the other so that it is difficult to draw conclusions as to which is responsible for an observed response. In other words, a correlation between a response and a variable A may mean that A is in fact the stimulus; but could equally be due to a second correlation between variable A and the true stimulus B. This kind of difficulty often turns up in the study of biological control systems, where we are often unable to open the control loop in order to take the stimulus–response measurements we would like. The solution which was adopted in the case of thermoregulation was to destroy the natural correlation between internal and skin temperatures either by maintaining steady work conditions at various levels, or by generating sudden changes in one of the temperatures (by taking hot or cold drinks or baths), and observing which of the two possible stimulus variations was reflected in the response.

The third interesting feature of Benzinger's study is the unsuccessful use of rectal temperature as an indicator of the controlled internal temperature. The comparison between rectal and sphenoid temperatures, as described by Benzinger and Taylor (1963), shows that the rectal temperature bears to the sphenoid temperature a relationship qualitatively similar to that discussed in Chapters 2 and 3 as the first order, or single time lag, response. This is not really surprising because there must inevitably be a thermal inertia effect in the system, but the magnitude of this effect, with a time constant of 5–10 minutes, serves as a reminder that similar time lags are likely to be present in the regulatory system itself, since heat is mainly generated and lost peripherally, whereas the controlled temperature is central. With time lags of this magnitude, instability would seem to be almost inevitable unless special stabilizing measures were incorporated. The cross coupling of information from central and peripheral sources can be seen, qualitatively at least, to be a possible means of providing the necessary stabilizing influence.

As a final comment on temperature regulation, it is worth mentioning two attempts to elucidate the system from a completely different point of view. Wyndham and Atkins (1961) made a series of approximations to describe the thermal conditions of the body by a model that was reasonably tractable mathematically (basically a series of concentric cylindrical shells), and calculated the heat exchange equations using thermodynamic principles. This thermal model was then applied to an analogue computer so that solutions to the equations could be obtained for various conditions. Unfortunately the reference cited is only a preliminary report on the project, and no results were given of comparisons between model solutions and experimental evidence. Nevertheless this kind of approach has considerable interest. A 'hardware' model can be subjected to a wide variety of tests that are not permissible or even not possible with animal subjects, particularly man, and predictions derived from biologically based theories can be tested rigorously and not disastrously on the model.

The other analogue computer approach has been made by Crosby, Hall and Hardy (1961), who based their models directly on experimental findings rather than theoretical equations. Elaborate mathematical analyses based on experimental results have also been made by Wissler (1961, 1963).

Perhaps the most important thing that needs to be elucidated is the dynamics of the system, and this is one aspect which may well be best done on a computer model, since controlled manipulations of temperature, and particularly internal body temperature, are difficult to produce with human subjects. Before we can place confidence in such results, however, models must first be rigorously tested for conformity with all existing data.

Optomotor Responses

It has long been known that movement of the visual environment (or a prominent part of it) can elicit activity in many animals including man. In insects particularly, the turning tendency elicited when surrounded by a striped drum has been used as a convenient indicator for measurements of visual acuity, while the optokinetic nystagmus produced in man by the same test situation has proved to be a valuable diagnostic tool for certain forms of ear disease and for various brain lesions.

There are two main aspects of this phenomenon that have been investigated. The first is the way in which the visual field is analysed by the animal in order to obtain information on which to base the response, and the second is the way in which the information is used to produce the response. The work of Mittelstaedt (1949) on the second of these problems has already been mentioned (see p. 84). He was questioning the existing assumption of a simple stimulus–response relationship: that is to say, the assumption that the responses were controlled by an open loop control system, in which the occurrence of any particular change in excitation of the retina would result in a corresponding particular reaction. The demonstration that optomotor responses were in fact mediated by a closed loop system paved the way for a reexamination of a number of other systems which had hitherto also been assumed to be simple reflex systems.

In a later paper, Mittelstaedt (1951) developed the investigation of the optomotor response system further by considering in more

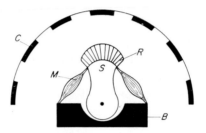

Figure 6. Generalized scheme showing the essential elements involved in optomotor responses. C: striped cylinder which acts as stimulus. R: receptor array. S: supporting structure. B: fixed base in which S pivots. M: musculature to move S relative to B. (Modified from Mittelstaedt, 1951.)

detail what patterns of interconnection between the various components of the system could be postulated in order to account for the observed behaviour. Figure 6 shows a generalized system in which the striped cylinder C is rotated about an animal consisting (for the purposes of this discussion) of four parts. The receptor system, R, is mounted on a supporting structure S. This structure is pivoted on a fixed base, B, and can be moved relative to the base by musculature, M. The precise physical equivalents of these parts will depend on the animal being considered: for the study of optokinetic nystagmus in a human for example, S is the eye, M the extraocular muscles and B the head; while for an insect's optomotor responses as studied by Mittelstaedt, S is the head, M the neck muscles and B the prothorax. Three parameters may be measured in this system:

$$\sigma = \text{angular speed of } C \text{ relative to } B$$
$$\mu = \text{angular speed of } S \text{ relative to } B$$
$$\phi = \text{angular speed of } S \text{ relative to } C$$

and it follows from the physical arrangement that

$$\phi = \sigma - \mu.$$

The system may therefore be represented in block diagram form as shown in Figure 7, where the rectangular block contains the internal part of the system whose function is to produce the output (μ) in response to the visual stimulation (σ). This diagram bears a close resemblance to that of a conventional control system (see Chapter 3) if the rectangular block is taken to be a linear amplifier with high

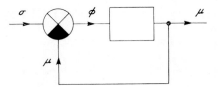

Figure 7. Block diagram illustrating the visual feedback loop involved in optomotor responses. The blank box represents the internal part of the system: if this were to consist of a linear amplifier then the system would be precisely analogous to a position control as discussed in Chapter 3.

gain: such a control system would make the output, μ, tend to follow the input, σ. If the gain of the amplifying box is V then

$$\mu = V\phi$$

and since $\phi = \sigma - \mu$, then it follows that

$$\frac{\mu}{\sigma} = \frac{1}{1 + 1/V}.$$

Thus the larger the value of V the closer we come to ideal following. In engineering systems where this kind of circuit is used, very large values of V can be obtained (of the order of tens or hundreds of millions) so that accurate tracking is feasible, but in a biological system a large value of V is less likely, and as a result we might expect an appreciable residual error under steady conditions.

   The first parameter of interest is the gain of the internal amplifying system; the only way of measuring this with reasonable accuracy is by opening the feedback loop. Since this loop is outside the animal, Mittelstaedt was able to open it by preventing the motor output from having normal feedback influence on the visual input. He did this by fixing a fly (*Eristalis*) with a piece of card glued to its thorax and held by a clamp, and by having the fixed animal, as shown in Figure

4*

8 (*a*), (*b*), standing on a platform which was mounted freely in a bearing so that it could rotate in a horizontal plane. Thus when the animal performed locomotory movements with its legs the platform rotated (enabling the 'output' speed to be measured) but the visual display was unaffected.

   This arrangement enabled measurements of the open loop gain to be made, but these measurements immediately revealed a major limitation of the simple forms of control system theory that have

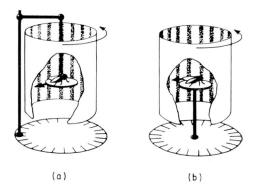

(a)   (b)

Figure 8. Method used by Mittelstaedt to measure the characteristics of the optomotor system in flies. (a) Open loop: the fly is fixed and his feet turn a disc. (b) Closed loop: the speed of rotation of the fly is measured without interfering with the feedback. (From Mittelstaedt, 1964.)

been discussed in earlier chapters, for it turned out that the amplifying system was grossly non-linear. Figure 9 shows the variation of gain with input magnitude; clearly there is not even any limited range over which linear operation could be taken as a reasonable approximation. Nevertheless, the analysis can be taken a step further by changing the parameters, and investigating the variation of the relative error, R, with the amplification V, where

$$R = \frac{\phi,}{\sigma}, \qquad V = \frac{\mu}{\phi}$$

and as before

$$\phi = \sigma - \mu$$

This gives

$$R = \frac{1}{1 + V}$$

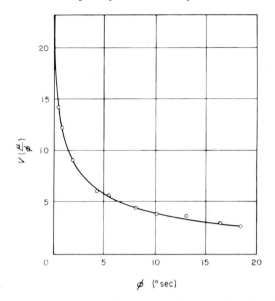

Figure 9. Variation of amplification, V, with input, ϕ, showing that the opto-motor system is grossly non-linear: for a linear system V would be a constant. (From Mittelstaedt, 1951.)

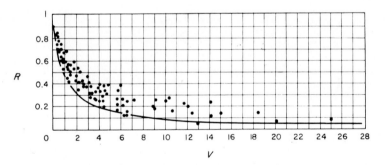

Figure 10. Variation of relative error, R, with amplification, V. The experimental points show a reasonable agreement with the theoretical curve (shown as a solid line), $R = 1/(1+V)$. (From Mittelstaedt, 1951.)

Mittelstaedt's results are shown in Figure 10, and display a reasonable agreement between the practical results and the theory.

Another main problem of optomotor reactions is to discover what information is extracted from the moving visual field; in other words to find out how the pattern of retinal excitation varying in time and space is processed in order to obtain a parameter or parameters to act as command signals to drive the optomotor control system. The earliest attempt to suggest a mathematical solution to this problem was made by Kalmus (1949), with specific reference to the optomotor response mechanism in flies (*Drosophila* and *Musca*). Disregarding the possibility of stereoscopic vision, the only information available to these animals is the pattern of light intensities impinging on the eyes from different directions, and the variation of this light pattern with time. The response of the receptor array to a varying spatial illumination pattern can therefore be denoted by a function J, which is a function both of direction and of time.

Consider now the situation illustrated in Figure 11, where the fly is at F, and the response of the ommatidium pointing in the direction FP is J. If, following Kalmus, we assume for simplicity that the fly

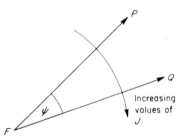

Figure 11. Situation considered by Kalmus in his analysis of optomotor responses.

has perfect visual acuity and that J is a continuous function of direction and time with continuous partial derivatives, then we may consider a plane through FPQ where PQ is the direction of increasing values of J. If angle PFQ is ψ, then the rate of change of J as we move in the direction PQ is $\partial J/\partial \psi$. If now in a short time interval δt the change in value of J at P is δJ, this change is to the insect indistinguishable from that which would be produced by a movement of the element of the incident illumination pattern by an amount $\delta \psi$ in the direction QP where $\delta \psi = \delta J/(\partial J/\partial \psi)$, and this gives rise to an apparent velocity in this same direction of

$$v = \frac{d\psi}{dt} = \frac{\partial J}{\partial t} \bigg/ \frac{\partial J}{\partial \psi}$$

Kalmus assumed that the effect on the animal of this apparent velocity of the visual element would be proportional to the velocity, to the area of the element ($\delta \Omega$) and to the spatial rate of change of J

$(\partial J/\partial \psi)$. The effect of this element of the visual field is then given by

$$\delta T = \frac{\partial J}{\partial t} \cdot \delta \Omega$$

and the total effect, T, is found by integrating over the whole visual field. To this quantity T was given the name 'apparent translation'. A further quantity derived from the visual stimulus pattern is the 'apparent rotation', R, of the pattern. The effect, δR, due to an element of the visual field, is numerically equal to δT, and represents an angular rotation of the pattern about an axis perpendicular to both the line of sight and the line of action of the apparent translation as shown in Figure 12. Both T and R result from the same changes

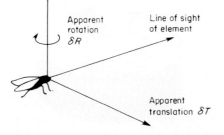

Figure 12. Relationship between the rotation and translation components of the visual environment as observed by a single receptor element. The total optomotor response will depend on the vector sum of all such elemental contributions.

in the visual pattern and accordingly the response of the fly might be expected to contain both translatory and rotatory components.

The theory advanced by Kalmus is based on these apparent movements of the pattern, and states that the fly will run as fast as it can in the direction opposite to that of T, and that it will turn around the axis of R in such a way as to reduce R to zero. Kalmus used a number of test situations in which various combinations of moving stimulus patterns were exposed in the visual field of a fly, and in which the fly was constrained in various ways so as to separate the postulated optomotor responses: the reactions obtained from the flies were in each case consistent with the theory. The main interest of this work lies less in the specific theory it produced (which, as its author admits, is only one of several that could fit the evidence) than in the approach towards making a mathematical analysis of the visual environment in an effort to find a mathematical quantity that could serve as a stimulus to the response.

A somewhat different attempt to specify mathematically the relationship between the visual environment and the induced optomotor
response has been made by Hassenstein and his colleagues, working
with beetles. His emphasis was on the mathematical relationship
between the motor responses and the changes of visual stimulation
of individual ommatidia of an insect. The isolation of the ommatidia
was achieved with an arrangement of three cylinders as shown in
Figure 13. The outer cylinder had broad black and white stripes,
while the inner cylinder was all black and contained pairs of slits so
spaced that the distance between the members of the pair corres-
ponded to the interommatidial angle for a beetle at the centre.
Between these cylinders (both of which were stationary) a third

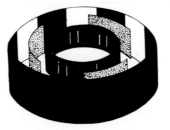

Figure 13. Hassenstein's cylinder arrangement. By moving the grey middle
cylinder, illumination changes in both senses can be directed at particular
ommatidia. (From Hassenstein, 1959a.)

cylinder, consisting of broad slits between grey areas, could be rotated
in either direction. By this means relative changes in light intensity
could be induced in neighbouring ommatidia at successive times.

The beetle was fixed in the same way as the flies in Mittelstaedt's
work, but the method of measuring the response was to use a Y-maze
globe as shown in Figure 14. This enables the beetle to make
repeated right–left choices without disturbing the visual input and
without the necessity for interference from the experimenter.

The situation considered by Hassenstein (1951) was of a series of
ommatidia arranged in a row, with a succession of illumination
changes passing along the row. He found that the simplest form of
stimulus that could elicit a response was a set of successive changes
of illumination in two neighbouring ommatidia A and B. The changes
of illumination take place at two successive times and may be denoted
with $+$ or $-$ depending on whether the light intensity increases or
decreases. An increase of illumination of A at time t_1 followed by an
increase of illumination of B at time t_2 causes the animal to turn in

the direction from A to B, and this situation may be denoted by succession $A_1^+ B_2^+$ leading to response $+R$. Similarly the succession $A_1^- B_2^-$ leads also to response $+R$. In addition to this situation, which corresponds to the natural stimulation afforded by contour displacement, Hassenstein also investigated the artificial successions $A_1^+ B_2^-$ and $A_1^- B_2^+$, both of which gave rise to the opposite motor response,

Figure 14. Y-maze globe used by Hassenstein. It is held by the insect and turns in response to locomotory movements, so that branchings repeatedly appear in the path and the animal has to make right–left decisions. (From Hassenstein, 1959b.)

which we may denote as $-R$. From the symmetry of the situation it is obvious that changing the order of stimulation from AB to BA reverses the direction of the responses. These results are summarized in Table 1; the pattern of signs in this table suggests that some form

Table 1. Relationship between the senses of the stimuli successively applied to two adjacent ommatidia, A and B, and the sense of the resulting response. The distribution of signs in the table suggests a multiplicative relationship.

	A_1^+	A_1^-
B_2^+	$+R$	$-R$
B_2^-	$-R$	$+R$

of multiplicative process is at work. A more elaborate mathematical model has been developed by Hassenstein and Reichardt (1956), (recapitulated in an English version by Reichardt, 1961), who described a theory that the analysis of the visual field is based on the principle of running autocorrelation, interactions occurring only between ommatidia that are adjacent and between those spaced two apart. This theory has been supported by a number of experimental tests in which specific manipulations of stimulus patterns produced

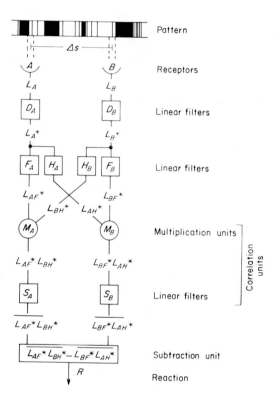

Figure 15. Reichardt's mathematical model for correlation processing of information from two ommatidia. (From Reichardt, 1961.)

predictable changes in response. The reaction produced as a result of stimulation of two ommatidia is described by the mathematical model shown in Figure 15, in which the spatio-temporal stimulus pattern is processed by a series of linear filters denoted by D, F and H, before being subjected to the multiplicative and averaging processes, which together form the autocorrelation function. From measurements of the variation of the strength of response with change of angular speed of the rotating cylinder (which in turn determines the temporal component of the stimulus pattern) Hassenstein (1958) was able to deduce that the F and H filters are simple linear components with the characteristics of first order lag circuits—that is to say, with transfer functions of the form $A/(1 + \tau s)$ (see p. 39)—while the D filter is of a more elaborate kind (not discussed in Chapter 3) having a transfer function of the form $A\sqrt{s\varepsilon}^{-\sqrt{B}s}$ where A and B are constants. By feeding adjacent ommatidia with stimulus patterns chosen to be mathematically convenient, the actual and the predicted responses were compared, and found to agree within experimental limits. In experiments using very similar apparatus, Bliss (1962) has also obtained results in general agreement with the predictions of Reichardt's model.

Working with crabs, Horridge and Sandeman (1964) have obtained results that are qualitatively in agreement with the work outlined above; these authors have attempted to deduce the general pattern of interconnection of the elements and they have advanced a theory of information processing by the eyes. These theories, however, still await rigorous quantitative tests.

Orientation

Until about two decades ago investigation of the orientation of animals was mostly concerned with correlating simplified stimulus situations with the postures and directed movements resulting from them. A vast amount of this kind of information has been summarized by Kühn (1919) and Fraenkel and Gunn (1940). The material was mostly observational, and the experimental work was largely based on varying combinations of external stimuli and some very simple surgical manipulations. The development of Cybernetics has made this approach obsolete; nowadays one wants to know considerably more about the organization of the systems controlling the processes involved, although in general we are still a long way from an attempt at full explanation derived from the properties and functions of the anatomical elements involved. Even in cybernetic terms, a comprehensive treatment of orientational reactions is at present quite impossible and only a few examples can be given here.

Orientational reactions are generally necessary for an organism to make controlled movements. These, in particular, enable an animal to exercise some choice over its immediate environment, rather than being forced to stay wherever it was born or hatched. A limited degree of control can be exercised without having orientational reactions: if an animal wanders haphazardly, for example, it has some chance of reaching a favourable environment; this is called *kinesis*, and is a slight advance on completely passive dispersal. Alternatively, an animal can attempt to reach a particular place simply by launching itself into a current of air or water. Either drift or movement may be released by a specific stimulus, for instance light, thermal radiation or chemicals, and the process may be terminated as a further controlled response to an external stimulus. If the control of habitat is to be more elaborate or more precise, however, the animal has need of some orientational mechanism.

Spatial orientation can be achieved in many different ways. It is possible for instance for an animal possessing only one receptor to be directed by the action of successive stimuli to this receptor. Thus the flagellate *Euglena* finds its direction either towards or away from a light source by the succession of light and shade reaching its eye spot as a consequence of its own spiral movement (Mast, 1911). More usually, more than one receptor is implicated in spatial orientation. Orientation to a directed stimulus, for instance a source of light like the sun, is called a *taxis*. This simply denotes that the animal maintains a constant angle with respect to the stimulus. In most situations the angle concerned is that between the horizontal projection of the stimulus direction and the direction of the motor response. We talk of *positive taxis* if the angle is zero, *negative taxis* if the angle is 180° and *transverse taxis* if the animal moves at right angles to the stimulus direction. Locomotion in any other direction relative to the stimulus is called *menotaxis*.

The idea of taxis was preceded by the idea of *tropism*, introduced by Loeb (1918), who tried to connect anatomical details of the animal's sensory and motor organization with their reactions towards stimuli, but it is now realized that the nervous system of most animals responds to the total stimulus situation and that the motor reactions are also organized as a whole. Thus the older distinction between *telotaxis* and *tropotaxis* has lost much of its validity. The idea behind this distinction was that a binocular animal could be led to a light either by fixing the light at a particular retinal area or by maintaining symmetrical stimulation of the two eyes. Thus if one eye was eliminated by cutting the optic nerve or by blackening, the telotactic animal can still maintain the angle to the light with the one functional eye, whereas the tropotactic animal cannot do so, but moves in circles, mostly in the direction of the non-functional eye.

A more useful classification has been proposed by Jander (1963a, b) mostly referring to the light and gravity reactions of insects. He distinguishes between switching the orientation mechanism on and off, so that the animal is either orientated or non-orientated, and changes between the basic directions of a taxis, which may happen either by a change of sign of the turning reaction (which he calls reversion), or by substitution of one orientating mechanism by another.

Jander also states that in a tactic reaction there is no rotation around the body axis, which will be orientated in the direction of the stimulus, but only angular rotation about the other two axes. A further important feature of the turning reactions of many animals is that when displaced from their equilibrium direction they will return to it by turning through the smallest possible angle.

Phototropotaxis can be subdivided into *archiphototaxis*, in which the intensity of the turning reaction depends only on the light intensity, *prophototaxis* (called *mesophototaxis* by Mittelstaedt, 1964), in which the reaction depends on both the intensity and the direction of the light, and *metaphototaxis*, in which the reaction depends only on light direction.

Prophototaxis can be accounted for by nervous mechanisms capable of computing the sine of the stimulus direction and of addition, subtraction and division. In prophotoactic animals, the menotactic angle varies with the stimulus intensity, so that it is difficult to conceive how animals could use such a mechanism in conjunction with time compensated sun or moon orientation (*astrotaxis*). This can be effected only by metaphototaxis (see p. 109).

Jander's conclusions about the working of the system for prophototaxis differ somewhat from those reached by Mittelstaedt (1962, 1963), who investigated a number of possible ways in which the response could depend on the input angle and on an internal command signal. The scheme proposed by Jander amounted to a relationship between turning reaction and stimulus of the form

$$r = w + F \sin \beta$$

where r is the turning reaction, w is the 'course order', β is the direction of the stimulus relative to the head axis and F is a constant.

If $w = 0$ (so that the 'course order' is to steer directly towards the stimulus), the functional relationship is a straightforward sine wave as shown in Figure 16 (*a*). The system will adopt a course that results in zero turning reaction; there are two such courses available, one at $\beta = 0$ and the other at $\beta = 180°$. From inspection of the curve, it can be seen that any displacement from the $\beta = 0$ position

produces a turning reaction that returns to zero, whereas a displacement from the $\beta = 180°$ position produces a turning reaction that takes
β further from 180°, and in fact reduces it to zero. Thus the $\beta = 0$
position is one of stable equilibrium, whereas the $\beta = 180°$ position
is one of unstable equilibrium. These two positions are distinguished
by Mittelstaedt (1962) as the *stability* and *lability* positions. Under
these conditions with $w = 0$, the return to the stability position after
a displacement is always by the shortest route. If $w \neq 0$, however,
the situation changes as shown in Figure 16 (b). The stable equilibrium position is altered so that the animal steers at some angle to
the stimulus instead of directly towards it: this is the desired result

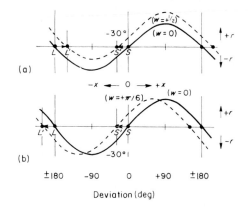

Figure 16. Possible forms of dependence of the turning reaction, r, on the
deviation for various course orders (w). (a) Theory of direct addition of course
order. (b) Bicomponent theory. (From Mittelstaedt, 1963.)

of making $w \neq 0$. The unstable equilibrium point, however, is now
no longer at 180° to the stability position, and this has the effect of
making the organism return by the longer route for some particular
deviations. A more startling prediction from this theory is that increasing w will increase the deviation at the stability position up to
90°, but that a further increase of w will produce continuous rotation
with no stability position at all. These conditions have never been
observed experimentally, and this leads to the conclusion that a
theory involving the addition of a course order to a sinusoidal deviation signal cannot account for the facts. Mittelstaedt considered
various alternatives, and finally proposed his bicomponent theory
(Mittelstaedt, 1962). This suggests that both the course order and the
deviation signal are divided into sinusoidal and cosinusoidal com-

ponents, and that these components are cross multiplied to produce a turning reaction given by

$$r = K_c \sin w \cos \beta + K_s \cos w \sin \beta$$

where K_c and K_s are constants.

A block diagram of the bicomponent theory is shown in Figure 17. In this case we see that when $w = 0$ the equation reduces to

$$r = K_s \sin \beta$$

as it did with the simpler theory. When $w \neq 0$, however, the bicomponent theory predicts that the functional relationship between

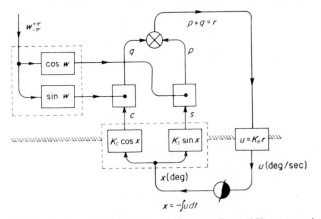

Figure 17. Block diagram of bicomponent theory of providing a turning reaction dependent both on the angle between the light source and the reference direction and on the course order. The line of hatching denotes the boundary between the organism and its environment. (From Mittelstaedt, 1963.)

r and β will be still sinusoidal and also still symmetrical about the axis, as shown in Figure 16 (c). This means that there is no limit to the deviation that can be maintained, that the unstable equilibrium position will always be 180° away from the stability point, and that return to the stability position after a displacement will always be by the shortest route. Mittelstaedt (1962) also claims that by making some additional assumptions about the relative magnitudes of the constants involved, virtually all the known properties of orientation systems in insects can be accounted for on the basis of this bicomponent theory.

Interaction of two modalities

In natural conditions, orientational behaviour depends on many kinds of stimuli impinging on one animal. The simplest situation, apart from considering single modalities, would be one in which two stimuli, for instance light and gravity, have effects of comparable magnitude. This situation has been investigated in several species of fish by von Holst (1950).

In one experiment the fish was in a centrifuge and could be illuminated with light of various intensities shining from various directions. When the apparatus was still and the light impinged vertically, the dorso-ventral axis of the animal was vertical. However, when a light shone on one eye only, from the side, the dorso-ventral axis and the sagittal plane inclined towards the side from which the light came and formed an angle α with the vertical. The magnitude of this angle depends on the light intensity and on the angle β between

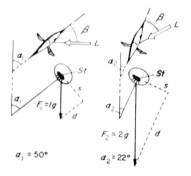

Figure 18. Behaviour of a fish under (a) normal conditions and (b) conditions in a centrifuge with increased gravitational force. (From von Holst, 1950.)

the direction of the light and the dorso-ventral axis. By keeping the gravitational force constant it is possible to determine the relationship between α, β and L, the light intensity. On the other hand, by operating the centrifuge and keeping β and L constant it is possible to investigate the dependence of α on gravitational force. This is detected in the fish by means of the otoliths, which are suspended on hairs, and it appears that the angle of inclination is so changed that the shearing force, s, acting perpendicular to the sensory hairs remains constant, while the pressure component, d, may increase considerably, as shown in Figure 18. The block diagram shown in Figure 19 represents the theory advanced by von Holst (1950) to account for these results, leaving out of account the possibility of imposing 'commanded' deviations into the system. This theory assumes that the turning reaction, r, is given by

$$r = L \sin \beta + F \sin \alpha$$

where L is the light intensity, β is the angle between the light and the vertical, α is the angle of deviation of the dorso-ventral axis from the vertical, and F is a constant.

Difficulties arise, as in the single orientating stimulus case discussed above, when consideration is extended to possible ways in which commanded deviations can be produced. Possibilities discussed by Mittelstaedt (1964) are basically similar to the single case,

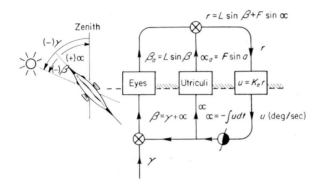

Figure 19. Block diagram of possible arrangement to account for orientational reactions of the fish. (From Mittelstaedt, 1964a.)

and an extension of the bicomponent theory has been proposed to cover this case also, the postulated relationship being

$$r = (F_s \sin \alpha + L_s \sin \beta) \cos w + (F_c \cos \alpha + L_c \cos \beta) \sin w$$

A function of this form can also account for the operation of the bee's 'sun-compass', as well as for the behaviour of other animals when exposed to light and gravity.

Prey capture

The term orientation is used when investigating the motor reactions of an animal with respect to such general physical stimuli as light, gravity, currents or odours. Direct orientation is probably more important, however, with respect to objects of biological significance to them, such as mates, young, predators and prey. One prerequisite for many of these reactions seems to be the possession of selection machinery in the nervous system to enable the animal to isolate the object that directs the orientational responses from the rest of the environment. In different animals this must happen to very different degrees and so far it has been imperfectly investigated. For our purpose we take such an ability for granted and will illustrate the large and complex field by just one example concerning prey capture; we will deal with the form of sudden pouncing as it occurs for instance when a frog or a chameleon stalks and catches a fly with its tongue, or a mantid catches an insect with its forelegs.

The problems associated with this form of activity have been investigated by Mittelstaedt (1952, 1957) who has applied to this system techniques similar to those already described with respect to orientation. The basic problem is one of coordination: information about the direction of the prey relative to the body must be used to determine the direction of strike of the forelegs. One obvious way of achieving this would be to have the forelegs under visual control, using the distance between leg and prey as an error signal that could be reduced to zero by a straightforward position control system. This possibility, however, must be excluded because of the speed of the strike: the duration of the strike is less than the inevitable synaptic and nerve conduction delays that would be involved in a nervous control loop. This therefore is a more complex information processing problem; information from the receptors must be converted into a complete set of strike orders for the effectors, the strike itself being controlled by an open loop system.

Information is available to the animal from the eyes, which can measure the angle between the head and the prey, and the proprioceptive apparatus of the neck, the sternocervical and tergocervical plates, which can measure the deviation of the head from the prothorax. The strike has to be directed at the appropriate angle to the prothorax; the relationship between the angles is shown in Figure 20. Two feedback loops are present in this arrangement, since a movement of the head affects the proprioceptive input and also the optical input. The block diagram shown in Figure 21 represents Mittelstaedt's hypothesis of the working of the system using the same symbols as in Figure 20.

The system can be interfered with in various ways in an attempt to gain insight into its working. The proprioceptive nerves can be sectioned either unilaterally or bilaterally; the head can be fixed in position by cementing a bridge of balsa wood between head and prothorax; the head can be loaded to alter the mechanical properties of the system; and combinations of these disturbances can be applied simultaneously. Mittelstaedt (1957) found that bilateral cutting of the proprioceptor nerves changed the animal's average strike performance (measured by the ratio of hits to strikes) from about 85% to about 25%. Fixing the position of the head of an unoperated animal had no effect if the head was fixed in the median position, but produced progressively larger deficits as the angle of deviation increased. A 20° deviation was sufficient to reduce the success ratio to about 25%, the misses being towards the right if the head was turned to the left and towards the left if the head was fixed to the right. The effects of head fixing and deafferentation were additive; if an animal

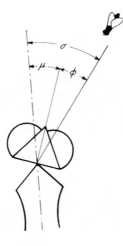

Figure 20. Relationships between the angles involved in the prey capture mechanism in mantids. (From Mittelstaedt, 1957.)

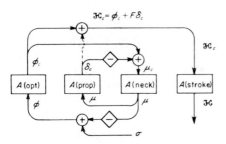

Figure 21. Block diagram proposed by Mittelstaedt to account for observed results of experimental interference with the prey capture mechanism. (From Mittelstaedt, 1957.)

with its head fixed at about 20° to the right of the median position was subjected to left nerve section, approximately normal performance was restored. Finally, the effect of adding large loads to the back of the head was investigated, and it was found that the animal was able to maintain almost normal performance until the additional torque exceeded the equivalent of twice the head weight at twice the head diameter. These results are all qualitatively in accordance with the flow diagram of Figure 21. It is interesting to observe that the overall system is open loop despite the existence of closed loop subsystems. This implies that the system must have a preset calibration that is sufficiently stable to operate reliably throughout the animal's many moultings.

Movements of Eye and Pupil

Where a biological system is concerned with the control of responses to the environment, there often exists a loop that is partly inside and partly outside the organism. Such loops are particularly amenable to experimental interference, since this interference can often be brought about without the use of surgery. Such experimental situations, being reversible and relatively easy to manipulate and control, also have the advantage that they produce results from normal intact animals, and this in turn means that work can be performed with human subjects. Two control systems connected with the eye have these external loops, and the workings of both in humans have been investigated in this way.

The pupillary reflex to light, whereby the opening of the iris controls the proportion of incident light that is allowed to reach the retina, is a particularly convenient system for experiment and analysis, as pointed out by Stark (1963). An attractive feature from the point of view of analysis is that the controlled quantity, the pupil area, has only one degree of freedom—a simplicity not common in biological systems.

It is tempting, on first looking at the pupil light reflex, to say that it is obviously mediated by a closed loop control system in which the pupil opening controls the light reaching the retina, and the light reaching the retina controls the pupil opening. This assumption, however, requires experimental justification, since it would be quite possible to postulate a system in which movements of the iris were unaffected by changes of retinal excitation. Such a system could then be either open or closed loop by virtue of its internal connections, but the particular loop involving the optical pathway would be absent. Such a system has long been known and has recently been demonstrated again in the toad by Campenhausen (1963), who has shown that the sphincter pupillae of this animal contracts when exposed to light even when the iris and muscle have been removed from the eye.

The human pupillary light reflex has, however, been shown to involve the retina, and Lowenstein (1955) has described the main pathways concerned, drawing upon experience derived from a large number of human case-histories and also from experimental work with animals. The sphincter muscle is innervated from the ciliary ganglion, which in turn is innervated from cells in the third nerve nucleus. These cells have inputs from both ipsilateral and contra-lateral pretectal regions—it is presumably these crossed connections that account for the consensual pupil reflex—and the pretectal regions are fed directly from the optic tracts.

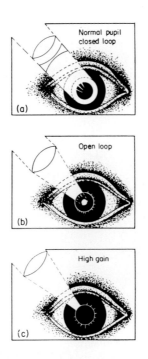

Figure 22. Method of making experimental analysis of pupillary reflex to light. (a) Normal closed loop operation. (b) Open loop working by focusing the beam so that pupil movements do not affect the incident light. (c) High gain working by exaggerating the effect of pupil movements on the incident light. (From Stark and Baker, 1959.)

The first systematic investigation of the pupillary light reflex using the techniques of control system analysis was made by Stark and Sherman (1957). They recognized that the system was likely to be non-linear, but nevertheless treated it as linear for the purposes of analysis. In order to make this approximation a reasonable one, they used very small driving signals. This 'small signal approximation' is standard practice in engineering systems since these too will never be truly linear over an indefinite range of signals (see p. 53).

The methods used by Stark and Sherman to modify the optical loop are illustrated in Figure 22. With the stimulus light forming a collimated beam the operation of the system is normal, with movements of the iris controlling the proportion of incident light reaching the retina. If the stimulus light is focused, as shown in Figure 22 (b), so that the diameter of the beam as it passes through the plane of the pupil is less than the smallest diameter that the pupil itself can assume, then it follows that changes in pupil size can have no effect

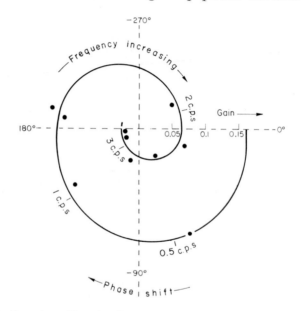

Figure 23. Open loop Nyquist diagram for pupil reflex system. The Nyquist criterion for stability is satisfied, showing that the system is stable. (From Stark and Sherman, 1957.)

on the amount of light reaching the retina, giving open loop working. By applying sinusoidal stimulus variations, Stark and Sherman were able to conduct a frequency analysis and to deduce that the open loop transfer function was given by

$$\frac{0 \cdot 16 \varepsilon^{-0 \cdot 18s}}{(1 + 0 \cdot 1s)^3}$$

This indicates a system with a reaction time, represented by the term $\varepsilon^{-0 \cdot 18s}$, and a triple lag (three lag components all with the same 'corner' frequency—see p. 50). The system is stable despite its three lags because of its low open loop gain (0·16); in other words, we may say there is an adequate gain margin (see p. 49). Alternatively, we may draw a Nyquist diagram as in Figure 23, and see that the Nyquist criterion of stability is satisfied since the point $(-1 + j0)$ is not encircled by the open loop curve (see p. 48).

A useful test of the analysis up to this point, and an indication of the extent to which the small signal linear approximation is justified, would be to increase the gain of the system until the Nyquist criterion was no longer satisfied and then to close the loop. The system would

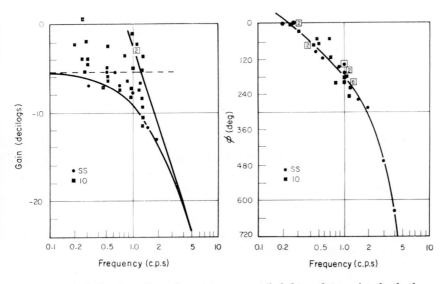

Figure 24. Bode plot of pupil response, compiled from data using both the normal open loop technique and the 'environmental clamp' method (see p. 119). The potential instability point, from the phase plot (the point at which the lag is 180°) is seen to occur at a little over 1 c.p.s. (From Stark, 1962a.)

then become unstable and if the increase of gain was achieved without altering the phase shift in the loop then the system should oscillate at the potential instability frequency as shown in Figure 24. A method of producing this increase in gain is shown in Figure 22 (c): it consists simply of adjusting the optical input apparatus as used by Stark and Sherman so that the beam of light instead of being focused so as to pass through the centre of the pupil and be unaffected by changes in pupil area, should now fall on the edge of the iris where a small change in pupil area will bring about disproportionately large changes in the amount of light reaching the retina. By using this technique, Stark and his colleagues (Stark and Cornsweet, 1958;

Stark and Baker, 1959) have shown that oscillation at the predicted frequency is in fact produced. Moreover, they applied a still more rigorous test to the analysis. By administering drugs to the eye, they were able to change the characteristics of the system so that the potential instability point was altered from 1·5 c.p.s. to 0·9 c.p.s. Increasing the gain of this altered system again produced instability at the predicted frequency. These results justify the use of linear analysis in the pupillary system, although it contains a considerable degree of non-linearity: by applying small signals to a non-linear system it is often, although not always, possible to obtain meaningful and useful results, but their validity must be verified before they can be used for prediction.

The fact that the system is far from linear has been demonstrated by Stark, Van der Tweel and Redhead (1962), by applying a super-position test. The response of a linear system to two superimposed inputs is the sum of its separate responses to each input acting alone (p. 29). Moreover the response of a linear system stimulated with short pulses takes a constant time to reach its peak value. The pupil system failed to satisfy these criteria. Other forms of non-linearity were discussed by Stark (1962b), who showed that the response to a step stimulus was different depending on whether it was a light–dark or a dark–light step.

One feature of the pupillary response was deliberately ignored in the analysis of the system described so far, but was considered later by Stark, Campbell and Atwood (1958). This is the problem of noise, which may be loosely defined as a component of output unrelated to the input. In fact the pupil area varies continuously and spontaneously even under conditions of constant illumination—a phenomenon known as pupil unrest. The origin of pupil unrest is unknown. Stark and coworkers (1958) were unable to find a generator of noise, but did make some deductions about its possible and impossible sources. It cannot arise from instability in the servomechanism, because of the ample gain margin in this system. Neither can it be due to spontaneous movements in the iris muscle, because pupil unrest has been shown to be coordinated between the eyes, and so must presumably be of central origin. Stark and coworkers (1958) concluded that pupillary unrest does not have any positive function, but is simply tolerated because it has no deleterious effect on visual acuity.

A slightly different experimental approach, allowing more flexibility in control of the system by the experimenter, has been used by Stark for some more recent investigations of the pupil light reflex system. He has called it an environmental clamp (Stark, 1961, 1962a) by analogy with the axon clamp technique used to study the transmission of nerve impulses. The device is illustrated in Figure 25, and consists of a 'clamping box' that introduces a variable gain and phase shift and that is connected into the open loop to form a new closed loop. This is easily achieved with the apparatus used in the earlier experiments, since a measure of the pupil area obtained by an infrared reflection method already existed in the form of an electrical signal, and the light stimulus was generated by a control system also operated by an electrical signal. The clamping box, therefore, needed only to be a straightforward electronic amplifier with variable gain and phase shift. With the gain of the clamping box set at zero, open

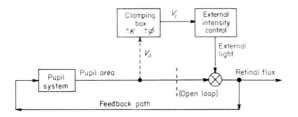

Figure 25. Clamping technique for obtaining the frequency response of a system. By inserting into the loop a box with variable gain and phase shift, the system can be made to oscillate over a range of frequencies. The gain and phase of the system can then be deduced from the fact that the net gain must be unity and net phase shift 180°. (From Stark, 1962a.)

loop operating could be obtained, but oscillation could now be obtained over a range of frequencies and with known increase of gain instead of using the empirical methods described previously. This is a powerful technique, and one that can be employed with advantage where a biological servomechanism can be driven by an external system having superior range and power. By its use, Stark (1962a) has further justified his linear analysis by obtaining oscillations over a range of frequencies in accordance with predictions made from the linear model, and has also been able to bring out certain non-linear features as well. These non-linear features have been investigated in greater detail by Baker (1963) who used an apparatus to stimulate both eyes with flashes of light and concluded that most of the non-linearity is introduced after the summation of the signals from the two eyes. It is therefore presumably cortical.

The second control system concerned with the eye is much more complicated than the pupil system. It is the system responsible for positional control of the eye: the determination of the direction of gaze. There are two main additional complications: first, the system has two degrees of freedom,* and secondly, it is much more markedly non-linear. There are three quite distinct kinds of eye movement: a slow drift, a fast tremor (physiological nystagmus) and irregularly occurring flicks (saccades).

When attempting to fixate on a stationary point the eyes are not kept still, but exhibit a combination of these three types of movement; when attempting to follow a moving stimulus the motion of the eyes can be analysed into a smooth following component with continuous tremor and irregularly spaced saccades superimposed.

The accurate measurement of eye movements presents a technically difficult problem because of their wide range both of amplitude and speed. This necessitates a high gain d.c. amplifying system with a wide bandwidth and very small drift. There are several methods that have been successfully used recently although they do not all satisfy all the ideal conditions: the extent to which they fail must be allowed for in drawing conclusions from the results.

The corneal reflection method used by Westheimer (1954) measures the distance between the temporal limbus and the image of a light source reflected from the cornea. This distance is obtained electronic- ally, using a scanning spot of light and measuring the time taken for the spot to pass from one point to the other. The method was im- proved on by Rashbass (1960) who employed a closed loop system. In this, a spot of light from an oscilloscope was focused on the limbus, and the output from a photomultiplier (which detected the reflected light) was used to position the spot so that it remained focused on the limbus independently of movements of the eye. Both these de- vices, however, could be used for detecting horizontal eye move- ments only. A more elaborate form of measurement using the corneal reflection was described by Mackworth and Mackworth (1958) who were able to make two dimensional measurements by using a tele- vision camera to detect the reflection, and the television picture as the stimulus; by mixing the eye position spot with the stimulus picture on a monitor screen a direct visual indication of direction of view can be obtained.

A second method of measuring eye movements is to make use of

---

* The direction of gaze relative to the head can be defined by two co- ordinates, and in addition the eyeball makes rolling movements about the antero-posterior axis. The rolling movements are largely compensatory in character and are not made independently of the other movements (Listing's and Donder's laws). It is a good approximation to say that there are only two degrees of freedom.

the standing potential that exists between the anterior and posterior poles of the eye. Electrodes placed on either side of the eye pick up a potential that varies with the direction of gaze; this method is known as electrooculography (EOG) and has been used by Ford, White and Lichtenstein (1959), and also by Shackel (1960) who has given a long and detailed account of the difficulties and limitations attached to it.

Other methods that have been used to measure eye movements involve the use of a contact lens: the most sophisticated of these methods is that used by Robinson (1963) using search coils embedded in the contact lens and exposing the head to alternating magnetic fields. By this means he was able to obtain simultaneous measurements of movements about three axes with an all electronic system having large bandwidth and high sensitivity. This method would seem to hold great promise for future detailed investigations of eye movements.

The other method requiring the use of a contact lens employs an optical lever principle, reflecting a beam of light from a mirror which is mounted either on the lens itself or on a stalk attached to the lens (Ratcliff and Riggs, 1950; Ditchburn and Ginsburg, 1953). It was by an extension of this method that the earliest experimental manipulations of the visual feedback system were achieved. A light was reflected from the mirror attached to the contact lens and focused through an optical system on to a screen where it could be used as the stimulus target. Thus target movements are produced by eye movements; by an appropriate optical arrangement the two movements can be made to compensate exactly producing the so-called stabilized retinal image (Ditchburn and Ginsburg, 1952; Riggs, Ratcliff, Cornsweet and Cornsweet, 1953; Ditchburn, 1955). The technique has been used by a number of workers interested in the effects of image stabilization on pattern perception, and many reports have appeared to the effect that pattern vision disappears after a short period of stabilization, and that it reappears intermittently thereafter (Ditchburn and Fender, 1955; Ditchburn, Fender and Mayne, 1959; Pritchard, Heron and Hebb, 1960). Some doubt has been cast on this result recently, however, by Cornsweet and Dwelley (1962) and Barlow (1963) both of whom claimed to have superior techniques of image stabilization; using these techniques, both sets of workers reported that the image disappeared permanently, suggesting that the reappearances previously observed were artifacts arising from imperfect stabilization. However, a counter-attack has been made by Bennet-Clarke and Evans (1963), whose stimuli were after-images impressed on the retina and so were certainly stabilized; they

5 + R.C.L.S.

obtained fragmentary reappearance of the image just as had been reported by the earlier workers.

The use of the stabilized image technique to make quantitative investigations of the eye movement control system was first reported by Fender and Nye (1961). They recognized that this procedure was in fact opening the visual feedback loop in the system, so that by applying sinusoidal disturbances of various frequencies to the stimulus object, the Nyquist and Bode diagrams could be obtained for the open loop system. The way in which a working model of the complete system has been derived by the use of a variety of analytical techniques has been described by Fender (1964a); it provides a good demonstration both of the difficulties of working with biological

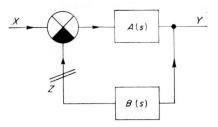

Figure 26. A practical limitation in investigating some biological feedback systems. The open loop response should be measured between X and Z, but the only points accessible are X and Y.

systems and of ways in which analytical methods can be modified to give useful information, even when the system to which they are applied does not strictly satisfy assumptions, such as linearity, on which the analysis is based.

The first complication we meet with is illustrated in Figure 26. The system has forward and feedback paths, with transfer functions $A(s)$ and $B(s)$, so that the open loop and closed loop transfer functions are given by

$$T_o(s) = A(s)B(s)$$

$$T_c(s) = \frac{A(s)}{1 + A(s)B(s)}$$

However, since the point Z of the diagram is at some completely unidentified point in the body (presumably somewhere in the cortex) it is not possible to follow the normal procedure of making open loop measurements between X and Z; instead, both open and closed loop measurements are made between X and Y. A mathematical manipulation (or in practical terms, a graphical construction) is necessary to convert the experimental data to the required form. This difficulty disappears if the transfer function of the feedback pathway, $B(s)$, is equal to unity (a situation commonly met in engineering systems— see p. 47). In this case, the measured open loop response, $A(s)$ is the same as $T_0(s)$ and the closed loop transfer function becomes

$$A(s)/[1 + A(s)]$$

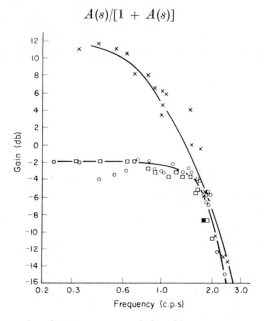

Figure 27. Comparison between open and closed loop measurements in the eye movement control system. Crosses: open loop measurements. Circles: closed loop measurements. Squares: closed loop points calculated from open loop measurements on the assumption that the feedback transfer function is unity. The agreement between calculated and measured points suggests that this assumption is justified. (From Fender, 1964a.)

By calculating this function from the open loop measurements and comparing results with the closed loop measurements, Fender (1964a) obtained the results shown in Figure 27, and concluded that the feedback transfer function is in fact unity.

A second, and more serious difficulty, is illustrated in Figure 28, which is a typical record of the response to a sinusoidal stimulus. The gross non-linearity of this response is evident, and in fact the results have to undergo considerable processing before they can be used. The response consists of two main parts: a following component

of about the same frequency as the input, and a succession of
saccades that are irregularly spaced and bear no simple relation to
the input. The work of Rashbass (1959, 1961) has shown that the two
components are generated separately, so it is reasonable in the first
analysis to regard the fundamental of the smooth following component at the input frequency as the response of the system, and to
disregard the saccades.

The eyeball and its associated musculature, considered as a purely
mechanical system, appears to have the dynamics of a lightly
damped second order system, as discussed in Chapter 3, with a natural
frequency of about 50 c.p.s. (Westheimer, 1954; Fender, 1964b). In

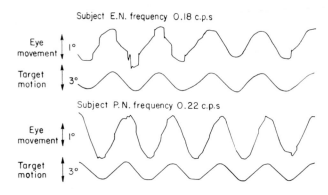

Figure 28. Two typical eye movement records, showing the gross non-
linearities in the response. In order to obtain useful results, the records have
to be analysed and the component of the response at the stimulus frequency
has to be extracted. (From Fender, 1964a.)

order to prevent excessive oscillation in such a circuit when subjected
to step changes (and it should be noted that step changes—saccades—
are a prominent feature of normal eye movements), we would expect
to find provision for additional damping by supplying a signal proportional to the rate of change of position (see p. 34): a suitable signal
exists in the output from the spindles in the extraocular muscles.
The use of these particular signals in this way has not been definitely
established, although pathways from the spindles have been traced
as far as the superior colliculus (Cooper, Daniel and Whitteridge,
1955).

The dynamics of the eyeball must figure in the final model that we
construct of the eye movement control system, although they do not
appear in most of the investigations that have been reported because
of the limited frequency response of the measuring apparatus. This

limitation on performance is one that has to be borne in mind when interpreting results of system analysis; it is brought out particularly in this case because the natural frequency of the system is so high compared with the normal operating frequencies (about 50 c.p.s. compared with about 1 c.p.s.).

In his analysis of the smooth following component of eye movements, Fender (1964a) first obtained the Bode diagrams and established linear approximations to the gain curve as shown in Figure 29.

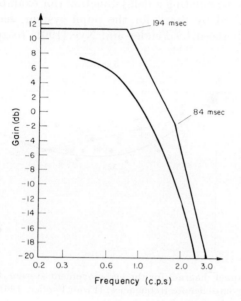

Figure 29. Bode plot of eye movement response, showing asymptotic approximations. (From Fender, 1964a.)

These approximations, as discussed in Chapter 3, were chosen to have slopes of multiples of 6 db/octave (see p. 50), and led to the conclusion that there were four simple lags involved, arranged in two pairs having corner frequencies of about 0·8 and 2 c.p.s. This would normally indicate a transfer function of the form

$$\frac{A}{(1 + \tau_1 s)^2 (1 + \tau_2 s)^2}$$

where

$$\tau_1 \doteqdot 85 \text{ msec}$$

and

$$\tau_2 \doteqdot 200 \text{ msec}$$

but unfortunately a further complication is encountered at this point when we come to consider the phase variation of the system. Figure 30 shows the Nyquist diagram, and we see that even for very high frequencies the phase lag does not exceed 90°, whereas the transfer function deduced from the gain characteristics would give a lag rising to 360°. It is common, especially in biological systems, to have a discrepancy between the gain and phase properties, but usually the difference is in the opposite direction—the actual lag is greater than that forecast from the gain curve—and in that case the difficulty can be resolved by postulating a delay function (for example the transfer function deduced by Stark for the pupil system: see p. 116). The hypothesis advanced by Fender and Nye (1961) to account for the

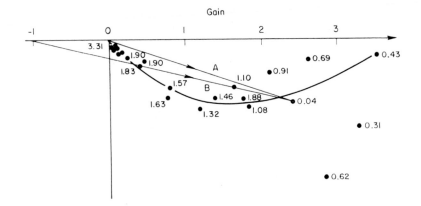

Figure 30. Nyquist diagram for the eye movement system. Figures alongside points denote frequencies. (From Fender, 1964a.)

absence of phase lag is that a positive feedback loop exists around part of the system, and they deduced from physiological evidence that such a positive feedback loop might exist in the oculomotor pathway. Fender (1964a) has also pointed out that a positive feedback pathway would be in line with the re-afference principle of von Holst and Mittelstaedt (see p. 85): the efference copy, which enables the organism to discriminate between its own movements and movements in the environment, is equivalent to a positive feedback pathway.

Yet another complication in the eye movement control system has been demonstrated by Stark, Vossius and Young (1962), who showed that the response to a regular, predictable stimulus was significantly different from the response to an unpredictable stimulus. This would imply that the responses obtained by straightforward sinusoidal stimulation include the predictor mechanism in the measurements, since a sinusoid is a predictable stimulus. To obtain comparable results for unpredictable stimuli, Stark, Vossius and Young presented

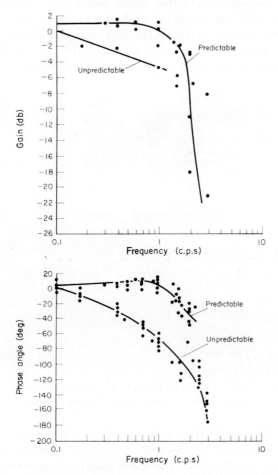

Figure 31. Dependence of the response of the eye movement system on the nature of the stimulus. Predictable inputs can be followed up to appreciably higher frequencies than can unpredictable ones. (From Stark and coworkers, 1962.)

a stimulus consisting of the sum of a number of sine waves of nonrelated frequencies, and analysed the response into components at each of the input frequencies. This analysis is complicated, and was conducted by a computer: the result of the analysis, compared with normal sinusoidal analysis, is shown in Figure 31. It is interesting to notice that the phase lag for unpredictable inputs is much more nearly in line with what would be expected from the gain plot.

The model finally proposed by Fender (1964a) is shown in Figure 32. This model is in part speculative, and is not as specific or quantitative as we would like, but nevertheless does give useful ideas about the possible form of organization of the system, and indicates some ways in which additional experiments might be devised to elucidate its workings further. Precise descriptions are in any case difficult for a

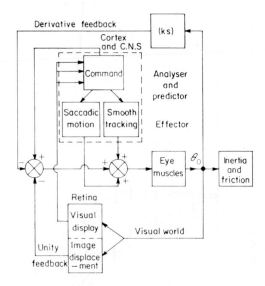

Figure 32. Schematic diagram of Fender's model for the complete eye movement control system. (From Fender, 1964a.)

system of this complexity, and may indeed be impossible without the use of some new mathematical language specially adapted for learning systems. However, the use of standard engineering methods of analysis has cast considerable light on the workings of the oculomotor system despite its gross departures from ideal engineering characteristics.

All of the work described so far, as pointed out at the beginning of this section, did not include the dynamics of the eyeball itself. Fender (1964b) examined this part of the system and found that the mechanical properties of the eyeball appeared to vary during the course of a saccade and also between different saccades, some of which displayed overshoot and some did not. The natural frequency

(see p. 35) varied between 20 and 100 c.p.s. and the damping coefficient between 0·1 and 0·8—at some times the damping coefficient
even appeared to be negative. The conclusion reached by Fender as
an explanation of this behaviour is that the muscles receive a complete pattern of excitation during a saccade rather than just a single
command signal related to the desired final position.

Recently, a quite different form of analysis of the eye movement
control system has been attempted by Young and Stark (1963a, b),
who have started from the assumption that the system operates in the
form of a sampled-data controller. This is a form of control in which
the state of the output is not monitored continuously, but is examined at intervals, the operation of the control system at any time
being dependent on information gathered at the last sampling instant.
The analysis of this kind of system is beyond the scope of this book;
suffice it to say that equations are operated on by using z-transforms
instead of Laplace transforms and that a form of transfer function
can be derived. The results of this analysis by Young and Stark are
impressive, and seem to take us nearer to a complete description of
the system than has been achieved by any other method, although
it was still confined to unpredictable inputs and to the gross following
and saccadic movements. The prediction mechanism still defies
analysis.

Conclusion

This chapter began with a description of early attempts to analyse
the responses of human operators while performing a tracking task.
It was said then that part of the complication arose because the situation is such an elaborate one, involving a considerable proportion of
the sensori-motor equipment of the subject. The foregoing description of the eye movement system illustrates this point, for this system
must play an important part in tracking, and we can expect the
significant features of the system to make a contribution to the tracking response. It should therefore come as no surprise to learn that a
sampled-data analysis of the tracking situation has been carried out
by Bekey (1962), using very much the same ideas and techniques as
Young and Stark (1963a) had done with the eye movement control
system. This analysis appears to provide a model that fits the experimental evidence more closely than any of the previously proposed
linear models, and, in view of the success of Young and Stark (1963b)
with a similar model for the eye movement system, seems to suggest
that the difficulties of the tracking task might reside entirely in the

eye movement system. Nearly all the features previously reported in tracking behaviour appear also in eye movements—it would be interesting to look for a range effect with unpredictable step inputs (see p. 82). It is certainly not necessary with the existing experimental evidence to postulate anything other than a straightforward coordinating system between hand and eye, all the elaborate information processing taking place in the eye system.

Having now come full circle in our discussions we can see that the contribution of mathematical and engineering techniques to the elucidation of biological control problems has already been considerable, and that further application of more sophisticated methods of analysis holds great promise for the future.

References

General

Bergeijk, W. A. van and David, E. E. (1959). 'Delayed handwriting.' *Percept. Mot. Skills*, **9**, 347.

Chapman, K. M. and Smith, R. S. (1963). 'A linear transfer function underlying impulse frequency modulation in a cockroach mechanoreceptor.' *Nature*, **197**, 699.

Craik, K. J. W. (1947). 'Theory of the human operator in control systems: I. The operator as an engineering system.' *Brit. J. Psychol.*, **38**, 56.

Craik, K. J. W. (1948). 'Theory of the human operator in control systems: II. Man as an element in a control system.' *Brit. J. Psychol.*, **38**, 142.

Ellson, D. G. (1949). 'The application of operational analysis to human motor behaviour.' *Psychol. Rev.*, **56**, 9.

Greene, J. (1954). 'Man as a servo component.' *Control Eng.*, **1**, 2, 58.

Hammond, P. H. (1960). 'An experimental study of servo-action in human muscular control.' *Proc. 3rd Int. Conf. Med. Electron.*, Inst. Elec. Engr., London, p. 190.

Hammond, P. H., Merton, P. A. and Sutton, C. G. (1956). 'Nervous gradation of muscular contraction.' *Brit. Med. Bull.*, **12**, 214.

Hassenstein, B. (1960). 'Die Bisherige Rolle der Kybernetik in der biologischen Forschung.' *Naturw. Rundschau*, **13**, 349, 373, 419 (in three parts).

Holst, E. von (1954). 'Relations between the central nervous system and the peripheral organs.' *Brit. J. Anim. Behav.*, **2**, 89.

Holst, E. von and Mittelstaedt, H. (1950). 'Das Reafferenzprinzip.' *Naturwissenschaften*, **37**, 464.

Kalmus, H., Fry, D. B. and Denes, P. (1960). 'Effects of delayed visual control on writing, drawing and tracing.' *Language Speech*, **3**, 96.

Machin, K. E. (1964). 'Feedback theory and its application to biological systems.' *Symp. Soc. Exp. Biol.*, **18**, 421.

Machin, K. E. and Pringle, J. W. S. (1960). 'The physiology of insect fibrillar muscle: III. The effect of sinusoidal changes of length on a beetle flight muscle.' *Proc. Roy. Soc. (London), Ser. B*, **152**, 311.

Maldonado, H. (1964). 'The control of attack by octopus.' *Z. Vergleich. Physiol.*, **47**, 656.

Mayne, R. (1951). 'Some engineering aspects of the mechanism of body control.' *Elec. Eng.*, **70**, 207.

Merton, P. A. (1953). 'Speculations on the servo control of movement.' *CIBA Found. Symp. Spinal Cord.*

Mittelstaedt, H. (1949). 'Telotaxis und Optomotorik von Eristalis bei Augeninversion.' *Naturwissenschaften*, **36**, 90.

Mittelstaedt, H. (1954). 'Regelung in der Biologie.' *Regelungstechnik.*, **2**, 177.

Pringle, J. W. S. (1962). 'The Input Element.' *Symp. Soc. Exp. Biol.*, **16**, 1.

Pringle, J. W. S. (1964). 'Input and output elements of behavioural models.' In Reiss, R. F. (Ed.), *Neural Theory and Modeling*, Stanford University Press, California, U.S.A.

Pringle, J. W. S. and Wilson, V. J. (1952). 'The response of a sense organ to a harmonic stimulus.' *J. Exp. Biol.*, **29**, 220.

Searle, L. V. and Taylor, F. V. (1948). 'Studies of tracking behaviour: I. Rate and time characteristics of simple corrective movements.' *J. Exp. Psychol.*, **38**, 615.

Smith, K. U. and Smith, W. M. (1962). *Perception and Motion*, Saunders, London.

Stark, L. and Herman, H. J. (1961). 'Transfer function of a biological photoreceptor.' *Nature*, **191**, 1173.

Stark, L., Iida, M. and Willis, P. A. (1961). 'Dynamic characteristics of the motor coordination system in man.' *Biophys. J.*, **1**, 279.

Stark, L., Payne, R. and Okabe, Y. (1962). 'Online digital computer for measurement of a neurological control system.' *Comm. Assoc. Comp. Mach.*, **5**, 567.

Taylor, F. V. (1949). 'Certain characteristics of the human servo.' *Elec. Eng.*, **68**, 235.

Taylor, F. V. and Birmingham, H. P. (1948). 'Studies of tracking behaviour: II. The acceleration pattern of quick manual corrective responses.' *J. Exp. Psychol.*, **38**, 783.

Taylor, W. K. (1957). 'Measurement and control in the human operator.' *Trans. Soc. Instr. Tech.*, **9**, 104.

Wiener, N. (1949). *Cybernetics*, Technology Press of M.I.T. and Wiley, New York.

Temperature Regulation

Bazett, H. C. (1949). 'The regulation of body temperature.' In Newburgh, L. (Ed.), *Physiology of Heat Regulation and the Science of Clothing*, Saunders, London.

Benzinger, T. H. (1959). 'On physical heat regulation and the sense of temperature in man.' *Proc. Nat. Acad. Sci. U.S.*, **45**, 645.

Benzinger, T. H. (1961a). 'The diminution of thermoregulatory sweating during cold reception at the skin.' *Proc. Nat. Acad. Sci. U.S.*, **47**, 1683.

Benzinger, T. H. (1961b). 'The human thermostat.' *Sci. Am.*, **204**, 134.

Benzinger, T. H. (1963). Peripheral cold- and central warm-reception, main origins of human thermal discomfort.' *Proc. Nat. Acad. Sci. U.S.*, **49**, 832.

Benzinger, T. H. (1964). 'The thermal homeostasis of man.' *Symp. Soc. Exp. Biol.*, **18**, 49.

Benzinger, T. H. and Kitzinger, C. (1949). 'Direct calorimetry by means of the gradient principle.' *Rev. Sci. Instr.*, **20**, 849.

Benzinger, T. H. and Kitzinger, C. (1963). 'Gradient layer calorimetry and human calorimetry.' In *Temperature—its Measurement and Control in Science and Industry*, Vol. 3, Reinhold, New York.

Benzinger, T. H., Kitzinger, C. and Pratt, A. W. (1963). 'The human thermostat.' In *Temperature—its Measurement and Control in Science and Industry*, Vol. 3, Reinhold, New York.

Benzinger, T. H., Pratt, A. W. and Kitzinger, C. (1961). 'The thermostatic control of human metabolic heat production.' *Proc. Nat. Acad. Sci. U.S.*, **47**, 730.

Benzinger, T. H. and Taylor, G. W. (1963). 'Cranial measurement of internal temperature in man.' In *Temperature—its Measurement and Control in Science and Industry*, Vol. 3, Reinhold, New York.

Brebner, D. F. and Kerslake, D. McK. (1960). 'The effect of cooling the legs on the rate of sweat production from the forearm when the rest of the body is not exchanging heat.' *J. Physiol.*, **152**, 65P.

Crosby, R. S., Hall, R. A. and Hardy, J. D. (1961). 'Electrical analogue simulation of temperature regulation in man.' *Trans. Inst. Radio Engrs.*, **BME 8**, 245.

Euler, C. von (1964). 'The gain of the hypothalamic temperature regulating mechanism.' *Progr. Brain Res.*, **5**, 127.

Hardy, J. D. (1961). 'Physiology of temperature regulation.' *Physiol. Rev.*, **41**, 521.

Hertig, B. A. and Belding, H. S. (1961). 'Rates of adjustment of sweating and body temperatures following exposure to heat.' *Federation Proc.*, **20**, 211.

Magoun, H. W., Harrison, F., Brobeck, J. R. and Ranson, S. W. (1938). 'Activation of the heat loss mechanism by local heating of the brain.' *J. Neurophysiol.*, **1**, 101.

Randall, W. C. and Hertzmann, A. B. (1952). 'Dermatomal recruitment of sweating.' *J. Appl. Physiol.*, **5**, 399.

Selle, W. A. (1952). *Body Temperature*, Thomas, Springfield, Illinois.

Wissler, E. H. (1961). 'Steady state temperature distribution in man.' *J. Appl. Physiol.*, **16**, 734.

Wissler, E. H. (1963). 'An analysis of factors affecting temperature levels in the nude human.' In *Temperature—its Measurement and Control in Science and Industry*, Vol. 3, Part 3, Reinhold, New York, p. 603.

Wyndham, C. H. and Atkins, A. R. (1961). 'An approach to the solution of the human biothermal problem with the aid of an analogue computer.' *Proc. 3rd Int. Conf. Med. Electron.*, Inst. Elec. Engr., London.

Optomotor Responses

Bliss, J. C. (1962). 'Visual information processing in the eye of the beetle *Lixus*.' In Pollock, D. K., Koesler, C. J. and Tippet, J. T. (Eds.), *Optical Processing of Information*, Spartan, Baltimore.

Hassenstein, B. (1951). 'Ommatidienraster und Afferente Bewegungsintegration.' *Z. Vergleich. Physiol.*, **33**, 301.

Hassenstein, B. (1958). 'Die Stärke von optokinetischen Reaktionen auf verschiedene Mustergeschwindigkeiten.' *Z. Naturforsch.*, **13b**, 1.

Hassenstein, B. (1959a). 'Wie sehen Insekten Bewegungen?' *Umschau*, **10**, 302.

Hassenstein, B. (1959b). 'Optokinetische Wirksamkeit bewegter periodischer Muster.' *Z. Naturforsch.*, **14b**, 659.

Hassenstein, B. and Reichardt, W. (1956). 'Systemtheoretische Analyse der Zeit-Reihenfolgen und Vorzeichenauswertung bei der Bewegungsperzeption des Rüsselkafers Chlorophanus.' *Z. Naturforsch.*, **11b**, 513.

Horridge, G. A. and Sandeman, D. C. (1964). 'Nervous control of optokinetic responses in the crab *Carcinus*.' *Proc. Roy. Soc. (London)*, *Ser. B*, **161**, 216.

Kalmus, H. (1948). 'Optomotor responses in *Drosophila* and *Musca*.' *Physiol. Comp. Oecol.*, **1**, 127.

Mittelstaedt, H. (1949). 'Telotaxis und Optomotorik von Eristalis bei Augeninversion.' *Naturwissenschaften.*, **36**, 90.

Mittelstaedt, H. (1951). 'Zur Analyse physiologischer Regelungssysteme.' *Verhandl. Deut. Zool. Ges.*, p. 151.

Reichardt, W. (1961). 'Autocorrelation: a principle for the evaluation of sensory information by the central nervous system.' In Rosenblith, W. A. (Ed.), *Sensory Communication*, M.I.T. Press, Massachusetts, and Wiley, New York.

Orientation and Prey Capture

Fraenkel, G. and Gunn, D. L. (1940). *The Orientation of Animals*, Oxford University Press, London.

Holst, E. von (1950). 'Die Arbeitweise des Statolithenapparates bei Fischen.' *Z. Vergleich. Physiol.*, **47**, 381.

Jander, R. (1963a). 'Insect Orientation.' *Ann. Rev. Entomol.*, **8**, 95.

Jander, R. (1963b). 'Grundleistungen der Licht-und Schwereorientierung von Insekten.' *Z. Vergleich. Physiol.*, **47**, 381.

Kühn, A. (1919). *Die Orientierung der Tiere im Raum*, Fischer Jena.

Loeb, J. (1918). *Forced Movements, Tropisms, and Animal Conduct,* Lippincett, Philadelphia and London.

Mast, S. O. (1911). *Light and the Behaviour of Organisms,* Wiley, New York.

Mittelstaedt, H. (1952). 'Über den Beutefangmechanismus der Mantiden.' *Verhandl. Deut. Zool. Ges.,* p. 102.

Mittelstaedt, H. (1957). 'Prey Capture in Mantids.' *Recent Advan. Invertebrate Physiol., Symp. Eugene, Ore,* p. 51.

Mittelstaedt, H. (1962). 'Control Systems of Orientation in Insects.' *Ann. Rev. Entomol.,* **7**, 177.

Mittelstaedt, H. (1963). 'Bikomponenten—Theorie der Orienierung.' *Ergeb. Biol.,* **26**, 253.

Mittelstaedt, H. (1964a). 'Basic solutions to a problem of angular orientation.' In Reiss, R. F. (Ed.), *Neural Theory and Modeling,* Stanford University Press, California, U.S.A.

Mittelstaedt, H. (1964b). 'Basic control patterns of orientational homeostasis.' *Symp. Soc. Exp. Biol.,* **18**, 365.

Movements of Eye and Pupil

Baker, F. H. (1963). 'Pupillary response to double pulse stimulation: a study of non-linearity in the human pupil system.' *J. Opt. Soc. Am.,* **53**, 1430.

Barlow, H. B. (1963). 'Slippage of contact lenses and other artifacts in relation to fading and regeneration of supposedly stable retinal images.' *Quart. J. Exp. Psychol.,* **15**, 36.

Bennet-Clarke, H. C. and Evans, C. R. (1963) 'Fragmentation of patterned targets when viewed as prolonged after-images.' *Nature,* **199**, 1215.

Campenhausen, C. von. (1963). 'Quantitative Beziehungen zwischen Lichtreiz und Kontraktion des Musculus Sphincter pupillae vorn Scheibenzunglen.' *Kybernetik,* **1**, 6, 249.

Cooper, S., Daniel, P. M. and Whitteridge, D. (1955). 'Muscle spindles and other sensory endings in the extrinsic eye muscles: The physiology and anatomy of these receptors and of their connections with the brain stem.' *Brain,* **78**, 564.

Cornsweet, T. N. and Dwelley, D. (1962). 'Perfectly stabilized retinal images.' *J. Opt. Soc. Am.,* **52**, 598.

Ditchburn, R. W. (1955). 'Eye movements in relation to retinal action.' *Opt. Acta,* **1**, 4, 171.

Ditchburn, R. W. and Fender, D. H. (1955). 'The stabilized retinal image.' *Opt. Acta,* **2**, 3, 128.

Ditchburn, R. W., Fender, D. H., and Mayne, S. (1959). 'Vision with controlled movements of the retinal image.' *J. Physiol.,* **145**, 98.

Ditchburn, R. W. and Ginsborg, B. L. (1952). 'Vision with a stabilized retinal image.' *Nature,* **170**, 36.

Ditchburn, R. W. and Ginsborg, B. L. (1953). 'Involuntary eye movements during fixation.' *J. Physiol.,* **119**, 1.

Fender, D. H. (1964a). 'The eye movement control system: evolution of a model.' In Reiss, R. F. (Ed.), *Neural Theory and Modeling*, Stanford University Press, California, U.S.A.

Fender, D. H. (1964b). 'Techniques of systems analysis applied to feedback pathways in the control of eye movements.' *Symp. Soc. Exp. Biol.*, **18**, 401.

Fender, D. H. and Nye, P. W. (1961). 'An investigation of the mechanisms of eye movement control.' *Kybernetik*, **1**, 81.

Ford, A., White, C. T. and Lichtenstein, M. (1959). 'Analysis of eye movements during free search.' *J. Opt. Soc. Am.*, **49**, 287.

Lowenstein, O. (1955). 'Pupillary reflex shapes and topical clinical diagnosis.' *Neurology*, **5**, 631.

Mackworth, J. F. and Mackworth, N. H. (1958). 'Eye fixations recorded on changing visual scenes by the television eye marker.' *J. Opt. Soc. Am.*, **48**, 439.

Pritchard, R. M., Heron, W. and Hebb, D. O. (1960). 'Visual perception approached by the method of stabilized images.' *Can. J. Psychol.*, **14**, 67.

Rashbass, C. (1959). 'Barbiturate nystagmus and the mechanisms of visual fixation.' *Nature*, **183**, 897.

Rashbass, C. (1960). 'New method of recording eye movements.' *J. Opt. Soc. Am.*, **50**, 642.

Rashbass, C. (1961). 'The Relationship between saccadic and smooth tracking eye movements.' *J. Physiol.*, **159**, 326.

Ratcliff, F. and Riggs, L. A. (1950). 'Involuntary movements of the eye during monocular fixation.' *J. Exp. Psychol.*, **40**, 687.

Riggs, L. A., Ratcliff, F., Cornsweet, J. C. and Cornsweet, T. N. (1953). 'The disappearance of steadily fixated test objects.' *J. Opt. Soc. Am.*, **43**, 495.

Robinson, D. A. (1963). 'A method of measuring eye movements using a scleral search coil in a magnetic field.' *Trans. Inst. Elec. Electron. Engrs.*, **BME 10**, 137.

Shackel, B. (1960). 'Electro-oculography: the electrical recording of eye position.' *Proc. 3rd Int. Conf. Med. Electron*, Inst. Elec. Engr., London.

Stark, L. (1961). 'An artificial clamp for pupil oscillations.' *Federation Proc.*, **20**, 347.

Stark, L. (1962a). 'Environmental clamping of biological systems: pupil servomechanism.' *J. Opt. Soc. Am.*, **52**, 925.

Stark, L. (1962b). 'Biological rhythms, noise and asymmetry in the pupil–retinal control system.' *Ann. N.Y. Acad. Sci.*, **98**, 1096.

Stark, L. (1963). 'Stability, oscillation and noise in the human pupil servomechanism.' *Bol. Inst. Estud. Med. Biol. Mex.*, **21**, 201.

Stark, L. and Baker, F. (1959), 'Stability and oscillations in a neurological servomechanism.' *J. Neurophysiol.*, **22**, 156.

Stark, L., Campbell, F. W. and Atwood, J. (1958). 'Pupil unrest: an example of noise in a biological servomechanism.' *Nature*, **182**, 857.

Stark, L. and Cornsweet, T. N. (1958). 'Testing a servoanalytic hypothesis for pupil oscillations.' *Science*, **127**, 588.

Stark, L. and Sherman, P. M. (1957). 'A servoanalytical study of the consensual pupil reflex to light.' *J. Neurophysiol.*, **20**, 17.

Stark, L., Tweel, H. van der and Redhead, J. (1962). 'Pulse response of the pupil.' *Acta Physiol. Pharmacol. Neerl.*, **11**, 235.

Stark, L., Vossius, G. and Young, L. R. (1962). 'Predictive control of eye tracking movements.' *Trans. Inst. Radio Engrs.*, **HFE 3**, 52.

Westheimer, G. (1954). 'Mechanism of saccadic eye movements.' *Arch. Ophthalmol.*, **52**, 710.

Young, L. R. and Stark, L. (1963a). 'Variable feedback experiments testing a sampled data model for eye tracking movements.' *Trans. Inst. Radio Engrs.*, **HFE 4**, 38.

Young, L. R. and Stark, L. (1963b). 'A discrete model for eye tracking movements.' *Trans. Inst. Radio Engrs.*, **MIL 7**, 113.

Conclusion

Bekey, G. A. (1962). 'The human operator as a sampled data system.' *Trans. Inst. Radio Engrs.*, **HFE 3**, 43.

6

The Control of Speech and Voice

D. B. FRY

Department of Phonetics, University College, London

The production of speech sounds involves broadly three kinds of muscular activity, those concerned with respiration, phonation and articulation. Speech represents a secondary use of muscular mechanisms which have for the human being a more vital, primary use. The breathing muscles which help to keep man alive by providing for oxygen and carbon dioxide exchange, provide the power supply for speech; the larynx which acts as an inlet–outlet valve to the respiratory system, works as a vibrator to provide the carrier wave for speech; the muscles of the pharynx, tongue, soft palate and lips used for taking in and swallowing food are used in speech for articulatory movements. The control of these three main systems which is sufficient to maintain life has to be modified and very much supplemented when they are used for speech, since the movements of speech are the most complex and finely coordinated movements that a human being makes. During speech, the muscles of the chest, the larynx and the head perform miracles of accurate timing which are made possible only by a very complex control mechanism.

Although we think of speech as a unitary activity, it incorporates several different levels of operation. Its function in normal circumstances is largely communicative but in all speech there is to a greater or less degree an expressive element. In a lecture, for example, we generally expect the communicative function to be predominant; in very emotional speech, the expressive element gains the upper hand and may even interfere with communication; in singing, the two functions almost coincide since the musical effect of the singing is very largely dependent on the expressive capabilities of the singer. Breathing activity is the basis of speech for both functions, though we shall see later that it is influenced by the proportion of communicative

137

to expressive element in the speech. Phonation gives rise to the basic sound in speech and is also the main vehicle for the expressive function. Although, therefore, the control of all three activities— respiration, phonation and articulation—is determined by both the communicative and the expressive functions, it is larynx action that is most affected by the expressive element. It is particularly the voice as distinct from speech that is the vehicle for expression and it is predominantly articulation that is the medium for communication.

In speech, as in all other movements, there is the direct control of the muscles through motor pathways, what we may think of as a 'feed-forward' control (open loop), and there is also control through feedback loops. In the following sections we shall deal first in some detail with the operation of the feed-forward control and then consider how feedback control is used to obtain the very complex coordination of muscle activity which speech involves.

The Linguistic Basis of Speech

The control of speech originates in the speech centres of the brain whose most important function is to organize in a language form what the speaker wishes to say (Fry, 1963; for a discussion of brain mechanisms in speech, see Penfield and Roberts, 1959). We can take as the unit of this organization the sentence, but this means a sentence in the spoken language, not in the written language. In the spoken language a sentence is any complete remark and it does not necessarily have the full grammatical form which we expect in a written sentence: often enough it may consist of a single word or a short phrase. Spoken sentences can, of course, also have a quite complex grammatical structure involving a number of subordinate clauses. Whether the sentence be simple or complex, its organization requires the same kinds of operation on the part of the speech centres —first, the selection from memory stores of the content words which will convey the line of thought and second, the selection of a suitable syntactic framework. Both types of selection form a continuing process during the production of speech. It probably never happens in spontaneous speech that a whole sentence is determined as to content words and form before the speaker embarks on the sentence. Work on hesitations in speech suggests that only the first part of an utterance is organized before speech begins and the modifications in both line of thought and in structure commonly take place as the utterance proceeds.

Control of speech in a given utterance begins then with sentence selection or more strictly with the selection of part of a sentence. This in turn entails word selection from a store, not only the selection of content words but also of form words which determine the grammatical structure of the sentence. At this level too the order of the words is decided. The next selection operation deals with the morphemes, that is to say with linguistic units smaller than words which have grammatical function—in English, for example, such affixes as *-ing*, *-ed*, *-ly*, *-ness* and so on. Finally there is the selection of appropriate phonemes, that is of the sounds which must succeed each other in the actual utterance if the listener is to be able to decode the spoken message. Whilst these selections are being made, other control circuits are determining the rhythmic and intonation patterns which are to be incorporated in the utterance.

In all these selection operations, the choice actually made by the speaker is determined to a very large extent by the laws of the language he is using. The structure of the sentence must conform to the sentence-forming rules of the language, words must be chosen from the vocabulary, the morpheme sequence must be acceptable, the phoneme sequence must be a series of choices from the phoneme inventory (in English, for example, choices from a total list of about forty units). Equally, the rhythm and intonation patterns must be selected from those which make up the particular language system.

Every utterance includes in addition a considerable personal component supplied by the individual speaker rather than by the language system. The choice of words, for instance, will be influenced by this factor since each speaker has his personal vocabulary and his favourite choices in given contexts. The personal contribution is however even stronger on the expressive side, and such matters as emphasis (i.e. rhythm), intonation and voice quality reflect this component to a great extent.

To summarize this level of speech control, then, we can say that the speaker's brain selects first certain content words and an outline sentence form; next it selects form words and the necessary morphemes, and then the phoneme string which constitutes these units; to these are added appropriate rhythm and intonation patterns. The output of each selection operation is a linear arrangement of the appropriate units, that is a string of words, morphemes or phonemes. The selections constitute a forward going, continuous process as a result of which the brain provides a flow of operating instructions to the muscles of the speech mechanism. In the next section we shall consider the effects of these neural control signals on the activity of the muscles of respiration, phonation and articulation.

The Control of Respiration in Speech

In speech the rhythm of quiet breathing is drastically altered; the more or less even rhythm in which the time taken up in inspiration is about equal to that used in expiration is replaced by a time scheme in which inspiration is very rapid and expiration is very long drawn out. This change in ratio during speech is such that the expiratory phase generally lasts five to ten times as long as the inspiratory, and may in extreme cases last as much as forty to fifty times as long (Peterson, 1957). The amount of time that expiration lasts and also the amount of speech produced in one expiration are matters of considerable individual variation. Within the speech of one individual there is also variation in the frequency of inspiration and in the number of words or syllables per breath. Although these two factors naturally tend to be connected they are not entirely interdependent, since the speaker may change his rate of speaking without altering his breathing rate. A speaker who is dealing with a topic with which he is emotionally involved tends to take in breath more often, and hence the frequency of inspiration is largely dependent on the proportions of the communicative and the expressive elements (Goldman-Eisler, 1955).

The need for inspiration at irregular intervals causes the occurrence of *breath pauses* in the stream of speech. The points at which such breath pauses may occur is dictated by the structure of the sentence: they are inserted only at the end of a sentence, a clause or a phrase and are thus subject to the grammar of the language. In organizing the flow of speech movements the brain, therefore, includes in the programme the instruction for inspiration only at permissible moments in the sequence. The individual who breathes during speech more frequently than another only uses more of these possible breathing points. The incidence of breath pauses can be observed most easily in a speaker who is reading aloud. In all spontaneous speech there occur also *hesitation pauses* which are not directly connected with respiration but which may themselves affect respiration indirectly. It will be as well at this point to say something about the function of hesitations in the control of speech.

The Role of Hesitation Pauses in Speech

It was said earlier that word selection is a continuing process during speech; it is not, however, uniformly continuous for it goes forward in little spurts, as it were. In any sentence there are the form words, which determine the grammatical structure of the sentence, and there are the content words which mainly determine its content. During the encoding of the speech, the form words are more readily found and organized in sequence than content words, and so there is a tendency for encoding to proceed rapidly through strings of form words and to slow down or to stop at points where there are content words. It is the process of searching for content words that causes hesitation pauses in spontaneous speech; it has been shown by experiment that the occurrence of hesitation pauses and in fact the actual duration of each pause is a function of the flow of information in speech. A pause tends to occur before a word of high information content (i.e. of low predictability) and the less predictable the word is the longer the pause is likely to be. Here again there is great individual variation in the frequency and the length of pauses, but the relation between pause time and information generally holds good (Goldman-Eisler, 1958).

The amount of time spent in hesitation pauses reflects the difficulty of the encoding process. Speakers have been shown to spend more time in pausing when performing a difficult speech task such as summarizing or abstracting the meaning of a scene or event than when giving a concrete description of the same subject (Goldman-Eisler, 1961). It is clear therefore that a certain time is necessary for organizing the encoding of the message and that this time depends on the amount of information to be conveyed. On the evidence of pause time experiments, we might consider that the unit of organization at this level, in fluent speech as well as in hesitant speech, is probably one content word plus the ensuing form words.

The Control of Vocal Cord Action

We have seen so far that the control programme provided by the speech centres is probably written in sections equal to several words (possibly content word plus form words) and that it contains first instructions to take in breath for the beginning of the utterance and to pay out this breath gradually until a convenient syntactical point is reached where the next breath may be taken. The next important section of the speech apparatus is the larynx mechanism where the vibratory action of the vocal cords has to be controlled.

The action of the vocal cords has a number of functions in speech. The opening and closing of the cords supplies the carrier wave for speech, it produces the sound of voice which makes speech highly audible and provides most of the acoustic power; variation in the frequency of vocal cord vibration is the medium for many features of the intonation pattern; changes in the wave-form of the laryngeal vibrations give rise to changes of voice quality and last, the switching on and off of vocal cord vibration is an important factor in certain phonemic distinctions.

The amplitude of vocal cord vibrations, which is largely responsible for the loudness of the resulting speech, is very much under feedback control and will be considered later. Variation in vocal cord frequency is a very important feature of speech and is correlated with intonation pattern. In speech there is continuous variation in the period of vocal cord vibration; this is the main feature which differentiates speech from singing, for in singing the vocal cord vibrations have to remain at the same frequency (apart from the regular modulation due to vibrato) for an appreciable time. During speech, control is concerned rather with the direction, extent and rate of frequency change than with absolute frequency level. The pattern of frequency variation is programmed for a considerable stretch of speech at one time, for a unit not less than a clause and often for a whole sentence. This means that although the whole sequence of word choices may not be made before a sentence begins, a decision has usually been made as to whether the sentence is to have the intonation proper to an affirmation, an exclamation, a question and so on. The expressive side of the utterance is also already determined and this affects particularly the range of vocal cord frequency variation. If there are hesitation pauses, the pattern of larynx frequency bridges the gap and is continued after it according to the original programme.

Another dimension of variation in vocal cord action is concerned with the disposition of the vocal cords and their modes of vibration. The tension of the muscle fibres at different points along the cords, the shape of the cords themselves, the speed of closing and opening of the cords, the duration of the closed and the open phase are all factors which regulate the speaker's voice quality and are all to some degree under the control of the speech centres. In some languages, changes of voice quality form part of the linguistic system; certain African and oriental languages, for example, include differences between the meanings of words conveyed by a contrast between ordinary voice quality and breathy voice quality, produced by incomplete closure of the cords and an increase of breath force from the lungs during phonation. Differences of voice quality do not have a linguistic function in European languages generally and in English speakers voice quality is dependent on the personality of the speaker and on the expressive element in the speech. Thus moods of harshness, admiration, awe, pity and so on are realized in speech largely through the action of the larynx, that is through voice quality.

One aspect of vocal cord action has a linguistic function in every known language and that is the switching from larynx vibration to the open position of the cords. This switching is correlated with the contrast between voiced and voiceless sounds, in English, for example, between *s* and *z*, *p* and *b*, *ch* and *j*, etc. Hence the control of this switching, like other aspects of articulation which will be considered later, is linked to the selection of the phoneme string. It is interesting to note that one component in many cases of stuttering is connected with the switching of the larynx mechanism. The kind of stammerer who is liable to re-iterate initial consonants is experiencing difficulty in switching on the vocal cords for the following vowel, and not, as is sometimes popularly supposed, finding difficulty in the articulation of a consonant.

The Control of Articulation

It is mainly as a result of phonation that speech is easily audible; in order that it should be intelligible, it is necessary for the sound of voice coming from the larynx to be modulated by the movements of the articulators. The train of air pulses generated by the vocal cords is fed into a vocal tract which is varied continuously in shape and length with consequent modulations of the overall amplitude and the frequency content of the resulting acoustic wave. The main components of the articulatory mechanism are the pharynx, the tongue, the soft palate and the lips, and there is some contribution to articulation by the larynx, as we have seen, and also by the thoracic muscles which play a part in the grouping of sound sequences and in syllabification. All these elements work as a coordinated whole; the time course of successive actions by different groups of muscles is controlled within narrow limits so that it is possible not only to achieve synchronous action by several muscles but to vary the phase relations of a considerable number of muscle actions as required. Some idea of the complexity of these operations can perhaps be gained from a very much simplified description of the movements involved in a short English sequence. Let us suppose that a speaker is going to utter the word *branch*; the gross articulatory movements he makes will take place in somewhat the following way. As he begins the utterance, the vocal cords are held apart, the two lips are brought together and pressed upon each other and at the same time the soft palate is drawn up and back to shut off the air passage to the nose. Soon after the lips are brought together, the tip of the tongue rolls up and slightly back to form a depression in front of the main body of the tongue. After a short interval the lips begin to part and at this moment the vocal cords are brought together and made to vibrate by the pressure of air from the lungs. The tip of the tongue then turns down towards the floor of the mouth and as it does so, the whole body of the tongue is pulled back and down so that the space behind the root of the tongue, in the pharynx, becomes considerably smaller, whilst the space in the front of the mouth is made larger. The time taken over this movement is greater than for the previous ones as this part of the sequence corresponds to the vowel of the word. Before it is over, the soft palate begins to be lowered and by the time it reaches its lowest position, the tongue has risen again and is spread laterally to make contact with the upper gums at the sides and in the front. The next movement is the raising and pulling back of the soft palate again to close the air passage to the nose and at the moment when the closure is complete the larynx vibrations are switched off. After a short silence, the tip and blade of the tongue are brought away from the upper gums in the front, the middle of the blade of the tongue is depressed to form a wide groove in the upper surface of the tongue and at the same time, the amount of air issuing from the lungs is somewhat increased, and then finally dies away.

This is a very complex series of movements although it is associated with only a single syllable and does not include any reference to the action of the thoracic muscles or to variation in vocal cord frequency. There must at some level be a detailed control programme which takes care of the timing and succession of these muscle actions. It seems probable, as we shall see later, that this programme is drawn up for a stretch about the size of one syllable at a time. The inter-action of one part of a syllable on another, from the articulatory point of view, is so great that one cannot imagine that the programme could control less than a syllable at one time and it may conceivably be written for whole words or phrases. If the latter were the case, how-ever, one might expect that such a programme once set in train would not be easily upset or interfered with and the present evidence is that this is not at all the case.

The Feedback Control of Speech

So far we have been concerned only with control signals fed forward from the speech centres of the brain through motor nerves to the muscles used in speech. The very fine control that is needed to regulate the complex patterns of movement which speech involves cannot be obtained by this means alone, but requires the aid of several feedback loops. Whilst the speaker is sending out a stream of speech sounds, his brain is receiving a continuous flow of information concerning the progress of the associated movements through three paths. First, his hearing mechanism is being activated by the sound waves he is producing. A large proportion of what he hears reaches his inner ear by way of the bones and tissues of the skull and very little in the form of sound waves in air. The bone-conduction pathway has its own particular frequency characteristic and consequently the speaker hears a very different version of his own voice from that heard by other listeners who hear it through the air pathway, with attendant reflections from surrounding surfaces and objects. This fact explains a speaker's inability to recognize his own voice when he first hears it recorded. It is this bone-conduction path with the cochlear mechanism and its sensory nerve connections to the brain that constitute the auditory feedback loop in speech.

In addition, the speaker's brain is continually supplied through proprioception with information about changes in muscles and this constitutes the kinaesthetic feedback loop in speech. As each complex of movements unfolds, the brain receives news of the progress of every muscle action through this loop and partly by this means is able to check the congruence between the programme and the actual movement and also to prepare the pattern of timing for movements which are to follow.

Thirdly there is the tactual feedback loop. We get sensations of touch from certain parts of the vocal tract, in particular from the tip and front of the tongue, the gums and the lips, and these sensations probably play some part in the feedback control of speech, though their role is undoubtedly less dominant than that played by the auditory and kinaesthetic feedback.

In general the importance of feedback control in speech is related mainly to the timing of speech movements and to the regulation of the loudness and the pitch of speech sounds.

Feedback in the Control of Respiration

We have seen that the action of inspiration may take place only at certain points in a speech sequence. Hence the duration of successive stretches of speech uttered on one expiration may differ quite widely and therefore the rate at which the inspired air is to be 'paid out' will vary. Both auditory and kinaesthetic feedback help to keep the brain informed of the progress of an utterance, enable it to estimate how long it will be approximately before another breath may be taken and tell it when the moment arrives for inspiration. As the utterance proceeds, the speaker both hears and feels how fast he can afford to use the inspired air and when an appropriate moment arrives, the breathing mechanism is triggered to take in another breath rapidly before beginning the next utterance.

Not only the timing of respiration but also the amount of air taken in is dependent upon feedback control. As we shall see more fully in the next section, a speaker tries to maintain a certain relation between the loudness of his own speech and that of other sounds which may be occurring around him. The auditory feedback loop supplies him with the information upon which he bases this judgment. If the ambient noise level is high, he has to produce louder speech and for this he needs to take in more air at a time and to use it more lavishly; if the ambient noise level falls, he can then make do with a smaller supply of air and probably with less frequent inspirations.

Feedback in the Control of Phonation

The speaker's efforts to achieve some minimum signal-to-noise ratio have an even greater bearing upon the action of the larynx. If we place a speaker in quiet surroundings, get him to begin talking and then produce some masking noise of which the level is gradually increased, the speaker raises the level of his speech in order to overcome as far as he can the masking effect of the noise. When the intensity of the noise becomes very great, the speaker shouts as loudly as he can and even so he may find himself in the position of being 'unable to hear himself speak'. What he is trying to do throughout this process is to obtain a satisfactory relation between the loudness of the noise and the loudness of his own voice. The latter quantity is largely a function of the characteristics of the vocal cord tone that he is producing and particularly of its amplitude. This in turn is dependent upon the rate at which air is supplied from the lungs, the pressure level built up below the vocal cords during their closed phase and hence also on the duration of the closed phase.

Information about the loudness of the speaker's own voice is provided entirely through the auditory feedback loop and it thus plays a most important part in regulating the action of the larynx during speech. Changes in the level of ambient noise are very frequent in everyday conditions and consequently there are continual variations in voice level which are controlled with the aid of the auditory feed-back loop.

The second feature of larynx action which is dependent upon feed-back control is the frequency of vocal cord vibration. Precise control of this frequency has been shown experimentally to be impossible when auditory feedback is impaired by the presence of a high level of masking noise. The experimental subject wears a pair of telephone receivers into which can be fed noise of such an intensity that the bone-conducted sound of the subject's own voice can be completely masked. If the subject is then asked to sing some agreed pattern of pitch changes, such as a rising and falling major scale, the accuracy with which he can produce the required pitches is many times worse than it is when there is no masking noise. In the case of a scale, for example, what happens is that the subject covers in all a pitch change of considerably less than an octave, with the degrees of the scale crowded together especially at the top and the bottom of the range (Deutsch, 1956).

Further evidence of the importance of auditory feedback in the control of larynx frequency is given by the observation that when a subject tries to sing a steady tone, i.e. of constant vocal cord fre-quency, in a normal way without masking noise, there are periodic fluctuations in the frequency of the tone which take place about five or six times a second; if the side-tone in the subject's ears is now delayed (by a method which will be given in the next section) both the period and the extent of the frequency fluctuations increase pro-portionately to the delay over quite a range of delay times (Deutsch, 1959). This suggests that the moment-to-moment modifications in larynx frequency that are carried out are indeed controlled with the help of auditory feedback.

It was noted above that the presence of very loud masking noise has the effect of compressing the pitch range associated with an octave. In the experiment referred to, however, it is significant that the subject does make changes in vocal cord frequency and these changes are in fact in the desired direction, even though the degree of change is not what the subject intends. Clearly the subject still has some control over larynx frequency partly through preset signals and partly through the kinaesthetic feedback. What he seems to re-tain is some general notion whether he is producing a high note, a low note or some intermediate tone, in other words he is able generally to monitor the direction of change. In a previous section we saw that for the intonation patterns of speech, it is the direction of change of larynx frequency that is important and it is not therefore surprising that although it is impossible to sing notes in tune in the presence of loud masking noise, the intonation patterns of speech in the same conditions are not particularly abnormal. As the speaker is likely to speak very loudly in this situation, there is a tendency for the range of vocal cord frequency to be extended upwards but the contours of frequency variation are not materially altered. This means that the *direction* of larynx frequency variation *can* be controlled on the basis of kinaesthetic feedback but that auditory feedback is indispensable to the production of any exact frequency. In speech there is no doubt that control of vocal cord frequency is regulated by means of both auditory and kinaesthetic feedback.

Feedback in the Control of Articulation

Much of the evidence concerning the role of feedback in the control of articulation has been obtained by means of experiments with delayed auditory feedback, which has already been mentioned. It will be useful first of all, therefore, to give a brief indication of the way in which these experiments are carried out. We have said that the sound of his own speech comes to the speaker's ear by way of the bones of his skull and that it is possible to mask this sound completely by applying high intensity airborne noise to the speaker's ears. It is equally possible to pick up the speech and to feed it, considerably amplified, through telephone receivers to the speaker's ears at such a level as to mask the bone-conducted sound of the same speech. The feedback signal may be delayed by first recording it and then playing it back after some desired delay. Since the speech in the telephones is loud enough to mask the bone-conducted sound, the auditory feedback now consists only of the delayed signal.

The effects of delayed auditory feedback vary from one individual speaker to another and with the amount of the delay. A delay of about one-tenth of a second has a very disrupting effect on the speech of most subjects, generally creating something in the nature of an artificial stammer (Lee, 1950). In order to understand why this should be so, we must look more closely at the functions of the auditory feedback. We said earlier that one of the most important tasks of the speech control mechanisms generally was to plan and to carry out the timing of speech movements. This function is nowhere more important than in the control of articulation, where the control of the timing involves also the control of the extent of movements. In the utterance of the word *branch*, described above, the extent and the rate of the tongue movement in any given direction has to be regulated within narrow limits. In normal circumstances, it is with the help of the auditory feedback that this is done; the ear decides when the silence of the initial *b* has lasted the proper time, when the larynx mechanism shall be switched on, how far back in the mouth the tongue shall move for the vowel, how far from the gums the tongue tip shall move in *ch* and so on.

It is easy to see that the introduction of a delay of a fraction of a second in the auditory feedback will probably have an adverse effect on this fine control of times and distances. Auditory information that a given movement has been initiated or is in progress will be expected to arrive at a certain moment, on the basis of previous experience; since it does not arrive at that time, the speaker will prolong or repeat the action and this gives rise to stuttering and to very prolonged continuant sounds. Once this effect has begun, a kind of reverberation is set up between the outgoing activity and the feedback loop so that the normal rhythm of the speech is destroyed, the articulation becomes more and more disrupted and in extreme cases the speaker finds it impossible to continue speaking.

Observations with delayed auditory feedback demonstrate very clearly that the programming of articulation includes close specification of the timing of movements and if the auditory feedback mechanism reports that the programme is not being implemented correctly, there is gross interference with further programming. There may be some significance in the fact that a delay of about one-tenth of a second has the greatest effect on most speakers since this time is just a little less than the average time taken to produce a syllable. If the articulatory programme is in fact drafted for units of about one syllable, then the delayed auditory signal is arriving too late to confirm that the programme for the first syllable has been properly implemented and too early to allow the programme for the second syllable to get underway without interference (Fry, 1964a). It is important to notice that absence of auditory feedback information does not have the same effect. A subject speaking in a noise so loud as to make his speech inaudible to himself can continue to make the movements of articulation in accordance with the control programme. In this case, control through the kinaesthetic feedback loop keeps the movements going quite adequately.

This statement holds good, however, only for short term effects. A consideration of the effects of deafness shows clearly that continued absence of auditory feedback has a profound influence on speech. Thus an adult who goes very deaf in middle life will at first continue to carry out fairly normally the movements of articulation but after a time his articulation is affected by the loss of auditory information. Furthermore, the learning of speech movements is extremely difficult if not impossible without the help of some auditory feedback so that in the case of a congenitally deaf child, for example, the problem of teaching speech changes completely if he has some hearing that can be used.

Neural Control Circuits in Speech

We are now in a position to summarize the observations that have been made above and to review the control mechanisms which must be functionally operative in speech and which must therefore be represented at the neural level. The whole speech activity is a combination of preset and controlled actions which are interlocked at the level of the motor control unit. It would be possible in theory to draw a complete flow diagram for speech but such a diagram would necessarily be extremely complex; it would need to include, for example, the circuits concerned with the reception of speech (which lie outside the scope of this chapter; see Fry, 1964b) and to show the links between the reception and the generation of speech. Since a complete diagram would be unwieldy, it is probably more helpful to visualize the control circuits as forming three systems, which are of course interconnected: the linguistic control system, the physiological control system and the feedback control system. Figures 1, 2 and 3 show these systems in diagrammatic form.

Figure 1 presents the linguistic control mechanism. The material for speech arises partly from an intellectual source and partly from an emotional source in the speaker and these are denoted here as *propositional source* and *affective source*. Each of these influences to some extent both the content and the form of the speech. These two together determine sentence selection and as a result we get serial selections of word order, words, morphemes and phonemes. In parallel with this, there is the choice of rhythm and intonation pattern, the two features being so closely bound together that one cannot suppose that they are independently controlled. They are, however, controlled in part by the language system, that is to say in this instance by the sentence selection, and in part by affective forces in the speaker. On the same level, we have the control of voice quality, which is independent of the language system and wholly determined by the affective source. There are thus three factors which control the sequential activity in the motor control system: phoneme selection, rhythm and intonation selection and voice quality selection.

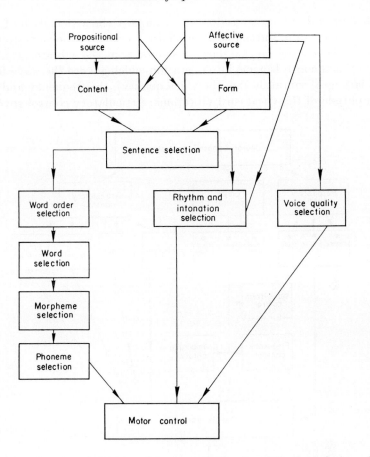

Figure 1. Diagram indicating the stages in the linguistic formulation of a spoken message down to the level of motor control patterns.

Turning now to the motor control system, in Figure 2, we must visualize this as not simply a 'stop–go' mechanism for ordering muscle action, but as a very complex circuit which is able to take care of the phasing of all muscle actions in speech. This is connected to three separate controls, respiratory, laryngeal and articulatory. The respiratory control operates first a three position switch providing for inspiration at appropriate moments, the control of expiration during speech sequences and a hold position which comes into operation during hesitation pauses. The larynx control also leads to a three position switch: hold open, during sounds which are not voiced, vibration, during all sounds that are voiced and hold closed, when glottal closure is needed to accompany the articulation of certain sounds, or in the course of emphatic speech or during a hesitation pause. When the larynx selector switch is in the vibrate position, additional controls come into operation which govern the wave-form of laryngeal vibration (i.e. the voice quality), the frequency and the amplitude of the vocal cord vibrations. Articulatory control governs

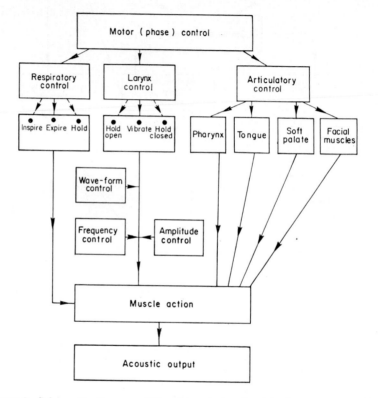

Figure 2. Schematic diagram of the elements involved in the control of motor speech activity.

four main muscle systems in parallel: the pharynx, the tongue, the soft palate and the facial muscles. All these various circuits together, their action very closely controlled by the motor phasing control, produce the extremely complex flow of muscle activity which constitutes the skilled movements of speech, and these movements in turn give rise to the sound waves.

The last of the three systems is shown in Figure 3. Here we see that the acoustic output of speech goes to the speaker's own hearing mechanism, which at the same time receives all ambient noise, and the combined information goes back to the motor control mechanism where, as we have seen, it is used in the control of timing, vocal cord amplitude and frequency, etc. Muscle action itself energizes the other important feedback loop through the proprioceptive mechanism and gives rise to kinaesthetic feedback. The movements that the muscles produce also excite tactual receptors in certain areas and information of this kind is also relayed to the motor control mechanism.

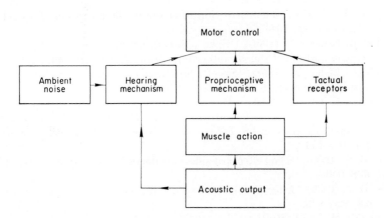

Figure 3. Feedback loops concerned in the control of speech movements.

This brief account of the control mechanisms in speech may be summarized as follows.

The origin of control lies in the linguistic organization of what is to be said.

The forward going control signals regulate the activity of muscles concerned in respiration, phonation and articulation.

Coordination of these three systems requires a control programme which supervises the continually changing phase relations between the systems and the different muscles which make up each system.

Adequate control of speech movements is possible only with the aid of auditory feedback, kinaesthetic feedback and a certain amount of tactual feedback.

Auditory feedback is particularly important in controlling:

1. the loudness level of speech;
2. precise vocal cord frequency;
3. the timing of articulatory movements.

References

Deutsch, J. A. (1956). *Stimulation and Behaviour*, unpublished D. Phil. thesis, Oxford.

Deutsch, J. A. and Clarkson, J. K. (1959). 'Nature of the vibrato and the control loop in singing.' *Nature*, **183**, 167–168.

Fry, D. B. (1963). 'Coding and decoding in speech.' In Mason, Stella E. (Ed.), *Signs, Signals and Symbols*, Methuen, London, pp. 65–82.

Fry, D. B. (1964a). 'The function of the syllable.' *Z. Phon. Sprachw. Kommun.*, **17**, 215–221.

Fry, D. B. (1964b). 'The correction of errors in the reception of speech.' *Phonetica*, **11**, 164–174.

Goldman-Eisler, F. (1955). 'Speech–breathing activity—a measure of tension and affect during interviews.' *Brit. J. Psychol.*, **46**, 53–63.

Goldman-Eisler, F. (1958). 'Speech production and the predictability of words in context.' *Quart. J. Exp. Psychol.*, **10**, 96–106.

Goldman-Eisler, F. (1961). 'Hesitation and information in speech.' In Cherry, C. (Ed.), *Information Theory: Proceedings of the Fourth London Symposium on Information Theory*, Butterworth, London, pp. 162–174.

Lee, B. S. (1950). 'Some effects of side-tone delay.' *J. Acoust. Soc. Am.*, **22**, 639–640.

Lee, B. S. (1950). 'Effects of delayed speech feedback.' *J. Acoust. Soc. Am.*, **22**, 824–826.

Penfield, W. and Roberts, L. (1959). *Speech and Brain-Mechanisms*. Princeton University Press, Princeton, New Jersey; Oxford University Press, London.

Peterson, G. E. (1957). 'Breath stream dynamics.' In Kaiser, L. (Ed.), *Manual of Phonetics*, North Holland Publ. Co., Amsterdam, pp. 139–148.

7

Circadian Regulation and Related Responses

H. KALMUS

Department of Human Genetics and Biometry
University College, London

Most people consider anticipation as a particularly characteristic attribute of man, even of civilized man and for all we know this point of view may be correct as far as conscious and deliberate activities are implied. There is however little doubt that *recurrent preparation*, of various kinds, for future contingencies is quite universal among organisms and that it occurs at the lowliest levels (see p. 166).

The oscillatory processes which are the carriers of this form of adaptation are commonly described under the heading of biological clocks (Bünning, 1964), but it is more useful in the context of this book to characterize them as circadian, anticipatory and regulatory.

The word circadian, derived from *circa diem* (about a day), was coined by Halberg in 1959. It describes endogenous arrangements and processes which, once entrained (Pittendrigh and Bruce, 1957) or released by a single cue or repeated cues ('Zeitgeber', Aschoff, 1958), continue oscillating (free running) with a period length of approximately 24 hours after the removal of the external cues in what is commonly called constant conditions.

The circadian organization of a living system often implies its anticipatory nature in so far as it produces daily recurrent adaptive changes prior to the activation by an external periodic source of energy. Seasonal and tidal preparedness may also be of such an anticipatory nature.

Regulation—whether resulting in some measurable constant value (homeostasis) or in some other 'regular' response—implies a change in time of some function of the organism to a temporal change in the environment. Thus temperature regulation, even if it results in a constant overall temperature, can only be achieved by increases in heat production or heat losses. If the external temperature changes in an environment are periodic, some thermic changes of the organism must also be periodic. These changes may be largely passive, as in a plant or a mycelium; or they may be active, as in a mammal or a bird (see p. 87). They may be sufficient to maintain a high body temperature, as in large mammals; or insufficient, as in the small humming birds which become poikilothermous at night.

Compensatory temperature responses in different organisms vary in their phase relative to the phases in the external course of temperature; endogenous temperature changes may in experimental situations appear as oscillations in their own right. The interactions of external and internal processes affecting temperature are usually complex and the results often appear imperfect. However, one ought to consider in any particular case whether the observed regulatory process and its results are indicative of the pursuit of the goal of constancy (true homeostasis) or of a changing goal (homeorhesis, Waddington, 1957), in particular of an oscillating one.

The universality and importance of clock arrangements in the living world has been pointed out by the author twenty-eight years ago (Kalmus, 1938) and there is now little doubt left concerning the formal similarity of the various time keeping devices in the plant and animal kingdoms. This similarity is quite comparable to the similarity of the various man-made chronometric devices ('clocks') e.g., watches, pendulum clocks, electronic clocks and others. Furthermore, these organizationally diverse biological clocks are less dissimilar among themselves than the chronometers of human technology, but like these they are of a limited number of types. One may thus inquire how many different kinds of biological clocks or chronometric principles exist, what they are and how they operate at the cellular, organismic and population levels. However, in this chapter only some general principles of a few clock types are being considered.

The change of day and night is undoubtedly one of the oldest synchronously variable complexes of ecological factors (light, temperature, moisture, currents, noise, prey or predator activities etc.) to which organisms have been subjected. If we assume that life on earth started some 2,000 million years ago, and that the rotation of the earth was not too different from that of today, most organisms have during their history experienced about 7×10^{11} dawns and an equal number of dusks. There is little doubt that the resulting convergent adaptations must have occurred frequently and often independently. Nevertheless, if we call non-homologous (but analogous) organs of different organisms by the same names, as for instance the eyes or legs of vertebrates, arthropods or molluscs, we may also call the different and independently arisen chronometric devices in the biological world by the same names.

Anticipatory actions occur widely in the organic world. A predator catching moving prey, a tennis player hitting a ball, a spider constructing a web, even a flower displaying its visual and olfactory attractions, all can be said to anticipate future events in their environments. This is also characteristic of circadian anticipation, but while the 'ultimate cause' of such 'expectations' is the 'confidence' in the future events, its immediate or physiological cause must be some signal ('cue') experienced during an earlier phase of the same or an earlier cycle. It is well known (Kalmus, 1964a) that the belief that 'the sun will rise tomorrow', contrary to what Laplace held, cannot be supported by the frequency theory of probability. Despite this, not only the clock industry, but also the totality of organisms act on this belief, by simple direct anticipatory responses, and also by such complex arrangements as the navigation of animals (Kalmus, 1964b) and the sexual photoperiodism of animals and plants (Pittendrigh, 1965).

Teleologically speaking biological clocks ensure that a particular object—be it a hormone particle or a mating partner—is at the 'right place' at the 'right time'. In nature such a state of affairs is usually safeguarded by several means. Mere observation cannot tell us what previous cues (changes in light, temperature, etc.) have entrained the underlying circadian oscillations, nor indeed whether any observed temporal adaptations are not merely passive responses to the situations prevailing immediately prior to observation. This question must be decided experimentally for every single behavioural, chemical or other cyclical change corresponding to an external periodicity (day, tides, season, etc.) and for every species separately. Some of the terms, methods and models used for such an analysis will be described later.

However, before entering into the more formal aspects of circadian phenomena, it should be pointed out that organisms, unlike the products of technology, do not have clearly separable parts for every function. Not only can single cells perform all the biological functions which are necessary for survival, but most organs of multicellular animals and plants have numerous functions. If one considers the nyctinastic leaf movements of a plant, it is by no means obvious which part of the plant is the originator of these movements nor what external or internal factors are responsible for them. It is, however, certain

that the frequency of the up and down movements of the leaves is not
the expression of their mechanical properties. Considered as mechanical pendulums, leaves would have a period length of a fraction of a
second and not one of 24 hours.

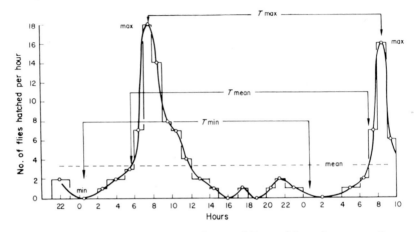

Figure 1. Hourly eclosion during 36 hours of *Drosophila melanogaster*, from a
culture bottle kept indoors near a window. The period lengths *T* max and
T min can be more accurately estimated than *T* mean (original).

Changes showing period lengths of about a day, a month or a year
can be observed at all levels or organization, in cells, organs, individuals and populations (see Figure 1). Changes at the higher levels
usually imply the existence of changes at some lower level.

Terminology and Symbols

Despite such uncertainties about the nature, localization and complexity of biological oscillations, phenomena in the field of circadian
periodicity are being increasingly described in terms of oscillation
theory. It is therefore useful to start our discussion of some of these
phenomena with an explanation of the most important of these terms
and the symbols representing them. In this we mostly follow the
recommendations distributed at a recent summer school on circadian
clocks (Aschoff, 1965). An *oscillating system* can be described in respect of either *internal* or *external* features. Viewed internally it may
appear as:

a. An *active* system, capable of *self-sustained oscillations* ('endogenous rhythms'). Such an active system can obtain the energy

necessary for maintaining the oscillation from a constant source of energy.

b. A *passive* system which is capable only of *forced* oscillations ('exogenous rhythms') or of free *damped* oscillations. The damping may be an increase in period length, a decrease in amplitude or both.

c. An *extremely passive* system which is not even capable of free damped oscillations.

Viewed externally an oscillating system may appear as:

a. *Autonomous*, that is not under the influence of a periodic source of energy, producing self-sustained or free damping oscillations.

b. *Non-autonomous*, when it is under the influence of a periodic source of energy. The resulting forced oscillation may occur in an active or a passive system.

Active systems have one (about 24 hour) or several limited (multiple hour about 48 hour, 72 hour, etc.; submultiple, about 8 hour, 12 hour, etc.) *ranges of entrainment*. That is they can be set into motion by external stimuli acting in the intervals specified, while passive systems have a more or less unlimited frequency range of forced oscillation.

Period is the time after which a specified *phase* of an oscillation recurs (for instance 24 hours). We shall denote the period length of external oscillations with T, those of biological rhythms with τ.

Frequency is the reciprocal of period (once in 24 hours). The terms period and frequency are strictly speaking only applicable if any selected phase leads to the same value. If different phases give different values, a special explanation becomes necessary. The phase used for the determination of period length should always be specified.

Amplitude denotes the difference between maximum (or minimum) value and *mean* value in a sinoidal oscillation. While more loosely applied to oscillations of a general shape, it is preferable to speak about positive (or negative) *extreme values*.

Range of oscillation is the difference between maximum and minimum values of the oscillating variable.

Phase (ϕ) is the total state of an oscillatory system at a particular instant within a period; it is usually represented by the value of a measured variable which, however, implies all its time derivations (mathematical) and corollaries (correlated material changes).

Phase angle ('argument' of the phase, also symbolized by ϕ) is the abscissa of a point on a curve representing the oscillation, given in units of a full period (degrees or time units). The phase angle of an external time cue is described as Φ.

Phase shift ($\Delta\phi$, $\Delta\Phi$) is a single displacement of an oscillation along the time axis. When in a succession of oscillations one or a few periods are shortened, we call it an *advancing phase shift*. When periods are lengthened we speak of a *delaying phase* shift.

Synchronization is a state in which two or more oscillations have the same frequency due to mutual or unilateral influences.

Entrainment is the process by which a *forcing oscillation* (for instance an external time cue, synchronizer or 'Zeitgeber') synchronizes a self-sustained oscillation (for instance a biological clock). Entrainment is possible only within limited *ranges* of frequency.

Phase relation is the time relation (phase angle difference, $\Phi - \phi$) between a specified phase in an external oscillator (for instance dawn) and a specified phase in a second synchronized oscillation (for instance waking up). Names for individual phases of the external and internal cycles are suggested in Table 1. Phase relations can be expressed as *phase angle differences*, that is in degrees (for instance $+30°$) or period units (for instance 1/12th of a day or 2 hours).

Table 1. Approximately coinciding external and internal phases during a *LD* 12:12 regimen*. (Compiled after Pittendrigh, 1965.)

Arbitrary† external zeitgeber time (AZT)	Phases of the external light cycle (T)		Phases of the circadian rhythm (τ)		Subjective circadian time (SCT)
00	dawn		subj. dawn		00‡
	morning		subj. morning	photophil	
06	noon	day	subj. noon	phase	06
	afternoon		subj. afternoon		
12	dusk		subj. dusk		12
	early night		subj. early night		
18	midnight	night	subj. midnight	scotophil	18
	late night		subj. late night	phase	
24	dawn		subj. dawn		24§

* See p. 16.

† Experimental dawn and dusk can be produced at any local hour, including the normal one.

‡ SCT 00 is that point in the cycle, which occurs, when dawn would have fallen on the first day of a *DD* free run following *LD* 12:12.

§ SCT 24 occurs one full cycle later than SCT 00, which is not usually 24 hours later.

If a specified phase of a synchronized biological rhythm (e.g. waking up) occurs earlier than a specified phase of the assumed entraining external oscillation (e.g. dawn) we call it *leading*; if it comes later it is called *lagging*. As the beginning of any cycle is arbitrary, and as phases are repetitive, this distinction is only of limited value.

A self-sustained rhythm which in the absence of external periodic time cues produces a sufficient number of undamped oscillations, is called *free running* or spontaneous. When the system is in a steady state it has a characteristic free running period, which in the case of a circadian oscillation more or less approximates to 24 hours.

A *transient* is a temporary oscillatory state between two steady states. It is characterized by changes in the period length of a free running biological system, following either a phase shift or other interference (e.g. extra illumination) of the external forcing oscillation or some other general change (e.g. in temperature, Figure 4, p. 45).

Shifts of biological phases resulting from such procedures can be graphically represented in the form of a *response curve** (Figures 5 and 6) of the general formula

$$\Delta\phi = f(\phi)$$

The intensity and possibly the spectral composition of light are the most important single environmental entrainers of circadian rhythms and therefore the following symbolism should be noted.

A light–dark cycle LD consists of a light time (light fraction) L and a dark time (dark fraction) D.

LD 16:8 (100:0·1) denotes a light–dark regimen where an organism is kept for 16 hours at 100 lux light intensity followed by 8 hours of 0·1 lux light intensity (= darkness). $LL(y)$ denotes continuous illumination at light intensity y (in lux); DD signifies constant darkness, that is a light intensity below the possibility of measurement. Symbolisms concerning temperature regimens and other cyclical changes should be devised in analogy to the above.

Operation of Circadian Rhythm under 'Field' Conditions

As mentioned on p. 157, it is rarely possible to deduce or even to guess from field observation alone, whether and to what extent an endogenous rhythm is instrumental in the regulation of an organism's overt activity. In special circumstances however a *prima facie* case can be made for the anticipation by an animal or plant of a daily change of environment.

---

* This is a special use of the term response curve which in general means a graph representing the dependent variable as a function of one or several manipulated variables.

6*

An example is the spider *Nemesia caementaria*, which in the adult state inhabits an opaque nest closed by a hinged lid. In summer the animals sit inside their nests in the daylight hours, but during darkness they are actively moving about clearing their nests and capturing insect prey. Observations show that many of these spiders open their trap doors late in the afternoon, sometime before it is dark (Figure 2). They then close them again and are only really active

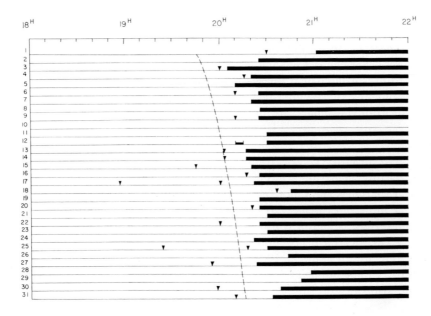

Figure 2. Activity in variable sometimes stormy weather of a trap door spider *Nemesia caementaria* during May 1961. The broken line indicates dusk. Sustained activity at night is preceded by isolated openings of the trap door. (From Büchli, 1964.)

later, when the light has gone (Buchli, 1964). As it can be shown that the light regimen in which these spiders live provides the timing cues for their activity, there is little doubt that the spider sitting in the dark must somehow 'know' that it will be dark outside quite soon and that it is prepared to spring into activity. Similar field observations on cave dwelling bats which get active before the outside dusk has fallen, or of birds which wake a little earlier than dawn are well known. Honeybees trained to find honey at a particular hour of the day will also appear at the feeding dish, slightly before the hour at which feeding begins (Kalmus, 1934).

Laboratory Demonstration of an Endogenous Rhythm

The above examples provide a *prima facie* case for the existence of endogenous rhythms, but they are by no means conclusive. In principle any phase of the light or temperature cycles or even an entirely unknown geophysical factor may exert instantaneous or delayed control over any of the described activities and no endogenous oscillation need be involved (Stoppel, 1926; Brown, 1960).

The first objection can be refuted by laboratory observations if these show that a biological oscillation can be generated (Figure 3) or can persist either in constant darkness, or sometimes under constant illumination. This has been shown for the eclosion of *Drosophila* and other insects, the locomotor activities of many vertebrates, insects

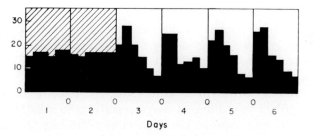

Figure 3. Initiation of periodic eclosion of *Drosophila melanogaster* by one transition from continuous darkness (shaded) to continuous light. (After Bünning, 1935, English translation, 1964.)

and lower, even unicellular organisms and for the leaf and petal movements of flowering plants. These all persist sometimes for considerable periods under 'constant' conditions.

The second objection to endogenicity, that based on the assumption of unknown external geophysical factors, is capable of refutation in three different ways, as follows. (i) By showing that a biological oscillation, which by manipulated cues has been entrained with an arbitrary phase angle to local time, will in constant conditions for several cycles, more or less persist in this abnormal phase relation, while parallel controls persist in their normal phase relation (see Kalmus, 1935). (ii) By showing that an oscillation, regardless of the way in which it has been entrained, will in constant conditions sustain a free running period. This generally will either be somewhat longer or somewhat shorter than 24 hours—hence the word circadian, coined by Halberg (1959, see p. 155). (iii) By showing that various treatments (e.g. changes in temperature or asphyxia), after entrainment, temporarily or permanently alter the period length of a free running oscillation (Kalmus, 1934, 1935, 1938).

Temperature Independence*

The selective advantage to an organism of possessing a timing mechanism independent of the vagaries of daily temperature changes would seem obvious, and a number of such temperature independent oscillations have in fact been found.

As the processes involved in these circadian oscillations are almost certainly of a biochemical nature which mostly have Q_{10}'s of 2–3, it is at first glance surprising that many free running systems would be temperature independent or only very slightly speeded up by a rise in temperature.†

However, it has been possible to devise a mathematical model for temperature independent oscillations by gearing two antagonistic biochemical processes, each having a high Q_{10} in such a way that the resulting oscillation has an almost constant frequency (Kalmus and Wigglesworth, 1960).

Free running circadian rhythms display relative temperature independence in two different ways according to whether entrainment and free run occur at the same or at different temperature. If, for instance, one entrains (by LD 12:12) side by side several *Drosophila* cultures at considerably different temperatures and subsequently observes their free running periods in DD, one finds a high degree of temperature independence (Kalmus, 1935; Pittendrigh, 1954). If, on the other hand, one similarly entrains several cultures at a particular constant temperature and subsequently observes their free run in DD at a different temperature, more complex events follow (Figure 4). A transfer of the *Drosophila* culture to darkness and low temperature induces a delay of the first two eclosion peaks while a change to darkness and warmer conditions advances those peaks. This was interpreted as a slowing down or speeding up of the oscillation (Kalmus, 1935). However, Pittendrigh (1954) has subsequently shown (see Figure 4) that these changes of period length are not permanent and has ascribed this to a generally transient state of the system, which after two cycles gives way to a new steady state, free

---

* Athermochrony (Kalmus, 1935).

† Bühnemann (1955) has even shown that the circadian sporulation rhythm in the alga *Oedogonium cardiacum* is slowed down by an increase in temperature.

running with more or less its original period length, though with shifted phases.* It is probable, but by no means certain, that the adaptative changes, which compensate for temperature changes after

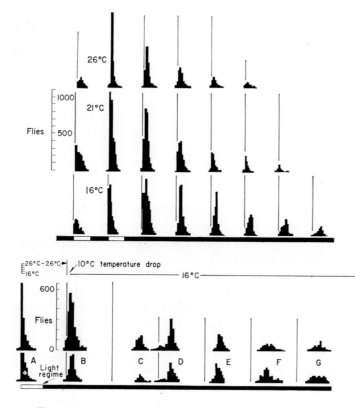

Figure 4. Temperature independence in darkness of persistent eclosion rhythm of *Drosophila pseudoobscura* entrained at different temperatures. At 16°c the period is about 24·5 hours (Top). Entrainment of 2 cultures at 26°c, followed by transfer to 16° and constant darkness result in a transient disturbance (peak C is delayed) and subsequent resumption of a free running period of approximately 24 hours. (After Pittendrigh, 1954.)

entrainment are also responsible for the faculty of being entrained with roughly identical frequency at different temperatures. These adaptations may in fact just be part of very general faculties which enable poikilothermous organisms to function at different temperatures (Stier and Wolf, 1932). Kalmus could show that 8 hours

* Considering the events during one or two cycles only, it is of course not possible to distinguish between phase shifts and changes of period length.

anoxia, caused by a vacuum of 18 mm in moist conditions, postpones the first eclosion peak under free running conditions by about 8 hours. The time sense of bees is also temperature independent within wide limits (5–38°C) but the visiting hour of time-trained bees can be postponed for several hours by cooling trained workers for several hours below 2–7°C (Kalmus, 1934).

The nature, localization and kinds of circadian mechanisms are still unknown in spite of much research, but a few properties in addition to their frequent temperature independence may be mentioned. They seem to occur at all levels of organization above the cellular one, being observed in fungi (Jerebzoff, 1961) and protists, but not as yet in bacteria or viruses, or in parts of cells such as mitochrondria or chromosomes. Most interestingly the entrained circadian rhythm of phototactic response persists in the flagellate *Euglena* even when, under favourable conditions, it multiplies more rapidly than once every 24 hours (Pittendrigh and Bruce, 1957). In these circumstances we must either assume that every cell contains several clocks, which are handed on functioning to the daughter cells or that we are dealing with chronometric devices, which go on keeping time while at the same time multiplying.

It is very likely that photosensitive pigments are implicated in many cellular clocks, and it can be shown that light of specific and sometimes narrow bands of wavelength is exclusively active in either releasing or stopping biological rhythms (action spectra—Ehret, 1960). In addition there is ample evidence from plant experiments that a number of enzyme systems are implicated (Hastings, 1960). However, different circadian systems react differently to individual enzyme poisons and, in view of the probably convergent origins of biological timing devices (see p. 157), it is not profitable to formulate generalizations.

The Uses of Chronometric Devices

The opportunist character of organic evolution leads one to suspect, that circadian mechanisms have, in addition to their direct uses, also been incorporated in adaptative and regulatory processes not directly concerned with the hours of the day. This seems indeed to have occurred quite frequently and in particular in respect of two classes of regulatory phenomena, namely the geographical orientation (sun navigation, astrotoxis) of terrestrial and marine animals and the seasonal maturation (sexual photoperiodism) of the higher plants and animals.

Celestial navigation

The first observations demonstrating that bees (von Frisch, 1950) and starlings (Kramer and St. Paul, 1950) could be trained to turn to known geographical directions were published almost simultaneously and it was soon accepted that the azimuth of the sun was the 'compass' of these animals. Since then several hundred species of arthropods, molluscs and vertebrates have been shown to possess some degree of sun-navigational faculty (Kalmus, 1964).

The word 'navigation', as used by biologists and indeed by seafarers and airmen, refers not so much to locomotor machinery as to the sensory and computational faculties involved in long distance transport, particularly in direction finding. Navigation is usually distinguished from pilotage, the use of landmarks, and sometimes used synonymously with celestial 'orientation'. However, some students of this field want to restrict the use of the words sun navigation, moon navigation and star navigation to procedures involving the determination of the longitude and latitude of two points (bicoordinate navigation) namely to that of a vessel's or animal's momentary position and that of its goal. Sun, moon or star orientation would involve only the assumption or maintenance of a particular compass direction (astrotaxis) of locomotion by means of the sun, moon or the fixed stars. Bicoordinate navigation over short distances can be excluded by the limitations imposed on an animal by its visual acuity (Adler, 1963) and it is controversial even for animals migrating over great distances. Sun and moon orientation—in the more general sense—are well established phenomena; we shall here only deal with the former.

Sun orientation is most commonly observed when animals change their position regularly—from the sea to a nesting place, from the hive to a flower bed, from summer quarters to winter quarters. One particular geographical direction is as a rule maintained during an outward flight and the opposite direction during the return.

The role of circadian clocks in these manoeuvres may be visualized as a simplified version of the well-known use of a watch in determining south. In the temperate and arctic zones of the northern hemisphere the approximate south direction is found by pointing the small hand of the watch towards the direction (azimuth) of the sun and then halving the angle between the small hand and the direction of 12 o'clock. The compensatory mechanisms of animals are simpler because their 'handle' turns 360° once only in 24 hours and not twice as the small handle of the watch. Consequently, while the sun is shining, the direction south and in fact any geographical direction can be simply determined by pointing the 'handle' at the sun's azimuth. South is then approximately at 0° (12 o'clock), West at 90° (3 o'clock) etc.

The maintenance of geographical direction is best described as an extension of menotaxis (Jander, 1963). This is the faculty in an animal of maintaining an arbitrary angle usually in the horizontal plane between its body axes and a directed stimulus, for instance the sun or an artificial light. The concept of a taxis does not necessarily imply translational locomotion, but need only refer to directional orientation (steering). It can be understood as a tendency for turning and is an angular vector.

Two rival theories of menotaxis at present being discussed (Mittelstaedt, 1962; Jander, 1963) need not be explained here. Both make the common assumption that *menotaxis* can be understood as the sum of the addition of a fixed angle to a position of a simpler relation between the axes of an animal and the stimulus direction, a *prototaxis* (e.g. positive phototaxis, or transverse geotaxis). If to any fixed angle between the light vector and the locomotor vector yet another variable (the hour dependent component is added) which compensates for the azimuth movement of the sun, the animal's orientation is maintained in its geographical direction. In experiments, in which a stationary lamp takes the place of the moving sun, the orientating animal will of course change the direction of its locomotion relative to the lamp.

Sun navigation can be inferred when an orientated locomotion of an animal depends on the momentary visibility of the sun.*

The daily movements of the sun's azimuth are a function of geographical latitude. At the North Pole the sun, which is visible during the summer half of the year only, moves almost parallel to the horizon with a constant speed of 15°/hour and in a clockwise direction. In the northern temperate zone the sun's azimuth moves during the hours of the day also clockwise, but more rapidly around noon than in the morning or evening. On the equator the sun rises—at the time of the equinox—due east and progresses along a vertical circle through the Zenith to the west. Its azimuth therefore stays constant at east (270°) until noon when it instantaneously changes to west (90°). Progressing towards the South Pole the obverse changes in the

* It is important to bear in mind, that the sun plays a dual role in the sun navigation of animals. Firstly it entrains the circadian clockwork, which is analogous to a ship's chronometer and secondly it provides a bearing point for the animal's analogue of a sextant. Experimentally these two 'uses' of the sun have been separated.

azimuth movements of the sun are observed from those on the northern hemisphere, but with the sense of direction reversed to counterclockwise.

From this situation several questions arise, concerning the angle and direction of compensation generated in an animal's brain when sun orientating in various latitudes, whether in its natural habitat or after it has been transported under various conditions of light regimen and previous training. In the latter situation astrotaxis may be greatly affected by transient conditions in the animal. A few facts relevant to these considerations may be mentioned. Honey bees must learn the direction and extent of the sun's apparent daily movement, according to the Hemisphere they are situated in (Lindauer, 1959), though there are traces of an innate component (Kalmus, 1956; New and New, 1962). Lizards show similar plasticity (Fischer and Birukow, 1962). On the other hand, pond skaters of the genus *Velia* find the south direction when sun orientating for the first time even if they have never previously seen the sun or its movements and only have experienced the change of day and night (Birukow and Busch, 1957). The 'light-compass' of most animals appears to be closely associated with other manifestations of the circadian clocks, e.g. motor activity, and as a consequence in common with these one can reset it by artificial manipulation of the *LD* regimen. Birds (Hoffman, 1960) will set off in a predictable but erroneous compass direction, when one has previously tampered with their hours of dusk and dawn. Although many other observations have been described, which shed some light on these problems, there is as yet little hope of a comprehensive theory for these most complex orientational mechanisms. But the role of circadian rhythms in them is unquestioned.

Photoperiodism

The great variability of some climates makes any strict dependence of seasonal sexual maturation or reactivation on such factors as momentary temperature or light intensity precarious. Anticipatory mechanisms, depending on less fickle factors, such as day length (light fraction) or its changes during the time of preparation for the sexual activities provide much better safeguards for correct timing, especially if they depend on integration over many days.

Until recently many people questioned the relevance of the circadian organization to photoperiodicity, but it is getting ever more clear that, while there might be argument on details, the explicit contention of Bünning (1960) that circadian rhythmicity underlies the photoperiodic timing mechanism is essentially correct.

We shall approach the problems of sexual photoperiodism as special examples for the seasonal modulation through photoperiod (as the entraining agent) of the state of circadian systems (Pittendrigh, 1964).

Photoperiodism occurs in many plants and animals but by no means in all of them. According to the phase relation between season and sexual activity or maturation, one can distinguish between long day, short day and day length-indifferent plants and animals. The responses to seasonal changes of day length in many cultivated plants and domesticated animals are demonstrably under genetical control and it is reasonable to suppose that the occurrence and frequency of the various genes responsible for any local population's photoperiodic behaviour are the outcome of local selection. It may be useful to start a discussion of the presumed relation between circadian events and photoperiodism by an anthropomorphic approach. We may ask ourselves the following question. If stranded on a desert island and ignorant of the time of year, but in possession of a watch how would the individual find out whether summer or winter was approaching? He could obviously do this by noting whether on successive days the sun was rising earlier and setting later or vice versa.

If then we postulate that seasonal preparations, such as sexual development or activity, are regulated by an analogous procedure, the following four propositions become plausible. (i) If it serves such a purpose, the circadian clock must be accurate, because, depending on locality, the time changes of light fraction may be as small as fractions of minutes per day and—apart from barren arctic regions—never exceed a few minutes per day. (ii) The effects of these daily increments or decrements must be capable of accumulation. (iii) The effects of addition or subtraction of a few minutes illumination depend on their timing relative to the phases of the internal circadian rhythm, characterized by subjective dawn* and subjective dusk. It is likely that a short additional time of illumination during the very end of the scotophil phase of the circadian cycle and possibly darkening at the beginning of the photophil period—both occurring around dawn— may have quite a different effect from similar additions of light or dark fractions at different circadian phases. (iv) The onset or cessation of illumination may have two distinct effects namely the *entrainment*,

---

* In natural conditions subjective and external dusk and dawn more or less coincide, but this would not be so during experiments (see also Table 1 p. 160).

which keeps the circadian oscillation synchronized (Figure 5) and the *induction* of the photoperiodic changes.

These considerations suggest experimental approaches to the problem of how exactly light fraction affects such sexual phenomena as the flowering of plants or the gonadal growth of vertebrates, together

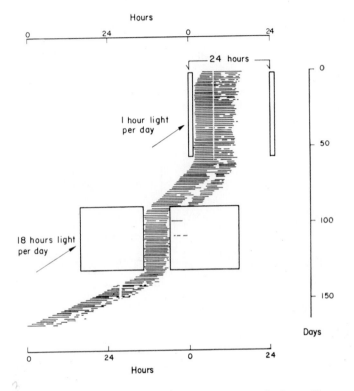

Figure 5. Motor activity in the rodent *Peromyscus maniculatus*. Free running periods (of differing lengths) between days 1–6, 60–92 and after day 132. In between entrainment (capture) of the oscillation by two different *LD* regimens. (After Pittendrigh and Minis, 1964.)

with their many associated changes. One such way is to observe the effects of short light pulses applied at various phases of the circadian rhythms under suitably manipulated *LD* regimens. Particularly interesting are experiments where the stepwise changes from light to dark and from dark to light—by which the circadian rhythm had been entrained—are replaced by two rather short light pulses at the beginning and the end of the light phase. The original overt rhythm may, during this regimen of skeleton illumination, appear maintained for a number of days.

If on the other hand one pulse (the entraining one) is daily applied at the usual hour, while the other (the inducer) is applied at various phases, a most interesting discontinuity of effect can be observed, namely a change of the roles between entrainer and inducer. Figure 6 illustrates the situation.

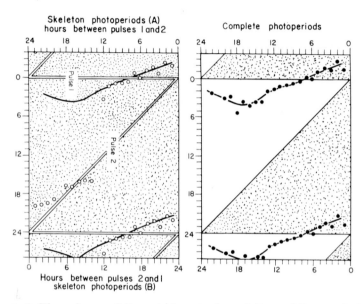

Figure 6. Phase jump of the eclosion rhythm of *Drosophila pseudoobscura* under the influence of skeleton and complete photofractions. The plotted points are medians of steady state experiments. (After Pittendrigh and Minis, 1964.)

In the upper panel the plotted points are 21 medians of steady state distributions of eclosion times of *Drosophila pseudoobscura* having been subjected for at least 6 days to complete photofractions (photoperiods) of between 1 and 23 hours. In the lower panel the complete photofractions have been displaced by 2 skeleton photofractions (light pulses) of 15 minutes duration at the beginning and the end of the corresponding complete photofraction. The figure shows the agreement between complete and skeleton photoperiod as nearly perfect, for up to 11 hours, not so good at 12 hours and unstable at 13 hours. At the 14 hours interval the *LD* 14:10 is 'interpreted' as *LD* 10:14 and the eclosion phase 'jumps' (Pittendrigh and Minis,

1964) by about 12 hours as compared with that observed with a complete 14 hours photofraction. One might say that at this point the two light pulses exchange their significance as either the initiator or the terminator of the photoperiod. For all skeleton regimens greater than 14 the photophil phase is that of the shorter interval between the light pulses.

These results go a long way in explaining the sometimes spectacular effects of night interruption by light flashes on flowering, and in particular the dependence of these effects on the hour of application. More important they make it possible to relate in numerical terms phase responses to single light pulses with the entrainment phenomena (Pittendrigh and Minis, 1964). Whether the effects of the inducer light pulses are in fact identical with those responsible for the phenomena of photoperiodicity remains to be seen. The question whether entrainment and induction are mediated by two different light sensitive substances must also remain open.

References

Adler, H. E. (1963). 'Sensory factors in migration.' *Anim. Behav.*, **11**, 566–577.

Aschoff, J. (1958). 'Tierische Periodik unter dem Einfluss von Zeitgebern.' *Z. Tierpsychol.*, **15**, 1–30.

Aschoff, J. (1965). *Circadian Clocks, Proceedings of the Summer School 1964*, North Holland Publ. Co., Amsterdam.

Birukow, G. and Busch, Elizabeth (1957). 'Lichtkompassorientierung beim Wasserläufer *Velia currens*.' *Z. Tierpsychol.*, **14**, 184–203.

Brown, F. A. (1960). 'Response to pervasive geophysical factors and the biological clock problem.' *Cold Spring Harbor Symp. Quant. Biol.*, **24**, 57–71.

Buchli, H. (1964). Personal communication.

Bühnemann, F. (1955). 'Das endodiurnal System der Oedogonium Zelle, III. Über den Temperatureinfluss.' *Z. Naturforsch.*, **10**, 305–310.

Bünning, E. (1960). 'Circadian rhythms and the time measurement in photoperiodism.' *Cold Spring Harbor Symp. Quant. Biol.*, **25**, 249–256.

Bünning, E. (1964). *The Physiological Clock* (trans. from 2nd German ed.), Springer, Berlin.

Ehret, C. S. (1960). 'Action spectra and nucleic acid metabolism in circadian rhythms at the cellular level.' *Cold Spring Harbor Symp. Quant. Biol.*, **25**, 149–158.

Fischer, K. and Birukow, G. (1962). 'Die Lichtkompassorientierung sonnenlos aufgezogener Smaragdeidechsen (*Lacerta viridis*).' *Verhandl. Deut. Zool. Ges.*, **1962**, 316–321.

Frisch, K. von (1950). 'Die Sonne als Kompass im Leben der Bienen.' *Experientia*, **6**, 210–221.

Halberg, F. (1959). 'Physiologic 24 hour periodicity; general and procedural considerations with reference to the adrenal cycle.' *Z. Vitamin-Hormon-Fermentforsch.*, **10**, 225–296.

Hastings, J. W. (1960). 'Biochemical aspects of rhythm: phase shifting by chemicals.' *Cold Spring Harbor Symp. Quant. Biol.*, **25**, 131–144.

Hoffmann, K. (1960). 'Experimental manipulation of the orientational clock in birds.' *Cold Spring Harbor Symp. Quant. Biol.*, **25**, 379–387.

Jander, R. (1963). 'Insect orientation.' *Ann. Rev. Entomol.*, **8**, 95–114.

Jerebzoff, S. (1961). 'Étude phénomènes périodiques provoqués par des facteurs physiques et chimiques chez quelques champignons.' Thesis, Toulouse, France.

Kalmus, H. (1934). 'Über die Natur des Zeitgedaechtnisses der Bienen.' *Z. Vergleich. Physiol.*, **20**, 405–419.

Kalmus, H. (1935). 'Periodizität und Autochronie (= Idiochronie) als Zeitregelnde Eigenschaften der Organismen.' *Biol. Gen. (Vienna)*, **9**, 93–114.

Kalmus, H. (1938). 'Tagesperiodisch verkaufende Vorgaenge an der Stabhheuschrecke (Dixippus neurosus) und ihre experimentelle Beeinflussung.' *Z. Vergleich. Physiol.*, **25**, 494–508.

Kalmus, H. (1938). 'Ueber das Problem der sogenannten exogenen und endogenen, sowie der erblichen Rhythmik und ueber organische Periodizität ueberhaupt.' *Rivista di Biologia (Florence)*, **24**, 191–225.

Kalmus, H. (1956). 'Sun navigation of *Apis mellifica* L. in the Southern Hemisphere.' *J. Exp. Biol.*, **33**, 554–565.

Kalmus, H. (1964a). 'Animals as Mathematicians.' *Nature*, **202**, 1156–1160.

Kalmus, H. (1964b). 'Navigation by Animals.' *Ann. Rev. Physiol.*, **26**, 109–130.

Kalmus, H. and Wigglesworth, L. A. (1960). 'Shock excited systems as models for biological rhythms.' *Cold Spring Harbor Symp. Quant. Biol.*, **25**, 211–216.

Kramer, G. and St. Paul, U. (1950). 'Stare (*Sturnus vulgaris*) lassen sich auf Himmels-richtung dressieren.' *Naturwissenschaften*, **37**, 526–527.

Lindauer, M. (1959). 'Angeborene und erworbene Kompoten in der Sonnenorientierung der Beinen.' *Z. Vergleich. Physiol.*, **42**, 43–62.

Mittelstaedt, H. (1962). 'Control systems of orientation in insects.' *Ann. Rev. Entomol.*, **7**, 177–198.

New, D. A. T. and New, J. K. (1962), 'The dances of honeybees at small zenith distances of the sun.' *J. Exp. Biol.*, **39**, 271–91.

Pittendrigh, C. S. (1954). 'On temperature independence in the clock system controlling emergence in *Drosophila*.' *Proc. Nat. Acad. Sci. U.S.*, **40**, 1018–1029.

Pittendrigh, C. S. (1965). 'On the mechanism of the entrainment of a circadian rhythm by light cycles.' In Aschoff, J. (Ed.), *Circadian Clocks*,

Proceedings of the Summer School 1964, North Holland Publ. Co., Amsterdam.

Pittendrigh, C. S. and Bruce, V. G. (1957). 'An oscillator model for biological clocks.' In Rudnick, B. (Ed.), *Rhythmic and Synthetic Processes in Growth*, University Press, Princeton, New Jersey, U.S.A.

Pittendrigh, C. S. and Minis, D. H. (1964). 'The entrainment of circadian oscillations by light and their role as photoperiodic clocks.' *Am. Naturalist*, **98**, 277–297.

Stier, I. B. and Wolf, E. (1932). 'On the temperature characteristics for different processes in the same organism.' *J. Gen. Physiol.*, **16**, 367–374.

Stoppel R. (1926). 'Die Beziehungen der tagesperiodischen Erscheinungen beim Tier und bei der Pflanze zu den tagesperiodischen Intensitaetsschwankungen der elektrischen Athmosphaere.' *Planta*, **2**, 342–356.

Waddington, C. H. (1957). *The Strategy of the Genes*, George Allen and Unwin, London.

8

Regulation and Control in the Endocrine System

K. BROWN-GRANT

Department of Human Anatomy, University of Oxford

The concepts of regulation and of homeostasis are implicit in many and explicit in some sections of the first edition of Bayliss' Principles of General Physiology (1915). However, in the chapter on the chemical messengers for which he and Starling had proposed the term hormones, the actions described and discussed are purely effector, with little mention of any possibility of their integrative function or coordination. In 1931, however, Aron and his coworkers (Aron, Van Caulert and Stahl, 1931) could outline the reciprocal, mutually regulating interrelationship of the anterior pituitary and the thyroid via the levels of circulating thyrotrophin and thyroid hormone. The concept of the endocrine system as a homeostatic and regulating system has since come to dominate physiological thinking in this field (see for example the introductory chapters of Cameron, 1945; Selye and Rosch, 1954; Gorbman and Bern, 1962). It is a striking example of the far reaching effects of the development of a simple, reliable technique for the performance of a single essential experimental manoeuvre that much of the development in this field after 1915 was made possible by the advent of Smith's method of hypophysectomy in the rat (Smith, 1927). This chapter will be concerned with the various types of control mechanism observed in the endocrine system, an enquiry into the purpose of these systems and how they may operate, and the question as to whether it is possible, or profitable, to consider them as feedback mechanisms and how far any such analytical treatment is valid or useful. A reasonable familiarity with endocrine concepts and terminology in the reader has

176

been assumed. For the non-biologist (as yet no adjective comparable to illiterate or anumerate has been suggested to describe members of this subculture) a good general popular account is that by Stuart-Mason (1960) and the textbook of Gorbman and Bern (1962) is a lucid and remarkably complete account of the current state of knowledge in the field for readers with some biological training. The glands of internal secretion that will be considered are the adrenal medulla and posterior pituitary gland; the endocrine pancreas and parathyroids; the thyroid and adrenal cortex, and the anterior pituitary in connection with these two and with the gonads. No attempt has been made to deal comprehensively with any of these, but rather to illustrate types of control mechanisms. A further necessary limitation, because of lack of any personal acquaintance with the field, is the omission of all but incidental reference to poikilotherms and the total exclusion of any consideration of invertebrates. Much of the available information on the control of endocrine function in these animals is assembled in a book by Jenkin (1962) and more detailed reviews can be found in the volume edited by Pincus, Thimann and Astwood (1964).

Mechanisms of Regulation of the Activity of Endocrine Glands

1. Direct neural control

Two organs with well established endocrine function, the adrenal medulla and the neurohypophysis (posterior pituitary), are derived embryologically from neural crest ectoderm and from the base of the diencephalon. The secreto-motor fibres to the adrenal medulla are preganglionic cholinergic sympathetic system fibres which, when stimulated, cause the release of noradrenaline or noradrenaline plus adrenaline from the chromaffin cells of the medulla. These cells appear to be of two types, one capable of the synthesis of noradrenaline only and the other of noradrenaline plus adrenaline, presumably by virtue of possessing the enzymes necessary for methylation of the amine group of noradrenaline. The proportion of these two types of cell varies in different species and adrenaline is absent from the foetus and newborn in some mammals, appearing later in life. The secretion of the two hormones can vary independently; predominantly adrenaline is secreted in response to the stimulus of hypoglycaemia and noradrenaline in response to hypotension. The anatomical basis of this differential secretion appears to be the presence of two types of neurones in the posterior hypothalamus which, on direct electrical stimulation, cause the release of predominantly one or other hormone

from the medulla (von Euler, 1956). A multiplicity of afferent pathways must in turn reach these hypothalamic neurones, as medullary
secretion occurs in response to such varied stimuli as pain, fright,
cold, reduced blood pressure and lowered blood sugar levels. The release of adrenaline may indeed be an all or none emergency mechanism as originally suggested by Cannon. Secretion may be brief, but
intense, and self-limiting by exhaustion of the available gland supply.
Noradrenaline, in contrast, has a greater role in minute to minute
regulation of the cardiovascular system and its release is more continuous and dependent on appropriate signals from the baro-receptors
monitoring the systemic arterial blood pressure (Figure 1).

The neurohypophysis exhibits the phenomenon of neurosecretion
in its most striking form in vertebrates. The hormones of the gland
(arginine or lysine vasopressin and oxytocin in mammals, vasotocin
(8-arginine oxytocin) and oxytocin in birds and reptiles, and vasotocin and ichthyotocin (4-serine, 8-isoleucine oxytocin) or oxytocin-
like polypeptides in amphibia and fish (Sawyer, 1963; Heller, 1966)
are now thought to be manufactured by the nerve cells of the
supraoptic and paraventricular nuclei of the hypothalamus. They
are transported along their axons to the neural lobe of the pituitary
where they are released from the nerve endings into the closely
adjacent blood vessels. The functions of the two hormones in mammals are antidiuretic and to cause contraction of the smooth muscle
of the uterus and the myoepithelial cells of the mammary gland,
though in lower vertebrates both hormones may affect water and salt
metabolism by a variety of mechanisms (Heller and Bentley, 1963).
The stimulus for their secretion in lower animals is not known.

In mammals, the stimuli that elicit secretion of the posterior
pituitary hormones are quite varied. There appears to be a non-
specific release of vasopressin in response to pain or trauma; the release of oxytocin, on the contrary, appears to be inhibited by such
stimuli (Harris, 1950, 1958). Oxytocin secretion is stimulated by
afferent impulses from the lower genital tract in the female, e.g.
uterine cervix or by stimulation of the nipples by suckling young. The
two responses to oxytocin, uterine contraction and milk ejection may
therefore occur simultaneously and at times inappropriately. Whatever the physico–chemical state of the hormones in the neurohypophysis (Sawyer, 1963) the physiological evidence is that one or other
may be released independently of the other in response to natural
stimuli, though experimental stimuli may induce secretion of both
(Holland, Cross and Sawyer, 1959a; Harris, 1955a, 1958). The classic
stimulus to vasopressin secretion is a change in the osmotic pressure

of the blood reaching the hypothalamus. More recently it has been suggested that a tonic control of vasopressin may be exerted by the brain stem reticular system, affected in turn by impulses from 'blood volume receptors' (possibly atrial stretch receptors) and by impulses from arotic and carotid baro-receptors to inhibit secretion as required (Farrell and Taylor, 1962). It would be of interest to know if the postulated afferent stimuli from the vascular system affect the rate of firing of individual supraoptic and paraventricular neurones in the same way as osmotic stimuli have been shown to do (Cross and

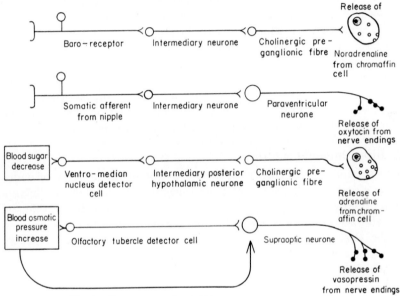

Figure 1. Possible pathways involved in the neural and humoral control of adreno-medullary and neurohypophysial secretion.

Green, 1959). Evidence for an additional tonic inhibition of supra-optic neurone activity and vasopressin release by afferents from the mid-brain in the cat has been presented recently by Suda, Koizumi and Brooks (1963).

The control of the secretion of the adrenal medulla and neuro-hypophysis is neural and many of the afferent pathways discussed above are also exclusively neural. At least two situations exist, how-ever, in which there appears to be a humoral link in the regulatory system (Figure 1). Increased adrenaline secretion in response to hypoglycaemia is dependent upon the hypothalamus and there must be somewhere neural elements capable of monitoring the blood sugar level. It seems possible that they may be in the region of the ventro-medial nucleus of the hypothalamus where the 'satiety centre' and the central 'glucostat' thought to be concerned with the regulation of food intake have been tentatively identified (Anand, 1961). Impulses from these neurones may affect secondarily the activity of cells in the posterior hypothalamus regulating the discharge of adrenaline from the adrenal medulla. Experiments demonstrating increased secretion of vasopressin in response to intracarotid injection of hypertonic saline localized the site of the osmoreceptors to the anterior hypothalamus and the vesiculated neurones of the supra-optic nucleus were thought to be both receptor and effector cells (Verney, 1947). More recent electrophysiological work (Holland, Cross and Sawyer, 1959b) has suggested that the actual osmotic pressure detectors may be located in the deep portion of the olfactory tubercle rather than in the effector cells of the supraoptic nucleus themselves. A neural link between the cells detecting changes in the blood and the effector cells may exist in both cases.

2. Direct humoral control

This section is concerned with the regulation of the activity of endo-crine glands which are not of neural origin and which do not have a functionally important nerve supply, nor is their activity affected by the secretion of some other endocrine gland. Control appears to be exerted by a direct influence upon the gland itself of the blood content of the metabolically important component, the level of which is influenced by the hormone secreted by the gland.

In most mammals other than ruminants the level of glucose in the blood is normally maintained within quite narrow limits (70–120 mg/ 100 ml). The hypoglycaemia-producing hormone is insulin, secreted by the beta islet cells of the pancreas which reduces the blood glucose level mainly by promoting the uptake of glucose by the skeletal muscles. A rise in the level of glucose in the blood reaching the pan-creas stimulates the secretion of insulin; the rate of release falls or ceases when the blood glucose level is subnormal (Young, 1963). Hyperglycaemia is not, in itself, harmful to the body; hypoglycaemia may result in rapid loss of consciousness or death probably because the central nervous system is dependent on glucose as its major or sole metabolic fuel and dependent, moreover, on a constant supply of glucose from the blood as there are no carbohydrate stores in the brain.

There are two major hormonal mechanisms that are brought

into action during hypoglycaemia apart from a decrease in insulin secretion. One is the neurally mediated release of adrenaline discussed above which acts largely by accelerating the breakdown of muscle glycogen. The other is the secretion of the second pancreatic hormone, glucagon, which raises the blood sugar level by an action on the liver to promote glycogen breakdown. The hormonal status of glucagon, a polypeptide produced by the alpha islet cells of the pancreas and first detected as a hyperglycaemic factor contaminating commercial insulin preparations, was at first debatable. It now seems established that it is indeed a true hormone, secreted by the pancreas in direct response to a fall in the glucose content of the blood reaching the organ (Foà and Galansino, 1962; Berthet, 1963; Foà, 1964). Hormonal regulation of the blood sugar involves the action of at least two other hormones, the glucocorticoids of the adrenal cortex and the growth hormone of the anterior pituitary. Both act to raise the blood sugar level but in different ways. The secretion of the first is certainly, and of the second probably, stimulated by a fall in blood glucose level, but the mechanisms involved are complicated and involve the intervention of the hypothalamus.

The second example of an endocrine gland, whose activity is controlled directly by the blood level of the component its secretions influence, is the parathyroid. Parathormone raises the blood calcium level largely by an action on bone and the stimulus to its secretion is a fall in the ionizable calcium level of the blood reaching the glands (Copp, Moghadam, Mensen and McPherson, 1961). This hormone has now been prepared in pure form and has both the effects on bone and on the renal handling of phosphate previously described in response to parathyroid extracts. It may not be the sole hormone of the parathyroids. There is indirect evidence (see review by Munson, Hirsch and Tashjian, 1963) that a second hormone, calcitonin, exists. Calcitonin is released in response to a *raised* level of ionizable calcium in the blood reaching the gland and acts in some way to reduce blood Ca^{2+} levels. Calcitonin, or a substance with similar actions, may be produced by the thyroid as well as by the parathyroid glands (Baghdiantz, Foster, Edwards, Kumar, Slack, Soliman and MacIntyre, 1964). The analogy with the actions of glucagon and insulin in regulating the blood sugar level is obvious. This type of direct humoral control via the blood level of some component which is influenced by the hormones of the gland is not common. It is possible, however, that the secretion of erythropoietin (a substance acting humorally to stimulate red cell production) by the kidney is controlled by the oxygen content of the blood reaching it, an indirect index under normal conditions of the red cell content of the blood (Naets, 1963).

3. Indirect humoral control by the trophic hormones of the anterior pituitary

Suggestive findings from clinical cases and the effects of injections of pituitary extracts into intact animals played a part in the recognition of the role of the anterior pituitary as an endocrine gland in its own right and also as influencing the thyroid, adrenal cortex and the gonads. The first successful experimental removals with prolonged survival, an essential step in the analysis of endocrine function, were performed by Smith (1916) on tadpoles. Later, he devised a method for performing the operation in rats (Smith, 1927), and, as discussed in the introduction, this was an essential step in the analysis of pituitary function. The possibility of prolonged survival after this operation, when performed by a skilled experimentalist, is demonstrated in a recent paper (Smith, 1963) on the effects of pituitary transplants made a year after hypophysectomy in rats.

The concept of the anterior pituitary as the major factor regulating the activity of other endocrine glands is now generally accepted. The gland contains and can secrete independently five separate protein hormones influencing other endocrine glands. These are often called trophic hormones (a suggested alternative generic name—endocrinokinetic hormones (Carlisle and Jenkins, 1959)—is more accurately descriptive) and are known as the thyrotrophic, adrenocorticotrophic, follicle stimulating, luteinizing, and luteotrophic hormones. The sixth pituitary hormone, which has general metabolic rather than endocrinokinetic action, is the growth hormone, somatotrophin. The following abbreviations will be used in this chapter: TSH and ACTH; for the gonadotrophins, FSH, LH, and the luteotrophic hormone will be referred to as prolactin. The unique feature of the regulation of the activity of the target endocrine glands by the pituitary trophic hormones is that this introduces the possibility of a homeostatic control of the level of the target gland hormone in the blood. In a sense, this is a highly specialized example of a regulation by a blood constituent whose level is affected by the hormone secreted, but the intermediary is not calcium as in the case of the parathyroid, but the hormonal product of the target gland. In view of the importance accorded to the concept of homeostasis in physiological thinking, it seems appropriate to deal with pituitary-target organ relationships first as examples of self-regulating feedback systems and to see how far their working in the intact animals can be explained in these terms. The pituitary-thyroid system will be discussed first.

The Control of TSH Secretion by a Humoral Feedback Mechanism

The concept of the anterior pituitary-thyroid relationship as a self-regulating feedback system through the action of thyroxine on TSH secretion and of TSH on the thyroid is now of respectable antiquity (Aron and coworkers, 1931; Hoskins, 1949). The effects of exogenous thyroid hormone in depressing thyroid gland activity and the response of the thyroid when the production of hormone is reduced by partial thyroidectomy, shortage of iodine from reduced dietary intake (Stanbury and coworkers, 1954) or other causes or by interference with the synthesis of thyroid hormones by drugs (Pitt-Rivers and Tata, 1959) or as a consequence of an inborn error of metabolism affecting some stage of hormone biosynthesis (Stanbury, 1963) can be explained on this basis. The speed at which the pituitary can react to an increase or a decrease in circulating thyroid hormone level is surprisingly high. Lang and Reichlin (1961) showed indirectly that TSH secretion stopped within about an hour and possibly within minutes after the intravenous injection of thyroxine in the rabbit and that a fall in the blood level of hormone activated the pituitary within 6 hours; Bakke and Lawrence (1964) demonstrated a fall in the elevated serum TSH level of hypothyroid rats within an hour of thyroxine administration. Apart from evidence derived from experiments (which are discussed later) in which thyroxine was injected into the pituitary, these short time intervals indicate that the changes in TSH secretion cannot be due to alterations in the metabolic activity of the body tissues. These do not occur as soon after injection as this (Pitt-Rivers and Tata, 1959) and suggest that the changes in TSH secretion are directly related to changes in blood thyroid hormone content.

In these examples, the regulation of thyroid gland activity by the action of the circulating hormone in modifying the rate of release of TSH has been considered as it operates under situations where the level of thyroxine rises due to the introduction of exogenous hormone or falls because of failure of thyroidal secretion from some cause. An increase in the peripheral rate of metabolism of thyroxine when the rate of secretion is constant will lead to a fall in the level of hormone in the blood and a decrease to a rise. The importance of such changes is obvious; the difficulty in the past has been to quantitate them. The major pathways of peripheral metabolism of thyroxine and triiodo-thyronine are now known to be by *deiodination*, with subsequent partitioning of the iodide between the thyroid and the kidney, or by *faecal excretion* of the intact molecule or a derivative modified in the alanine side chain or some conjugate such as the glucuronide following hepatic excretion into the bile and a greater or lesser degree of entero-hepatic recirculation (Pitt-Rivers and Tata, 1959; Tata, 1964). Variations in the rate of metabolism by either of these routes can now be studied quantitatively by the use of thyroid hormones of high specific activity labelled with radioactive iodine.

An interesting example of the effect of changes in peripheral metabolism is provided by a study of the thyroid changes during lactation in rats maintained on a diet of high iodine content. Grosvenor and Turner (1958) first showed that the level of thyroid activity was greater than that of non-lactating female control rats; the cause of the increased activity was not known at that time; various possibilities could be envisaged, such as an association of prolactin and TSH secretion. Van Middlesworth (1960) drew attention to the fact that the rate of faecal excretion of thyroxine in the rat may be related to faecal mass, high residue diets resulting in a greater loss than low residue diets of equal calorific value. Grosvenor (1962a) reexamined the thyroxine metabolism of lactating rats in the light of these observations and showed that the greater food intake and resulting greater faecal mass had resulted in increased rate of loss of thyroxine by this route. The raised level of thyroid gland activity is necessary to maintain the normal blood level of hormone in the face of an increased rate of peripheral metabolism. Alterations in the rate of loss of hormone consequent upon changes in food intake may be most pronounced in the rat and other small rodents where the faecal loss of thyroxine is a dominant feature of the thyroxine economy of the body (Grossie and Turner, 1962). Another situation in which such changes may play an important part is in the response of the pituitary-thyroid system during exposure to cold over a period of days or weeks. In these circumstances, food intake is increased, the rate of faecal excretion of thyroxine is greatly increased and the level of thyroid hormone in the blood falls in both intact and thyroxine-maintained thyroidectomized rats (Intoccia and Van Middlesworth, 1959; Kassenaar, Lameyer and Querido, 1956). The part played by metabolic changes acting through a lowered blood level of thyroid hormone in the response to prolonged cold exposure has been considered in more detail elsewhere (Brown-Grant, 1966a).

Changes in the rate of metabolism of thyroid hormones by the peripheral tissues or of their excretion may also follow alterations in the level of circulating hormone and in a sense act as regulatory mechanisms. The details of these changes vary from species to species. In the rat, for instance, Maclagan and Wilkinson (1954) showed that a large dose (50–100 μg) of thyroxine was excreted more quickly than a tracer dose (4 μg) and that the increase was due largely to a marked rise in faecal excretion. No comparable decrease in the biological half-life or increase in faecal excretion was observed after the administration of increasing doses of triiodothyronine. Myant (1957) has provided an explanation of these findings. At normal plasma levels, thyroxine is bound by non-covalent linkages to specific thyroxine-binding protein(s). The affinity of these proteins for thyroxine is very high, but they are present only in minute amounts in serum and their capacity is limited. As the level of thyroxine is raised, the hormone is distributed between the binding proteins and the albumin fraction of plasma, increasingly in favour of the latter. The affinity of albumin for thyroxine is lower, though the binding capacity is very high, and the level of free (non-protein bound) thyroxine in the plasma therefore rises. Myant showed that as the plasma level of the hormone was raised, the rate of hepatic clearance of thyroxine increased from 2 ml plasma/hour at physiological levels to 35–40 ml/hour at high levels. The increased faecal excretion of hormone is probably a consequence of this increased rate of secretion into the gut. Triiodothyronine is much less firmly bound to the specific binding protein than is thyroxine and the biliary clearance rate was 30–70 ml/hour at all plasma levels but was not related to the absolute plasma level. Myant (1956) has also measured the hepatic clearance of thyroxine in man. Similar increases in clearance after the injection of large doses of thyroxine were observed but quantitatively the limited data available suggest that this self-regulating mechanism, though present, is less important in man than in the rat.

The rabbit differs from the rat and from the mouse (Brown-Grant and Gibson, 1955; Brown-Grant and Tata, 1961; Brown-Grant, 1963) in that an excess of thyroxine does not result in an increase in faecal excretion rate or a reduction of the biological half-life. An increase in the renal excretion of iodide derived from the injected hormone suggested that in the presence of a raised blood level an increase in the rate of deiodination occurred. This may represent an alternative method for controlling the level of circulating thyroxine; similar changes have been observed in guinea-pigs (Brown-Grant, 1963). The mechanism responsible for the change in the

pattern of thyroxine and triiodothyronine metabolism when an excess
is present in the rabbit and guinea-pig is not yet known. In more chronic
experiments, Tata (1961) showed that the tissue level of the enzyme,
thyroxine dehalogenase, assayed by the deiodination of thyroxine
by homogenates of muscle *in vitro*, was reduced following thyroidectomy and increased following L-thyroxine administration to
normal or thyroidectomized animals but not following D-thyroxine
or 2:4-dinitrophenol administration. Whether changes in enzyme
level can occur over the much shorter time periods involved in the
experiments *in vivo* is not known. In addition, the assay method used
has subsequently been subjected to considerable criticism as to its
specificity (Galton and Ingbar, 1963; Morreale de Escobar, Rodriguez,
Jolin and Escobar del Rey, 1963). However, although the exact
mechanisms involved are not yet clearly established in all cases, it
does appear that with increasingly accurate methods for assessing
the rate of extra-thyroidal degradation and excretion of thyroid
hormones, some degree of regulation at the tissue level involving increased removal when the blood level is high and a reduction when
the level is subnormal has to be considered (Hogness, Wong and
Williams, 1954; Sterling and Chodos, 1956).

The association between thyroxine and certain specific thyroxine
binding protein or proteins (TBP) has been mentioned briefly above.
This phenomenon has been the subject of intensive study in recent
years and the methodology is advancing rapidly. Several excellent
reviews have been published (Ingbar and Freinkel, 1960; Robbins and
Rall, 1960; Tata, 1960, 1964; Farer, Robbins, Blumberg and Rall,
1962), but inevitably all are now somewhat incomplete. The essence
of the situation is that many of the effects of thyroid hormones on
tissues, including the effect on the pituitary to depress TSH secretion
and also the rate of thyroid hormone metabolism and excretion,
appear to be determined not by the *total* hormone level in the plasma
but by the level of *free* hormone, that is thyroxine not 'bound' to
protein. The actual level of free thyroxine in the plasma of euthyroid
man was calculated to be only about 0·05% of the total thyroxine
content (Robbins and Rall, 1960). A value of this order has been
found experimentally by Lee, Henry and Golub (1964) and by
Oppenheimer and Surks (1964).

The presence of a binding protein in plasma does not affect
the concept of how the pituitary-thyroid system operates when the
animal is considered in the steady state if the level of free thyroxine
is the operationally important controlling factor. If the level of
TBP is abnormally high or if a normal amount of TBP with an

increased affinity for thyroxine is present, then the total thyroxine level of the plasma at which the *free* thyroxine level is normal will be raised; similarly if the TBP concentration is low or affinity is reduced, the total content for a normal *free* T4 level will be reduced. The total extra-thyroidal pool of thyroxine will also be increased or decreased following the change in plasma total T4 level, assuming that there has been no change in the thyroxine distribution space, but the fractional rate of turnover will also be decreased and increased respectively and the absolute rate of thyroxine disposal in µg/day will be the same. The thyroid gland will therefore be called on to supply exactly the same quantity of hormone per day to maintain the system in a steady state, despite a raised or lowered total blood level of hormone that would superficially be expected to depress or stimulate TSH secretion. The finding of a normal level of thyroid gland activity in euthyroid individuals with a raised or lowered total thyroxine level coupled with an idiopathic elevation or reduction of serum TBP has confirmed these predictions (Ingbar, 1961; Nicoloff, Dowling and Patton, 1964).

A possible regulating function for TBP can be envisaged, rather in the way that buffering by plasma proteins assists in the maintenance of a constant hydrogen ion concentration in the blood. However, as Lee and coworkers (1964) have pointed out, in the presence of an excess of hormone the 'buffering' is not at all effective, the rise in free thyroxine being a parabolic function of total thyroxine level, though for any given increase of total T4, the rise in free T4 was least in sera containing the highest level of TBP. When acute changes in TBP level occur, changes in thyroid activity may follow. A rise in TBP level has been demonstrated following the administration of oestrogenic hormones and during pregnancy in primates, though not so far in any other species, and a fall may occur following the administration of testosterone or anabolic steroids (see Robbins and Rall, 1960; Brown-Grant, 1966a, for references). Rising TBP levels lead to a fall in free T4 and stimulation of TSH secretion until the extra hormone secreted, together with a reduced rate of peripheral metabolism, lead to restoration of free T4 levels in the presence of an increased total extra-thyroidal pool. The converse changes occur when TBP levels fall.

The temporary variations induced by changes in TBP do not conflict with the general concept of a feedback mechanism and indeed the realization of the part played by these proteins has removed certain inconsistencies that previously existed in the application of this principle to the pituitary-thyroid interrelationship. Thus the simultaneous occurrence of a raised total thyroxine level and a euthyroid state of the peripheral tissues in human pregnancy has been shown to be related to a rise in TBP (principally the thyroxine binding globulin) with the maintenance of a normal free thyroxine level (Oppenheimer, Squef, Surks and Hauer, 1963; Lee and coworkers, 1964). The role of oestrogen in eliciting the rise of TBP level in pregnancy is still uncertain (Ingbar and Freinkel, 1960). The rise in the radioiodine indices of thyroid activity in human pregnancy may be related to an increase in the renal excretion of iodide leading to a relative iodine deficiency (Aboul-Khair, Crooks, Turnbull and Hytten, 1964).

Conversely, a reduced plasma level of thyroxine in the absence of any evidence of hypothyroidism or of activation of the pituitary-thyroid system following the administration of 2,4-dinitrophenol was reported by Wolff, Rubin and Chaikoff (1950), and Goldberg and Chaikoff (1951) commented that these results were not consistent with the classic idea of a feedback regulation, a view which was further elaborated by Goldberg, Wolff and Greep (1957). Subsequent studies have shown by a variety of methods that dinitrophenol interferes with the binding of thyroxine to TBP in both human and rat serum (Christensen, 1960a; Wolff, Standaert and Rall, 1961; Osorio, 1962). Indirect experimental evidence from experiments *in vivo*, suggesting that this results in a rise in the free thyroxine level, is the acute increase in the rate of biliary secretion of thyroxine in the rat after dinitrophenol administration (Escobar del Rey and Morreale de Escobar, 1958) which is comparable to that observed by Myant (1957) after the injection of large doses of thyroxine. In the presence of dinitrophenol a normal free thyroxine level will be obtained at a lower total thyroxine level, hence the euthyroid state of the tissues and the absence of any increase in TSH secretion. The changes observed are in fact completely consistent with the idea of feedback control by the level of *free* thyroxine in the plasma. Similar changes (reduced total thyroxine level, euthyroidism and a normal level of TSH secretion and thyroid gland activity) have also been observed following the administration of salicylates and related compounds and diphenyl hydantoinates (a group of drugs used in the treatment of epilepsy); these compounds have also been found to interfere with the binding of thyroxine to TBP *in vitro* and *in vivo* (Christensen, 1960b; Oppenheimer, Fisher, Nelson and Jailer, 1961; Oppenheimer and Tavernetti, 1962; Osorio and Myant, 1963; Woeber and Ingbar, 1963, 1964). Although most of the changes observed can be explained by an effect on thyroxine binding (Good, Hetzel and Hogg, 1965), other mechanisms may be involved following the chronic administration of 2,4-dinitrophenol (see for example, Goldberg, Wolff and Greep, 1957; Reichlin, 1960; Morreale de Escobar and Escobar del Rey, 1961a, 1961b).

The idea of 'feedback' regulation of TSH secretion by the level of thyroid hormone in the blood has been modified to take account of the binding of hormone by plasma proteins and certain apparent inconsistencies can be explained in this way. Something more is involved in the control of TSH secretion, however, than a mere bathing of the pituitary cells in thyroxine solution. Recent studies on the actions of a group of goitrogenic (antithyroid) drugs related to thiouracil have emphasized this. Thiouracil derivatives block the synthesis of thyroid hormones, possibly by an action on the peroxidase involved in the oxidation of iodide to iodine at the stage before iodotyrosine formation (Pitt-Rivers, 1950; Pitt-Rivers and Tata, 1959), and this was for a considerable period considered to be their only action. Some evidence was obtained as early as 1949 (Andik, Balogh and Donhoffer, 1949) that these drugs might also interfere with the actions of exogenous thyroxine and Van Arsdel and Williams (1956) showed that propylthiouracil inhibited the peripheral deiodination of thyroxine in the rat. Their findings have since been confirmed and extended to the other thiouracil derivatives in rats and in man (Escobar del Rey and Morreale de Escobar, 1961; Hershman and Van Middlesworth, 1962; Hershman, 1964) and an effect on the metabolism of other thyroxine analogues has been demonstrated (Van Middlesworth and Jones, 1961). This effect on peripheral deiodination results in a rise in the plasma PBI level in thyroxine-maintained thyroidectomized animals before the decrease in urinary iodide excretion and increase in faecal thyroxine excretion, consequent upon the inhibition of deiodination, are apparent.

In more prolonged experiments, it has been shown that a dose of thyroxine that maintains a normal plasma hormone level does not prevent growth of the thyroid in propylthiouracil-treated rats and that goitre prevention (a restoration of the TSH secretion rate to normal) is only achieved when a dose of thyroxine sufficient to maintain two to three times the normal blood level is administered (Van Middlesworth, Jagiello and Vanderlaan, 1959). In contrast, when goitre formation is induced by potassium perchlorate or methimazole, a dose of thyroxine that maintains a normal blood level prevents growth of the thyroid (Jones and Van Middlesworth, 1960; Grosvenor, 1962b). Neither of these drugs affects thyroxine deiodination. It seems that in the presence of propylthiouracil the pituitary is much less responsive to a given level of circulating thyroxine. This change is quite rapid in onset; within 24 hours, at a time when the hormone level is still normal, TSH secretion is increased. At this time overall deiodination of the hormone has already been affected. These effects on TSH secretion can be prevented by large doses of hormone (Escobar del Rey, Morreale de Escobar, Garcia and Garcia, 1962; Morreale de Escobar and Escobar del Rey, 1962; Grosvenor, 1963) and the absolute quantity of thyroxine deiodinated is maintained under these conditions. During the administration of drugs of the thiouracil group it appears that there is a disruption of the usual relationship between plasma hormone level and TSH secretion which is not related to any effect on the binding of thyroxine by plasma proteins (Escobar del Rey and colleagues, 1962), but which may be related to an interference with the process of deiodination of thyroxine by the peripheral tissues. The actions of the hormone on the peripheral tissues are partially blocked as a consequence of this and it is tempting to suggest that the action on the pituitary is similarly modified but no direct evidence of this is at present available.

The Role of the Central Nervous System in the Control of TSH Secretion

General introduction

So far this discussion of the means by which the pituitary regulates and controls the thyroid has been confined to the interaction of these two glands via changes in the blood level of thyroid hormone and of TSH. Not to consider the role of the central nervous system would be to ignore the most important development in endocrine research in the past thirty years. H. M. Evans in his Harvey Lecture of 1924 dealt with the brilliant pioneer work from his laboratory on the demonstration of gonadotrophic, thyrotrophic, adrenocorticotrophic and growth promoting protein hormones in extracts of the anterior pituitary. It is ironical, in retrospect, that he should have dealt so successfully with the then current misinterpretation of changes in endocrine function following damage to the hypothalamus, as indicating direct neural control of the gonads and of the thyroid gland. This so emphasized the primary importance of the pituitary that the role of the central nervous system acting through the pituitary in the control of endocrine function was largely ignored for the next ten to fifteen years. The change in the intellectual climate may be judged by comparing the account of the control of pituitary function given by Marshall (1922) or Vincent (1922) and that of Rolleston (1936).

Various clues from reports of clinical cases and some scattered papers on experimental studies during this period can be seen, in the light of current knowledge, to have hinted at the role for the hypothalamus in the control of pituitary function, but as Schrieber (1963) pointed out in an excellent account of the historical background of present day views on neuroendocrine relationships, the implications were not appreciated at the time.

The studies of F. H. A. Marshall of the University of Cambridge played a central part in establishing a body of knowledge derived from both observation and experiment on the periodicity of sexual function and the role of the external environment ('exteroceptive factors') in initiating and sustaining sexual cycles. From this the need for a reevaluation of central nervous control of the pituitary became apparent and using the knowledge it was possible to devise suitable critical experiments. Marshall's writings illustrate the accumulation of data and show how his views developed to this stage (Marshall, 1910, 1922, 1936, 1942). In the first edition of his textbook in 1910 there is no mention of the pituitary gland in the index. In 1942 he wrote '. . . it would appear certain that many external factors which regulate the (sexual) cycle act through the intermediation of the central nervous system upon the anterior pituitary, this gland playing the part of liaison organ between the nervous system which is affected by stimuli from without and the endocrine system. . .'. Consequently, many of the early studies on neural control of the anterior pituitary gland were concerned with gonadotrophin secretion but the techniques involved (electrical and chemical stimulation of the hypothalamus and pituitary, placement of lesions in the hypothalamus, transplantation of the gland and section of the pituitary stalk) are equally applicable to the study of the regulation of the secretion of other tropic hormones.

A major advance occurred when Harris (1948) suggested that the crucial link between hypothalamus and pituitary was not neural but vascular, by way of the hypothalamo-hypophysial portal vessel system. Variable results following section of the pituitary stalk were explained when he demonstrated that the portal vessels were capable of regeneration and that the functional activity of the gland after stalk-section could be related to the degree of regeneration (Harris, 1950a). Another key experiment in support of his concept was the grafting of pituitary tissue from new born rats into the hypophysial capsule of hypophysectomized adults and the demonstration that when such grafts were vascularized by vessels from the primary plexus of the hypothalamo-hypophysial portal system they became capable of normal adult function including the cyclic release of gonadotrophins (Harris and Jacobsohn, 1952). The function of the portal vessel system was suggested by Harris (1955a) to be the transfer of neurohumoral transmitter agents secreted by nerves ending in the region of the primary plexus in the median eminence to the pituitary, where they act to stimulate and possibly also to inhibit the synthesis or release of the various trophic hormones. This view is now widely accepted (see for example the Chapter by Fraps (1962) in Zuckermann (1962) and Guillemin and Schally, 1963). Detailed discussion of the evidence on which these views are based can be found in successive reviews by Harris (1948, 1955a, b, 1959b, 1962). The neuroendocrine aspects of the control of TSH secretion can now be considered.

Neural control of TSH secretion

All the standard procedures (hypothalamic lesions, pituitary stalk-section, pituitary transplantation, electrical stimulation of the hypothalamus) have been applied to study the part played by the CNS in controlling TSH secretion. There are by now a considerable number of reviews devoted wholly or in part to this topic; there is also, perhaps more surprisingly, a considerable measure of agreement among different writers as to what has been established. What follows is a brief summary of the collective views of Bogdanove (1962), Brown-Grant (1960, 1966a), D'Angelo (1955, 1963), Greer, Yamada and Iino 1960), Harris (1955b, 1959a), Knigge (1960a), Reichlin (1964).

After effective pituitary stalk-section (i.e. if regeneration of the portal vessels is prevented) or after transplantation of the gland to a site where it is not vascularized by portal vessels (anterior chamber of the eye or beneath the kidney capsule) the basal level of TSH secretion is reduced though the level of thyroid activity is maintained at a higher level than after hypophysectomy, indicating a continued secretion of TSH. Reimplantation of the pituitary in the hypophysial capsule and vascularization by portal vessels restores TSH secretion to near normal values. Destructive lesions of the hypothalamus involving the median eminence reduce the secretion of all pituitary trophic hormones. However, bilateral ventral lesions anterior to the ventromedian nucleus but behind the optic chiasma and usually involving part of the supraoptico-hypophysial tract reduce TSH secretion specifically. Electrical stimulation in this region increases TSH secretion in rabbits and rats. The hypothalamus appears to exert a chronic stimulatory control over TSH secretion, probably by the secretion of specific thyrotrophin releasing factor (TRF) into the portal vessels (Schrieber, 1963) though other possible mechanisms have been proposed (Brown-Grant, 1957; Purves, 1960), and possible changes in the rate of peripheral metabolism of thyroxine, such that a lower rate of thyroxine and TSH secretion would be adequate to maintain a normal circulating hormone level, have never been specifically excluded in critical experiments (Brown-Grant, 1966a). Although lingering doubts as to the existence of a tonic stimulation of TSH secretion in basal conditions persist in some quarters, there is no doubt that an intact hypothalamus in normal vascular relationship to the pituitary is essential for certain acute changes in TSH secretion to be elicited. A variety of non-specific procedures ('stresses') that have in common the ability to elicit an abrupt discharge of adrenocorticotrophic hormone (ACTH) also produce a decrease in TSH secretion (Harris, 1955c). In the rabbit, it was shown that immobilization or the administration of large doses of stilboestrol which normally cause a decrease in TSH secretion failed to do so after pituitary stalk-section (Brown-Grant, Harris and Reichlin, 1954, 1957). Conversely, the increased release of TSH that normally rapidly follows exposure to reduced environmental temperature is abolished in rabbits and rats after transplantation of the pituitary (von Euler and Holmgren, 1956b; Knigge and Florsheim, 1960) or in hamsters with transplanted pituitaries or lesions of the median emminence (Knigge and Bierman, 1958). Andersson, Ekman, Gale and Sundsten (1962b, 1963) have shown that the thyroid activation that normally follows cooling of the preoptic-anterior hypothalamic area is blocked by lesions of the median eminence and that conversely the thyroid response to cold exposure can be abolished by local warming of the anterior hypothalamus.

Integration of nervous and humoral mechanisms in the control of TSH secretion

With clear evidence available that the level of thyroid hormone in the blood affects the secretion of TSH and that the hypothalamus influences pituitary TSH secretion, the question of the site or sites of action of thyroxine becomes of considerable importance. Thyroxine administration reduces TSH secretion in the rabbit after section of the pituitary stalk (Brown-Grant and colleagues, 1957) or after tranplantation to the eye (von Euler and Holmgren, 1956a), in rats after stalk-section (Blanquet, 1962) or after placing lesions in the anterior hypothalamus (Florsheim, 1959; Averill, Purves and Sirett, 1960) or transplanting the pituitary (Khazin and Reichlin, 1961a). Direct injection of thyroxine into the pituitary or hypothalamus

7*

was first performed by von Euler and Holmgren (1956a) in the rabbit. They showed that a dose of thyroxine that was ineffective when given systemically inhibited the thyroid gland when injected into the pituitary. These results were confirmed in the rabbit by Harrison (1961) and by Yamada and Greer (1959) and Kendall (1962) in rats. Yamada and Greer (1959) and Yamada (1959), observed in addition a delayed (8–9 hours) response when thyroxine was injected into the anterior hypothalamus. The sensitivity of the hypothalamus appeared to be lower than that of the pituitary and diffusion of thyroxine from the hypothalamus to the pituitary via the portal vessels cannot be excluded in their work (Bogdanove and Crabill, 1961). A direct effect of thyroxine on the pituitary to suppress TSH secretion appears to be established. No negative result can exclude the existence of thyroxine-sensitive receptors in the hypothalamus but it does not appear necessary at the moment to postulate their existence.

The effects of reduced blood thyroxine level on TSH secretion after stalk-section, pituitary transplantation or anterior hypothalamic lesions are less clear cut. All workers agree that some evidence of thyroid activation follows the administration of goitrogenic drugs in such animals and that blood TSH levels are raised and pituitary TSH levels reduced. Failure of the thyroid to grow to the levels seen in normal animals has been the principal deficiency observed and has led some workers to suggest that the response to a reduced hormone level is impaired. However, Van der Werff ten Bosch and Swanson (1963) have shown that the relative thyroid weight increase is quite comparable to that seen in intact rats when lesioned rats treated with goitrogens are compared with lesioned rats on a normal diet. Increased TSH secretion in response to a lowered blood thyroxine content can be elicited at the pituitary level and the increase is comparable to that seen in intact animals when the reduced basal level of TSH secretion in the operated animals is taken into consideration.

A considerable degree of adaptive change in TSH secretion in response to changes in blood thyroxine level exists in stalk-sectioned or lesioned animals or after pituitary transplantation. The absolute deficiency in such animals is in the response to such stressful stimuli as restraint or large doses of stillboestrol (in the rabbit) and in the acute (1–3 hours) response to cold exposure. These responses appear to involve the hypothalamus (probably via TRF release) in an obligatory manner. The response to cold exposure provides some very interesting information as to the interaction of a neural 'drive' and the feedback mechanism operating at the pituitary level. First, in the intact animal, the available evidence suggests that although there may be a transient rise in the level of circulating hormone immediately after cold exposure this does not persist; a falling off in the response despite continued cold exposure is evident in the experiments of Brown-Grant, von Euler, Harris and Reichlin (1954), Knigge, Goodman and Solomon (1957) and Andersson, Ekman, Gale and Sundsten (1962a) which may be due to the raised blood hormone level. The acute neural response to cold exposure can be suppressed by the simultaneous injection of thyroxine (Knigge, 1960b; Andersson, Gale and Ohga, 1963). Prolonged cold exposure generally results in a near normal or slightly reduced blood thyroxine level, an increased rate of peripheral thyroxine metabolism and evidence of increased TSH secretion (see Brown-Grant, 1966a, for references). It is likely that the sequence of events under these conditions is an increased rate of metabolism of thyroxine leading to reduced blood levels and hence pituitary activation, a sequence of events not involving the hypothalamus in contrast to the changes during acute cold exposure. This would explain the findings of Barrnett and Greep (1951) and Van Beugen and Van der Werff ten Bosch (1961) that the response to chronic cold exposure is not abolished in stalk-sectioned or lesioned rats.

Although neural mechanisms may produce either a transient rise or a transient fall in blood thyroxine level consequent upon a stimulation or depression of TSH secretion, e.g. in cold exposure (Andersson and coworkers, 1962) or immobilization stress (Brown-Grant, 1957), it appears that under normal conditions the feedback mechanism acting at the pituitary level is capable of overriding them. Indeed there appear to be only two situations in which a persistently raised blood thyroxine level can be maintained for any length of time other than by the injection of thyroxine. One is during prolonged electrical stimulation of the hypothalamus (Harris and Woods, 1958) and the other is in clinical cases of hyperthyroidism. These latter may be due to the uncontrolled activity of a small portion of the thyroid acting autonomously—a toxic adenoma—or to a more or less uniformly hyperactive gland (Graves' disease). The cause of the hyperactivity is not established; it does not appear to be due to an excessive secretion of TSH (Werner, Hamilton and Nemeth, 1952; Trotter, 1962) caused by some abnormality in the feedback control mechanism. Indeed, TSH secretion may be already maximally suppressed, as would be expected, in such patients. The currently favoured hypothesis is that some abnormal thyroid stimulating substance, not of pituitary origin, may be present in the serum of the Graves' disease patient (McKenzie, 1960, 1961, 1965).

This discussion of the regulation of TSH secretion has introduced several of the concepts fundamental to the consideration of the control of the pituitary-target organ systems. In subsequent sections the extent to which the same mechanisms operate in the control of ACTH and gonadotrophin secretion will be considered. In summary, the regulation of TSH secretion is illustrated in Figure 2. The central role is played by a 'feedback' mechanism operating at the pituitary

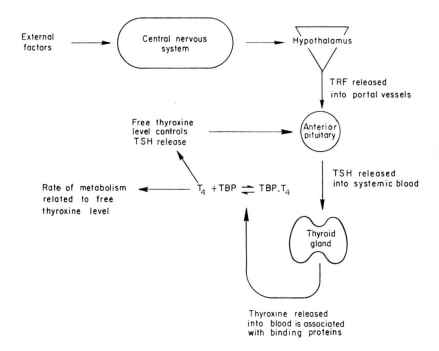

Figure 2. The interaction of neural and humoral control mechanisms in the regulation of the anterior pituitary-thyroid gland system.

level to stimulate or depress TSH secretion and maintain a constant blood thyroxine level. Recent studies have suggested that it is the free thyroxine of plasma rather than the total (free plus protein bound) whose constancy is maintained. If the tissue (or possibly more precisely, the pituitary) metabolism of thyroxine is interfered with by drugs such as methylthiouracil, the free thyroxine level may have to be higher to maintain the system in a steady state; perhaps a more precise description would be that a level of free thyroxine in the blood is maintained such that the rate of pituitary metabolism of thyroxine is kept constant. When the pituitary is supplied with portal vessel blood that has passed through the primary capillary plexus in an intact hypothalamus it appears that either in some way the blood exerts a reduced effect on TSH secretion or that some factor stimulating the release of TSH has been added to it. This appears to be a function of the hypothalamus *per se* and not dependent upon impulses from other parts of the brain. An isolated island of hypothalmic tissue connected to the pituitary maintains a normal basal level of thyroid activity in the cat (Woods and Bard, 1959).

An intact nervous system enables certain acute changes in TSH secretion, inhibition following emotional stress and acute increases following cold exposure, to occur. This second action may be of fundamental significance. The relation of the centre responsible for thyroid activation in the cold to the somatic heat control mechanisms is discussed by Andersson, Gale and Hokfelt (1964). It seems to be characteristic of homoiotherms that they show thyroid activation in the cold, initially by a purely nervous mechanism and subsequently following an increased rate of peripheral metabolism of thyroid hormones and a fall in blood level. Poikilotherms, in contrast, show a decrease in thyroid activity at reduced environmental temperatures (Clements-Merlini, 1962; Leloup and Fontaine, 1960; Fortune, 1955; Talmage, Doty and Yates, 1962; Shellabarger, Gorbman, Schatzlein and McGill, 1956). In the eel and the frog, it has been shown that the rate of thyroxine metabolism is reduced at low temperatures and that the thyroid gland depression is a result of the operation of a feedback mechanism as the blood level of thyroxine rises (Leloup and Fontaine, 1960; Dowling, Razevska and Goodner, 1964).

The Control of ACTH Secretion

The secretion of ACTH will be considered here solely in connection with the control of the rate of production of glucocorticoids by the adrenal cortex. The steroid hormones concerned are hydrocortisone (Compound F) and corticosterone (Compound B) see Figure 3, which are present in varying proportions in the blood of vertebrates from primitive cyclostomes to man (Jones and coworkers, 1962). The other major corticoid secreted by the adrenal is the potent mineralocorticoid, aldosterone (Figure 3); its secretion is influenced by ACTH but also by many other factors and will be considered later.

A reciprocal relationship between ACTH secretion and the glucocorticoid level in the blood is well established for such situations as the chronic administration of glucocorticoids or following the

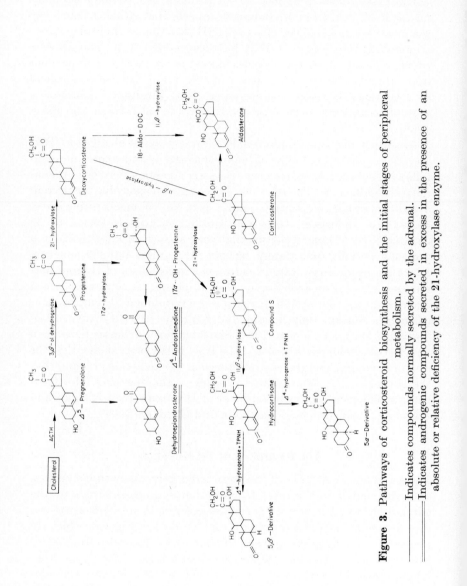

Figure 3. Pathways of corticosteroid biosynthesis and the initial stages of peripheral
metabolism.

——— Indicates compounds normally secreted by the adrenal.

=== Indicates androgenic compounds secreted in excess in the presence of an
absolute or relative deficiency of the 21-hydroxylase enzyme.

removal of one adrenal (see for instance Sayers, 1950). However, the concept of the pituitary-adrenal system as a homeostatic mechanism has not influenced experimental studies of this relationship to the same extent as it has the study of the pituitary-thyroid system. This is because the pituitary-adrenal system, if it can ever be said to be in a steady state, is characterized by the ease into which this precarious equilibrium can be disturbed. The term 'stress' was introduced by Hans Selye (see Selye, 1950) to cover the multitude of stimuli that could be shown experimentally to bring about a rapid increase in the rate of secretion of ACTH and thus to increase glucocorticoid release by the adrenal cortex. This concept has been criticized as being so all embracing as to be meaningless, but the fact remains that the existence of a stereotyped response, an abrupt increase in ACTH secretion following any noxious or disturbing stimulus, is a generalization of major biological importance. The nature of the stimuli capable of inducing ACTH discharge in the absence of any physical or chemical damage to the animal (flashing lights, loud noises, anxiety about examination, or interviews) and the speed of the response (see for instance Fortier, de Groot and Hartfield, 1959) suggested the involvement of the nervous system in the acute stress response. The whole range of neuroendocrine methods has been used in the study of this response coupled with increasingly precise biochemical methods for assessing the response of the adrenal (see discussion of techniques in the reviews by Ganong, 1963 and Fortier, 1963, 1966). Various theories have been proposed to explain the results available at different stages of these investigations and it is of interest to review some of them to see how, although they may have been superseded, they stimulated and contributed to later work.

The suggestion that stress results in an increased tissue utilization of corticosteroids, a fall in the blood level and thus a stimulation of ACTH release, was put forward by Sayers (1950) and was fully consistent with the information then available to him. Subsequent advances in biochemical methods for the measurement of corticoid levels in blood have demonstrated that the level in peripheral blood is in fact increased in the acutely stressed animal. This is in itself a surprising finding and is additional evidence of some added stimulus to the pituitary-adrenal system under these conditions. Yates and Urquhart (1962) have suggested that the modification at the hypothalamic level induced by stress is a change in the sensitivity of some receptor mechanism for monitoring the blood corticoid level so that following a given stress stimulus ACTH secretion is regulated so as to maintain the blood corticoid level for a time at a different, higher level. They base their suggestion on experiments in which graded doses of corticosterone, given immediately before the application of a standardized stress in rats, appeared to modify ACTH release so that the same corticosteroid level in the blood was attained 15 minutes later whether the stress or stress plus corticosterone was used (Yates, Leeman, Glenister and Dallman, 1961). These results have not been confirmed by other workers; Hodges and Jones (1963) gave doses of corticosterone that would produce equal or greater rises in blood level to those observed 30 minutes after their stress procedure (sham adrenalectomy) and showed that in each case the effect of stress and corticosterone were additive. Stockham (1964) has obtained similar results. Smelik (1963) gave a single, rather higher dose of corticosterone and measured both plasma levels and the rate of corticoid production by the adrenal *in vitro*. He showed that the rise in the level of corticosterone in the plasma preceded any inhibition of ACTH release and that inhibition was still maximal at a time when plasma levels had returned to normal values.

In patients undergoing surgery, it has been shown that large doses of synthetic glucocorticoid which reduce the basal level of endogenous glucocorticoid secretion to a very low value do not prevent the rise that normally follows surgery, nor does the maintenance of a supranormal level of hydrocortisone (the major glucocorticoid secreted by the human adrenal cortex) by intravenous infusions prevent a rise in plasma ACTH levels during surgery. A further elegant point was made in these studies by Estep, Island, Ney and Liddle (1963), as follows. When the 11β-hydroxylase activity of the adrenal was blocked by the administration of a specific inhibitor, so that the main adrenal secretion was the metabolically ineffective Compound S (11-deoxyhydrocortisone, see Figure 3), the plasma levels of Compound S also increased during surgery to the same extent in patients whose plasma hydrocortisone levels were 10–20 or 100–500 μg/100 ml.

The idea that ACTH release during stress is delicately controlled by the prevailing glucocorticoid level of the blood has not been confirmed. Very large doses, far outside the physiological range, will suppress the response to many minor procedures but severe stress will still provoke ACTH release. The later stages of the stress response over a period of hours or days are modified by the prevailing corticoid levels; the most detailed study of this particular point is probably the comparison of the responses to splenectomy and to adrenalectomy in the rat by Fortier (1959a, b). Neither the simple nor the modified feedback theory is adequate to explain the stress response.

While these studies on the feedback mechanism were being carried out, the possible role of the nervous system was being intensively investigated. A possible central role for adrenaline in the control of ACTH release was proposed but by 1955 the weight of the evidence was against this view (Harris, 1955a) and subsequent work has confirmed this (Ganong, 1963). Studies of the effect of stimulation of the central nervous system through implanted electrodes indicated that the posterior hypothalamus and, acting through this region, other parts of the limbic system could influence ACTH release (Harris, 1955a, 1959a, b; Ganong, 1963; Schrieber, 1963). The effects of destructive lesions in various parts of the hypothalamus, of pituitary stalk-section and of transplantation of the pituitary on the stress response to a variety of stimuli were also studied.

The outcome of these studies was general agreement that the response to certain types of stress stimuli could be blocked by these procedures, indicating an obligatory involvement of the hypothalamus and the vascular link to the pituitary provided by the portal vessels. Other forms of stress were still able to stimulate ACTH secretion, however, and Fortier proposed (1951) that 'neural' and 'systemic' stresses existed, differing in their ability to elicit ACTH release with or without the mediation of the central nervous system. This operational definition (a neural stress is one whose ACTH releasing effect is abolished by stalk-section or pituitary transplantation) appealed to some workers (Harris, 1955a) as an interim measure but not to others (Sayers, Redgate and Royce, 1958); a more plausible explanation is now available and is discussed later. Apart from the observations on the control of acute episodes of increased ACTH release, it was also observed that the basal level of ACTH secretion (as judged by adrenal weight) was also reduced in the stalk-sectioned animals but was restored when revascularization by portal vessels occurred (see for example Harris, 1959b). The mass of circumstantial evidence, suggesting the existence and importance of a neurohumoral transmitter of hypothalamic origin that caused ACTH release, stimulated an intensive search for this hypothetical corticotrophin releasing factor (CRF).

The background to this work is well summarized by Schrieber (1963) and the present state of our knowledge by Guillemin and Schally (1963) and by Harris and colleagues (1966). The agents isolated from median eminence tissue and also found as contaminants in some commercial preparations of posterior pituitary glands, are poly-peptides related in structure to and with some of the biological proper-ties of the intermediate lobe hormone (melanocyte-stimulating hormone, MSH), of ACTH, and of vasopressin. α-CRF is described as containing the amino acids found in α-MSH, plus threonine, alanine and leucine. It has some inherent ACTH-like activity and could be a precursor of ACTH. β-CRF is related to vasopressin and contains the amino acids of vasopressin plus serine and histidine and possibly valine and alanine. It is thought that this may be the physiologically important CRF and it is the most active compound with this type of biological activity so far isolated. It has no inherent ACTH or MSH activity but may have intrinsic antidiuretic properties in the way that vasopressin has slight ACTH releasing activity. The reasoning behind the design of experiments intended to demonstrate the exist-ence of a CRF in a particular extract and the criteria by which a CRF-like action (or for that matter thyrotropin or luteinizing hor-mone releasing activity) may be recognized are fairly obvious (e.g. that the extract should have no effect in the recently hypo-physectomized test animal) and have been restated in several publica-tions (Leeman, Glenister and Yates, 1962; Guillemin and Schally, 1963). By these criteria it appears that β-CRF may well be the long sought neurohumoral agent concerned in ACTH release. One rela-tively simple experiment that rather surprisingly remains to be carried out is the demonstration that the direct intrapituitary infusion of CRF causes ACTH release at a lower dose level than when given systemically; this technique has already been widely used by workers interested in the isolation of the LH releasing factor (Harris, Reed and Fawcett, 1966).

The existence of CRF satisfactorily explains many of the ob-servations concerned with the role of the nervous system and the portal vessels in the acute response to 'neurogenic' stresses and in the tonic maintenance of ACTH secretion. The persistence of ACTH release in response to systemic stress in animals with pituitary transplants or after effective stalk-section presents certain problems; is it necessary to invoke some other mechanism of control of ACTH secretion to explain these results and if so, does this additional mechanism play a part in the control of ACTH secretion in the in-tact animal? However, following large lesions of the median eminence the acute response to *all* stress stimuli tested is abolished and simi-lar findings have been reported for rats treated with large doses of pentobarbitone plus morphine. The exception to this statement is the effect of large doses of vasopressin in such preparations; as discussed above, this hormone has intrinsic CRF activity which probably explains this finding. Clearly, it is neither possible nor necessary to postulate the existence of two types of stress stimuli when discussing results obtained from this type of preparation.

These findings suggest that the persistence of a response to a systemic stress in an animal with a transplanted pituitary or after stalk-section may be related to the fact that the presumed source of CRF, the median eminence, has neither been destroyed nor rendered unresponsive by pharmacological blockade. The simplest explanation is that CRF, under these conditions, is released into the systemic circulation to act upon the transplanted or isolated gland. The failure of 'neurotropic' stresses could be due to the fact that they are, in fact, simply less severe and cause less CRF release, an explanation put forward in a slightly different context by Sayers and coworkers (1958). Ganong (1963) also inclines to this view and cites evidence for the detection of CRF in systemic blood. As he pointed out, the results of experiments in which the effects of stress on animals with transplanted pituitaries both with and without lesions of the median eminence are compared should provide fairly direct evidence on this point. Such experiments have been carried out in connection with the question of an influence of TRF secreted into the systemic blood on TSH secretion by pituitary transplants, but the results have been inconclusive (Florsheim and Knigge, 1961; Khazin and Reichlin, 1961b).

Two further indirect suggestions that systemic transmission of CRF is not an entirely unreasonable hypothesis may be mentioned. The first concerns the prodigality of the response of the normal animal. There is evidence, summarized by Ganong (1963), that circulating ACTH levels during stress may be the result of pituitary secretion of more than fifteen times the amount of ACTH necessary to produce a maximal adrenal response. If CRF secretion is on an equally lavish scale, then if even a small fraction of the total released reached the pituitary this might be enough to produce a detectable increase in ACTH secretion. The second concerns the persistence of CRF in the systemic blood; the considerable potency when the material is given as a single injection systemically (a detectable dose of 0·05 μg is quoted by Guillemin and Schally, 1963) suggests that the half-life in the circulation cannot be extremely short. Vasopressin and oxytocin have half-lives in the blood of the rat which are estimated at about 1 minute and 1–2 minutes respectively (Lauson, 1960). Although CRF polypeptide is of rather similar structure the half-life may be much longer than this; there is no direct information on this point. Florsheim, Austin and Velcoff (1963) suggested that the half-life of the TSH releasing factor in the rat may be measured in hours rather than minutes, but their conclusions have been criticized by Greer, Matsuda and Stott (1966).

The regulation of ACTH secretion appears to be dominated by the influence of the hypothalamus acting through one or more specific neurohumors released into the portal vessels to a greater extent than is the regulation of TSH secretion. At least in the conditions of acute stress under which it has been studied most intensively, blood corticoid levels appear not to play a central role in initiating or controlling ACTH secretion although in some way the exposure to different corticoid levels over the preceding few hours can modify the extent of the response (Hodges and Jones, 1964; Fortier, 1966). In contrast to the thyroid, the neural drive appears to be quite capable of maintaining sustained rises in blood corticoid levels. The response to adrenalectomy, at least in the rat, is not the simple sustained rise in blood ACTH levels that might be expected (Ganong and Forsham, 1960), but shows a biphasic pattern, blood ACTH rising initially for a few hours but later falling to low levels for several days. Ganong and Forsham concluded that the ACTH releasing effect of a decline in blood glucocorticoid levels is not a prompt dramatic response. None the less it does exist and although detailed studies of the adrenalectomized rat may have produced unexpected results (Hodges and Jones, 1963; 1964), long term studies, particularly of human cases of adrenal disorders, have at the same time served to reemphasize the importance of this mechanism over longer time periods. The adrenogenital syndrome (congenital bilateral adrenal hyperplasia in young children, presenting as pseudo-hermaphroditism in girl children at birth or precocious virilization in the absence of testicular abnormalities in male children 2–3 years of age) has been shown to be associated with a genetically determined complete or partial block of the process of 21-hydroxylation of corticoid precursors in the adrenal (Brooks, 1962). The 21-methyl compounds produced (Figure 3) are not biologically active glucocorticoids and pituitary ACTH secretion is increased due to the reduced blood corticoid level, leading to adrenal overactivity and an excessive production of adrenal androgens—both directly and by peripheral breakdown of the 21-methyl compounds produced in large quantities to C19 compounds. Administration of hydrocortisone or other potent glucocorticoids in adequate dosage protects the child from the harmful effects of a shortage of glucocorticoids and depresses ACTH secretion with a consequent decrease in the rate of production of unwanted androgenic compounds. This is a clear example of the importance of feedback mechanisms in the control of ACTH secretion.

A second abnormality of adrenal function may also, in some instances, be an example of an abnormality of the feedback mechanism. Cushing's syndrome is a condition of corticosteroid excess characterized by a peculiar form of obesity, high blood pressure, diabetes and characteristic skin changes. The cause may be a benign or malignant growth of the adrenal cortex acting independently of the level of ACTH secretion; indeed ACTH secretion may be depressed and the normal adrenal tissue atrophic. In some cases, however, the adrenals are uniformly enlarged and there is evidence that raised circulating levels of ACTH are responsible (Ney, Shimizu, Nicholson, Island and Liddle, 1963; Davies, 1964). Adrenal activity can be suppressed in such patients by very high doses of exogenous corticoids in contrast to the lack of effect of thyroxine in hyperthyroidism (Williams, Island, Oldfield and Liddle, 1961). It appears that ACTH secretion in such cases is held in check only by a level of circulating glucocorticoids that produces pathological changes in the peripheral tissues of the body. Further evidence of this has been found following bilateral adrenalectomy in such patients. Here the eucorticoid condition is maintained by the administration of a dose of steroid that is adequate to keep in good health an Addisonian, or a patient adrenalectomized for some other reason, such as advanced malignant disease of the breast. The Cushing's disease patient may have high blood ACTH levels and may develop tumours of the pituitary gland that secrete ACTH in large quantities despite the maintenance of normal blood corticoid levels (Nelson and coworkers, 1960). The hypothalamus-pituitary system of these patients appears to demand a pathologically raised blood corticoid level for ACTH secretion to be normal.

The site of action of corticosteroids on ACTH secretion is therefore of importance. There is evidence that compensatory hypertrophy of the remaining gland after unilateral adrenalectomy does not occur in animals with lesions of the median eminence but does occur after section of the pituitary stalk (see Ganong and Forsham, 1960; Fortier, 1963, 1966 and Ganong, 1963, for references). These findings suggest that a reduced level of corticosteroids may act through the hypothalamus. However, as Sayers and coworkers (1958) and more recently Yates and Urquhart (1962) have pointed out, such lesions do not only affect ACTH secretion and in none of these experiments has the possible effect of a reduced level of thyroid activity on adrenal function been excluded. Administration of corticoids still produces adrenal atrophy in these animals, suggesting that a raised blood level may act directly upon the pituitary. Kendall, Matsuda, Duyck and Greer (1964) in more acute experiments observed a depression of corticosteroid secretion in rats with an isolated median eminence connected to the pituitary, but obtained only equivocal results when the same dose of steroid was given to animals with a pituitary but no hypothalamic tissue.

More direct experiments have involved the injection or implantation of steroid into the hypothalamus or pituitary. Rose and Nelson (1956) claimed that continuous infusion of hydrocortisone into the pituitary capsule would present compensatory hypertrophy after unilateral adrenalectomy in the rat, at a dose level that was ineffective when given systemically, but Kendall (1962) found intrapituitary injection of steroid to be no more effective than systemic injection in suppressing adrenal secretion in rats. Davidson and Feldmann (1962, 1963) found that implants of hydrocortisone in the median eminence or mamillary body, but not in the pituitary itself, would prevent compensatory adrenal hypertrophy in rats. Chowers, Feldman and Davidson (1963) showed that hypothalamic, but not pituitary, corticosteroid implants in rats resulted in adrenal atrophy and blocked the acute response to unilateral adrenalectomy. Smelik and Sawyer (1962) observed a blockade of the stress response to immobilization in rabbits following implantation of hydrocortisone into the anterior hypothalamus or median eminence in rabbits but not following implants in the posterior hypothalamus or into the gland itself.

The ineffectiveness of intrapituitary implants in these studies is striking; it seems that the situation with regard to the feedback control of ACTH secretion is rather the reverse of that observed with TSH, in that there is good evidence that the effects of both increased and decreased levels of steroid act at the hypothalamic level to affect ACTH secretion, but that the evidence for a direct effect upon the pituitary is not convincing, though it cannot be excluded. This would be in keeping with the greater importance of neural mechanisms in the control of ACTH secretion. The precise site of action of steroid within the hypothalamus is not known; this is consistent with the findings from lesion and stimulation experiments; there is far less agreement about the precise areas involved in ACTH regulation than there is about TSH or gonadotrophin controlling 'centres' (see for example, discussion by Harris, 1959b, 1962).

It has been suggested (Brodish, 1963) that apart from lesions involving the primary plexus of the portal vessels in the median eminence, the effect of hypothalamic lesions on ACTH release is proportional to the mass of tissue destroyed rather than the destruction of any specific area. Such a diffuse system would explain the conflicting reports as to the position of critical lesions and indeed is not really unexpected in view of the multiplicity of afferent pathways potentially involved in transmitting stimuli concerned in ACTH release. The complexities of the neural control of ACTH are not restricted to stimulatory pathways; there have been several reports of inhibition of ACTH secretion following stimulation of parts of the limbic system outside the hypothalamus (see Ganong, 1963 for references). The interpretation of these findings is not made any easier by the fact that classical views of the connections of these areas with the hypothalamus are undergoing considerable revision following studies using modern neuroanatomical techniques (Raisman, 1966).

Effect of protein binding of corticosteroids

The role of hormone binding proteins has been discussed at length in connection with the thyroid hormones. Many of the ideas put forward there may also be applicable to the corticosteroids. In human plasma at physiological levels (10–15 µg/100 ml), hydrocortisone is reversibly bound to a specific plasma protein, transcortin or corticosteroid-binding globulin (CBG) which is distinct from the thyroxine-binding globulin. Most of the information about the binding of corticosteroids comes from studies in the human (see for instance, reviews by Daughaday, 1960; Yates and Urquhart, 1962; Mills, 1962). The level of binding protein in canine and bovine plasma is very much lower (Plager, Knopp, Slaunwhite and Sandberg, 1963; Lindner, 1964). Some comparisons with the thyroxine-binding proteins are of interest; it has been estimated that 5–10% of the hydrocortisone in human plasma is free (i.e. not bound to protein) at normal plasma levels, in contrast to some 0·05% of thyroxine. The capacity of CBG is about 25 µg/100 ml in normal human serum; this is about the same level as the capacity of TBG for thyroxine but whereas the thyroxine level never reaches these heights, even in extreme hyperthyroidism, hydrocortisone levels of 20 µg may occur daily in normal individuals and in stressed individuals often exceed 25 µg/100 ml. The far greater amounts of free hormone may contribute to the much more rapid turnover of hydrocortisone than of thyroxine; the half-life of the steroid in the blood is about 100 minutes; that of thyroxine is 6–8 days in normal individuals.

Some of the consequences of protein binding of corticosteroids are examined in detail by Yates and Urquhart (1962) and by Tait

and Burstein (1964). The general principles involved are much
the same as those discussed in connection with thyroxine; on
the whole, rather less quantitive information is available for the
corticosteroids. As with thyroxine, one of the most interesting
findings is that during pregnancy or following oestrogen administration in man the level of binding protein rises but there is little
evidence of an excess supply of steroids to the tissues or of a depression of adrenal function. A possible solution of this apparent
paradox is that the level of free steroid is unchanged but the direct
evidence suggests that, at least after oestrogen administration, the
level of free steroid is in fact increased (Plager, Schmidt and Staubitz,
1964). The pituitary or hypothalamus may act to maintain a constant rate of metabolism of steroid (compare the situation with regard
to thyroxine after the administration of methyl thiouracil).

Effect of alterations in steroid metabolism

The pathways of corticosteroid metabolism are reasonably well
known and are illustrated in Figure 3. The first and probably most
important stage, as it is irreversible and leads to the production of
biologically inactive compounds, is reduction of the double bond in
ring A of the molecule. The major site where this occurs is the liver;
the enzyme systems involved have been studied in some detail
(Forchielli and Dorfman, 1956). There appear to be two groups of
enzymes, a group associated with the microsomal fraction obtained
by high speed centrifugation of homogenates of liver tissue which
catalyse the transfer of hydrogen to produce only the 5α-stereoiso-
mers and a group associated with the soluble protein fraction of the
homogenate which produce only the 5β-stereoisomer as reduction
product. Both groups are obligatorily dependent upon TPNH
(NADPH) as hydrogen donor. Assays of the activity of these enzymes
in vitro is relatively simple as the reduction of the 4–5 double bond
removes the conjugated ketone group from the molecule and the
characteristic absorption band in the ultraviolet at 2400 Å is lost.
Measurement of the optical density at this wavelength, after extraction of the steroid with organic solvents following incubation under
anaerobic conditions with the enzyme preparation, enables the extent
of reduction to be estimated. Assays in the presence of a TPNH
generating system allow the enzyme level to be estimated; when
assays of unfortified homogenates or of liver slices are made, the
availability of reduced TPN may be the limiting factor; assays of the
second type may be more closely related to the activity of the liver
in vivo.

The relatively short half-life of corticosteroids in the blood (20 minutes in male rats and about 100 minutes in man) implies that acute changes in the rate of hepatic metabolism will have rapid effects upon the blood level of hormones and that sustained differences in the rate of hepatic inactivation will have marked effects on the rate of steroid secretion necessary to maintain a given plasma concentration of corticosteroid. In animals such as the rat, it may be no exaggeration to talk of the pituitary–adrenal–liver triangle (Urquhart, Yates and Herbst, 1959). The greater size of the adrenal gland in the adult female as compared to the male rat has been known for many years. Although the liver of female rats contains only one group of enzymes (the 5α-microsomal enzyme) the Δ^4-reducing activity of whole homogenates is several times greater than that of homogenates from the liver of male rats containing both the 5α and the 5β soluble enzymes (Forchielli, Brown-Grant and Dorfman, 1958; Yates, Herbst and Urquhart, 1958; Hagen and Troop, 1960). These differences in enzyme concentration are reflected in differences observed *in vivo*. Glenister and Yates (1961) showed that the half-life of injected radioactive corticosterone was 22 minutes in male rats but only 13 minutes in female rats. The estimated secretion rate (µg Compound B/100 grams body weight minute) in the female was about three times as high as in the male. The reverse situation is found in the hamster where the activity of liver from males is higher than that from females and the size of the adrenal is greater in the male (Yates and coworkers, 1958; Brown-Grant, Forchielli and Dorfman, 1962). Acute changes in the liver capacity to carry out ring A reduction were studied by Herbst, Yates, Glenister and Urquhart (1960). They showed that a variety of stresses which had in common the effect of reducing food intake reduced the capacity of rat liver slices in inactivate corticosterone. The activity of homogenates to which a TPNH generating system was added was unaltered, which suggests that a reduction in available TPNH rather than a change in enzyme level was involved. The half-life of corticosterone *in vivo* was found to be increased in these experiments. The effect of a reduced rate of removal will be to produce a rise in the blood level of steroid which will eventually depress ACTH secretion.

More direct evidence for an affect of reduced liver inactivation on ACTH secretion was obtained by Urquhart, Yates and Herbst (1959) who showed that adrenal enlargement was significantly less 24 and 48 hours after subtotal hepatectomy in male rats than after a comparable control operation (splenectomy). No differences were observed between adrenal weights in partially hepatectomized and splenectomized female rats. These results have been confirmed in the author's unpublished experiments in which adrenal weight in control, sham operated and partially hepatectomized male rats were compared Urquhart and coworkers suggest that the more rapid regeneration of the liver in female rats may be the explanation for the sex difference; alternatively the limiting factor in corticosterone inactivation in the female may be the rate at which steroid reaches the liver and the remnant (approximately one third of the original liver mass) may be capable of dealing with as much corticosterone as the whole organ normally does.

Changes in liver function in man have also been shown to affect the rate of metabolism of corticosteroids both in chronic liver diseases (Brown, Willardson, Samuels and Tyler, 1954; Englert, Brown, Wallach and Simons, 1957) and during surgical operations (Sandberg, Eik-Nes, Samuels and Tyler, 1954; Engell, Winkler, Tygstrup and Buus, 1961; Sandberg, 1960).

Finally, the influence of the level of thyroid function on the rate of hepatic inactivation of corticosteroids provides a clear example of the effect of changes in peripheral metabolism in influencing blood levels and hence ACTH secretion. It is well known that hypothyroidism is frequently associated with adrenal atrophy, although the gland shows a normal response to exogenous ACTH and, conversely, hyperthyroidism produced by the administration of large doses of thyroid hormone is associated with adrenal hypertrophy. The blood corticoid level however is not significantly altered in either situation. Various explanations involving shifts in pituitary trophic hormone production were originally proposed to explain these findings. More recent studies have shown that the rate of removal of corticosteroid from the circulation is decreased in hypo- and increased in hyperthyroidism, though not in cases of hypermetabolism of non-thyroidal original (Garren and Lipsett, 1961), and that this is probably the consequence of early changes in the level of available TPNH and later in the level of the 5α-microsomal Δ^4-hydrogenases in the liver (Yates and coworkers, 1958; McGuire and Tomkins, 1959). The pituitary is indirectly influenced by changes in the rate of peripheral metabolism of corticosteroids leading to transient alterations in the blood corticoid level.

The Control of Gonadotrophin Secretion

As discussed earlier, it was in connection with the role of external factors such as light in stimulating gonadal activity in seasonally breeding animals that the concept of an influence of the central nervous system on anterior pituitary function was first clearly formulated (see review by Marshall, 1942). Many of the techniques used in neuroendocrine research were first developed for the study of problems concerning the control of gonadotrophin secretion, for example, the remote control method for stimulation of the hypothalamus of conscious, unrestrained animals (Harris, 1955a, b). The basic hypothesis of a neurohumoral transmitter passing via the portal vessels to the pituitary was originally proposed in connection with studies on induced ovulation in the rabbit (Harris, 1937). Schrieber (1963) has pointed out that restriction of the breeding season to a suitable time of the year is characteristic of the less highly evolved vertebrates and that neural regulation of gonadotrophin secretion may be phylogenetically an ancient phenomenon, while regulation of ACTH and TSH secretion may be a relatively recent development in more highly evolved animals.

Despite the intensive efforts devoted to its study, control of gonadotrophin secretion is not well understood. The problem is necessarily a complex one; the secretion of not one but three pituitary hormones has to be considered, i.e. follicle stimulating hormone (FSH), luteinizing hormone or interstitial cell stimulating hormone (LH) and luteotrophic hormone or prolactin. The target glands upon which these hormones act are the testis and the ovary and in the case of prolactin the mammary gland also. The hormones secreted by the Leydig cells of the testis are testosterone and the closely related compound androst-4-ene-3,17-dione, which have similar biological properties and are interconvertible in the body (Lindner, 1961). The ovary produces a series of 18 carbon steroid compounds, the oestrogens, the characteristic secretion of the ovary containing ripening or mature follicles, and 21 carbon progestational compounds from the corpora lutea. These are formed from the follicles after they have ruptured and the ovum has been discharged. The hormone of the corpus luteum was thought to be a single compound, progesterone, but recent work has shown that a second biologically active steroid, pregn-4-ene-20α-ol-3-one, is also secreted by the ovary as well as being formed as a metabolite of progesterone in the peripheral tissues (Zander, 1959; Fotherby, 1964). The importance of the preovulatory secretion of progesterone, formed either by the follicles or ovarian interstitial tissue, has been recognized more recently (Rothchild, 1965).

The concentrations of the biologically active steroids in peripheral blood are very low and it is only in the last few years that biologically meaningful measurements have become possible. Migeon (1960) discussed the measurement of 17-ketosteroids in plasma; as a measure of androgenic hormone the usefulness for physiological purposes may be judged from the fact that there was no significant difference between males and females! More refined methods have since enabled testosterone levels to be measured in 5 or 10 ml samples of human peripheral plasma (Burger, Kent and Kellie, 1964; Brownie, Van der Molen, Nishizawa and Eik-Nes, 1964). The results agree well with clinical assessment of the degree of androgenic stimulation. Values for males were 0·70 μg/100 ml, and 0·18 μg/100 ml for females; a female child with signs of masculinization from congenital adrenal hyperplasia (see section on ACTH control) had a value of 0·59 μg/100 ml. Recently plasma free oestrone and oestradiol levels have also been measured in men (0·023 and 0·003 μg/100 ml) and in women (0·058 and 0·026 μg/ml at midcycle) by Ichii, Forchielli, Perloff and Dorfman, 1963.

These results represent very considerable advances in our knowledge but as yet the volumes of blood required and the expense in time and money have precluded their extended use in animals such as rats from which so much of the physiological knowledge of gonadotrophin regulation has been obtained. Any discussion of the 'feedback' control of gonadotrophin secretion must be based on very crude experiments involving extreme variation in blood levels or the administration of large and arbitrary doses of steroids. It may well be that many of the theories currently accepted may have to be abandoned when detailed biochemical studies on sex hormone levels in the plasma of animals like the rat at different stages of the cycle are available. Knowledge of different aspects of the control of gonadotrophin is most uneven; inevitably this discussion will be rather patchy.

General control of FSH and LH secretion

Hypophysectomy results in severe atrophy of both ovary and testis. After effective section of the pituitary stalk, transplantation of the pituitary or lesions of the median eminence or of the arcuate nucleus of the hypothalamus in the rat (see Lisk, 1960) the maturation and rupture of ovarian follicles ceases; atrophy of the testis and a marked decrease in androgen secretion is reported by some workers after stalk-section or transplantation of the pituitary but there are persistent reports that testicular function may be maintained at near normal levels under these circumstances, though not after hypothalamic lesions (see Everett, 1964 and Szentágothai, Flerkó, Mess and Halász, 1962 for references). Possible explanations for this difference include a greater sensitivity of the testis to a low level of gonadotrophin secretion or a difference in the pattern of secretion of neurohumoral transmitter by the hypothalamus (see discussion on the effects of androgenic hormones on the hypothalamus later in this section). The secretion of prolactin is not decreased by these procedures; on the contrary it appears that in the rat transplantation of the pituitary at a time when functional corpora lutea are present in the ovary will result in the maintenance of their function for a period of weeks or months. Similarly there is evidence that prolactin secretion adequate to sustain lactation continues, and also that the synthesis of prolactin continues when pituitary tissue is cultivated *in vitro* (Everett, 1964; Meites, Nicoll and Talwalker, 1963). These findings are compatible with the view that the pituitary is normally under some form of humoral stimulation by the hypothalamus in the adult animal to maintain basal levels of FSH and LH secretion and that, in contrast, the influence on prolactin secretion may be an inhibitory one.

The level of circulating androgen or oestrogen plays a part in regulating this basal level of FSH/LH secretion. The actions of the steroid hormones are not sex-specific; large doses of either testosterone or oestrogen will depress gonadotrophin secretion in either males or females leading to testicular or ovarian atrophy. Conversely, removal of one gonad leads to compensatory hypertrophy of the remaining gland. The effects of total gonadectomy have been studied in more detail. Enlargement of the pituitary following castration is an old observation (see Rolleston, 1936 for references). The recognition of typical changes in the basophilic cells of the pituitary and the development of methods for the parabiotic union of a castrate with a hypophysectomized partner (see review by Shipley, 1962) stimulated a great deal of work on the effect of various steroids in suppressing the excessive secretion of gonadotrophin that occurs after castration. It was soon established that both oestrogens and androgens could do this both in male and female castrates but that oestrogens were far more effective than androgens; progesterone was essentially inactive (Shipley, 1962).

The site of action of the hormones in producing these effects is of some interest. The development of castration cells was shown to be abolished following stalk-section, pituitary transplantation or lesions of the median eminence (Everett, 1964), suggesting that the effect of a greatly reduced level of circulating steroid was exerted via the hypothalamus. The site of action of steroid hormones has been investigated in more direct experiments; Szentágothai and

colleagues (1962) give an extensive account of this work and describe
many of their own ingenious experiments on rats in detail. The area
most intensively studied was the basal anterior hypothalamus between the optic chiasma and the anterior part of the median eminence. Bilateral lesions of this area have no consistent deleterious
effect on gonadal function in male rats but in female rats they produce a syndrome characterized by persistent vaginal cornification,
and small ovaries due to the absence of corpora lutea. The ovaries
contain many large, often cystic, ovarian follicles which neither rupture to release ova nor do they undergo follicular luteinization. The
picture is one of sustained oestrogen secretion brought about by an
unrestrained release of FSH plus some LH. The suggestion is that
the inhibitory action of physiological levels of oestrogen on FSH/LH
secretion was lost following destruction of this area. Donovan and
Van der Werff ten Bosch (1959a, b) obtained results in ferrets which
showed that lesions of the anterior basal hypothalamus caused
animals to come into oestrus during the normal winter anoestrus
period; prolonged electrical stimulation did not hasten the onset of
oestrus. The results were interpreted as the destruction of a centre
normally inhibiting gonadotrophin secretion which contains oestrogen sensitive elements whose threshold is affected by environmental
lighting conditions.

The same workers extended their studies to immature rats
and showed that precocious ovarian growth and vaginal opening
were produced by lesions situated basally behind the optic chiasma. Some of these animals displayed prolonged oestrus and did
not ovulate. From the published data, these lesions appear to involve
many of the structures whose destruction leads to the development of
the 'constant oestrus' syndrome in adult rats or guinea-pigs.
Horowitz and Van der Werff ten Bosch (1962) imply that because
relatively few of the animals showing precocious puberty fail to
ovulate, the exact extent of the effective lesions in the two cases
cannot be the same. It should be pointed out, however, that the
majority of animals were killed before they were 50 days old and all
before they were 100 days old. In other experiments it has been
shown (Swanson and Van der Werff ten Bosch, 1964) that failure of
ovulation may only develop some time after the attainment of sexual
maturity; perhaps all rats with precocious puberty following lesions
would eventually have become anovulatory. It has been known for
some time (see Shipley, 1962) that FSH secretion after castration can
be suppressed by lower doses of oestrogens or androgens in pre-
pubertal animals than in adults, and similar findings for the effect of

oestrogens on LH secretion in gonadectomized rats have been reported by Ramirez and McCann (1963a). It is tempting to relate the onset of puberty to a decrease in the sensitivity of the anterior hypothalamus to the inhibitory effect of oestrogens, and precocious puberty following lesions in this area in rats to destruction of oestrogen sensitive FSH inhibiting neurones.

It appears that release of sufficient FSH to develop ovarian follicles to the stage where ovulation can occur may be the last process in the chain of events leading to ovulation to mature in rats. The pituitaries of newborn rats are capable of FSH and LH release before the normal time of puberty if they are transplanted beneath the median eminence of an adult female (Harris and Jacobsohn, 1952). The precise age at which changes in the ovary and hypothalamus occur varies with the strain of animal and the rate of growth in different colonies (compare Zarrow and Quinn (1963) and Zarrow and Brown-Grant (1964). However, results obtained by administration of an FSH-like gonadotrophin, pregnant mares' serum gonadotrophin (PMS) either alone or in combination with an LH-like gonadotrophin (human chorionic gonadotrophin, HCG), to immature rats have shown that during development before puberty the rat passes through the following stages.

Initially (18–22 days) the ovary is relatively unresponsive to PMS and ovarian growth is minimal. Then for a few days rapid ovarian growth with the development of ripe follicles and oestrogen secretion is seen following the administration of a single dose of PMS but ovulation does not occur although injection of HCG in addition to PMS will cause rupture of the follicles and the formation of corpora lutea. Some days later ovulation is observed following the injection of PMS alone indicating a release of endogenous LH. In rather younger animals which do not normally release enough LH to cause ovulation the injection of progesterone facilitates LH release and this action of progesterone can be blocked by the administration of barbiturates (McCormak and Meyer, 1964). At the stage where LH is normally released without the injection of progesterone, this release can be blocked by a variety of drugs such as Nembutal, atropine or chlorpromazine (McCormak and Meyer, 1962; Zarrow and Quinn, 1963; Zarrow and Brown-Grant, 1964) and appears to be related in its timing to the environmental lighting schedule (Strauss and Meyer, 1962a, b; Wagner and Brown-Grant, 1965). It can also be blocked by the injection of a single dose of testosterone within a few days of birth (Schuetz and Meyer, 1963; Brown-Grant, Quinn and Zarrow, 1964). These are all characteristic findings for the adult cyclic rat (see later section). They suggest that the hypothalamo-hypophysial system for ovulatory LH release is present in these animals 1–2 weeks before follicular ripening would normally occur in response to FSH release, which thus seems to be the last element of the hormonal mechanism related to cyclic ovarian function to appear.

Szentágothai and colleagues (1962) also studied the effect of low doses of oestrogens in depressing FSH/LH secretion in parabiotic pairs of castrate and normal female rats. One μg of oestradiol to the castrate member reduced the weight of the hypertrophied uterus in the intact rat but this effect was abolished when the castrate partner had a lesion of the anterior hypothalamus. They also noted that the deleterious effect of 1 μg oestradiol per day on the corpora lutea present in the ovaries of pubertal rats was prevented by anterior hypothalamic lesions, as was the effect of large doses of testosterone on intrasplenic ovarian grafts in castrate rats. They further demonstrated that ovarian tissue grafted into the anterior hypothalamus resulted in a significant decrease in uterine weight in rats, whereas ovarian tissue in the posterior hypothalamus or anterior pituitary or grafts of liver tissue into the anterior hypothalamus did not.

Attempts to demonstrate the presence of oestrogen or testosterone-sensitive structures in the anterior hypothalamus of the rat by experiments in which the circulating level of oestrogen was reduced have given less consistent results. Compensatory hypertrophy of the remaining gonad after unilateral ovariectomy is reported to be abolished in such animals (see Szentágothai and colleagues, 1962, for references and also Barraclough, 1963) but increased gonadotrophin secretion was observed in the parabiosis experiments of Szentágothai and Flerkó, discussed above. In addition, Taleisnik and McCann (1961) showed that ovariectomy in rats with lesions producing the 'constant oestrus' syndrome but not in rats with lesions of the median eminence leading to constant dioestrus was followed by an increase in pituitary LH content similar to that seen in normal rats, although plasma LH levels did not rise.

Desclin, Flament-Durand and Gepts (1962) showed that in rats with anterior hypothalamic lesions, transplantation of the ovary to the spleen was followed by luteinization of the ovary which had previously contained only follicles and the development of castration cells in the pituitary. (The suggested mechanism in this type of experiment is a reduction of blood oestrogen level, evidenced here by a fall in uterine weight, consequent upon secretion into the hepatic portal circulation and consequent inactivation or removal by the liver.) Further evidence that oestrogen can affect gonadotrophin secretion after destruction of the anterior hypothalamic centres, thought to be sensitive to oestrogen, is the finding of Bogdanove (1963c) that pituitary enlargement and castration cell development occur in rats with large anterior hypothalamic lesions and these changes can be prevented by oestrogen administration. Furthermore, it appears (see pp. 218–219 in Szentágothai and colleagues, 1962) that earlier claims by the Hungarian workers that anterior hypothalamic lesions prevent ovarian atrophy after long treatment with large daily doses of oestrogen have now been withdrawn following more extensive experimental studies.

The direct application of hormones to the nervous system is a relatively new technique and probably the experiments of Kollross (1942) (see reviews by Harris, 1955b and by Brown-Grant, 1966b) on the direct action of thyroxine on the maturing nervous system of amphibian larvae were among the earliest of such experiments. The technique of 'hormones on sticks'—implantation by stereotaxic methods of needles or wires with small amounts of hormones at their tip in various areas of the hypothalamus or pituitary—enables discrete areas of the brain to be subjected to the action of hormones in the absence of a raised level in the systemic blood and hormonal stimulation of the rest of the CNS or body tissues in general. By this elegant method, Harris and Michael (1964) have, for instance, been able to induce oestrus behaviour in ovariectomized cats by implants in the mamillary region of the hypothalamus in the presence of a completely anoestrous genital tract.

The application of Harris's technique of 'hormones on sticks' to the study of the site of action of sex hormones in regulating pituitary secretion has supplemented the data obtained by the intrahypothalamic or intrapituitary grafting of ovarian tissue, but it cannot be said to have resolved the question of an action at one or other of these levels (Brown-Grant, 1966b). Littlejohn and De Groot (1963) reported that oestrogen implants in either the anterior hypothalamus or the mamillary body suppressed compensatory ovarian hypertrophy in the rat. Lisk (1960) noted a decrease in testis weight from implantation of oestradiol-loaded needles anywhere in the hypothalamus, but the effect was not statistically significant and the same degree of atrophy was produced by *subcutaneous* implantation of needles. Implantation in the region of the arcuate nucleus produced a significant decrease in testis and secondary sex organ weights in males. In female rats, implants in this region reduced uterine and ovarian weights and suppressed the cyclic vaginal changes; the ovaries contained few or no large follicles but the corpora lutea remained functional over the 30 day experimental period. The ovarian state is consistent with a depression of FSH and LH secretion and a persistent secretion of prolactin. Similar effects were seen after testosterone implantation (Lisk, 1962) in intact male and female rats. Intrapituitary implants were not studied and the lack of effect of implants in the preoptic area or anterior hypothalamus is quite striking.

Lisk (1963) studied the effects of oestradiol implants on the development of pituitary castration cells in ovariectomized rats. Only implants in the arcuate nucleus-median eminence region resulted in a generalized lack of castration cells throughout the gland; subcutaneous implants had no effect on the pituitary changes; when the needle tip was in the pituitary there was a *localized* effect, castration cells were not found in the vicinity of the needle tip but were well developed at more distant sites in the gland. Davidson and Sawyer (1961a, b) observed gonadal atrophy after testosterone implantation in the posterior median eminence of dogs but not after intrapituitary implantation and similar results from oestradiol implants in female rabbits. Kanematsu and Sawyer (1963b, 1964) described the effect of oestradiol implants in the posterior median eminence which prevented changes in pituitary cytology in ovariectomized rabbits. Although the authors do not emphasize the finding, the number of 'gonadotrophs' was also significantly reduced by intrapituitary implants as was the LH content. Bogdanove (1963a) studied the effects of intrapituitary ovarian grafts or oestrogen implants in the castrate rat in detail and presented clear evidence for local regression of the castration cell changes, indicating a direct effect upon the pituitary.

In this paper and also in more extensive reviews (Bogdanove, 1963b, 1964) he puts forward the following very cogent arguments to support his view that oestrogen can act directly on the pituitary. Effects produced by intrahypothalamic implants may be due to a very efficient spread of hormone throughout the gland via the portal vessels. (It is striking that the most effective site appears to be in or near the median eminence in the region of the primary plexus of the portal vessels, and that both in Lisk's and Kanematsu and Sawyer's work there was evidence of systemic effects of the steroid. The failure to detect peripheral changes after single intrapituitary implants may be because the steroid is only reaching and producing cytological and functional effects in a limited portion of the gland.) Ramirez, Abrams and McCann (1964) have shown that *bilateral* oestradiol implants in the pituitary in castrate rats will reduce the plasma LH level. It is difficult, however, to accept Bogdanove's arguments as explaining the reported effects of anterior hypothalamic implants on gonadotrophin secretion; however, these are so far available only as abstracts (Littlejohn and De Groot, 1963; Ramirez, Abrams and McCann, 1963). It is reasonable, though, to accept the view that oestrogen can affect gonadotrophin secretion after destruction of the anterior hypothalamus and can act by a direct effect on the pituitary; it does not follow from this that an action on either or both the anterior hypothalamus or the median eminence-arcuate nucleus region in the intact animal is excluded, but merely that the evidence at present available to support this view can be interpreted differently. It may be—and much of the evidence based on lesion experiments discussed above supports this view—that more subtle effects involving minor variations in hormone level act through the anterior hypothalamus whereas more gross changes act directly at the pituitary level.

It is unfortunate for the further analysis of this problem that the 'isolated' pituitary (following stalk-section, transplantation or median eminence lesions) has such a low level of FSH/LH secretion that experiments comparable to those undertaken, to analyse the effects of thyroxine and corticosteroids on TSH and ACTH secretion, are not feasible except by analysis of pituitary hormone content. Such information as is available supports the concept of a direct action upon the gland (Van Rees and Wolthuis, 1962). Attempted distinctions between the effects of lesions on changes in pituitary hormone level or production and on release of gonadotrophins from the pituitary are appearing in many publications (see for example, Van der Werff ten Bosch, Van Rees and Wolthuis, 1962), but in the absence of methods for measuring the rate of hormone synthesis these are necessarily indirect at present.

Control of ovulation

The discussion so far has been concerned with the general control or, as Everett (1964) terms it, the day-to-day control of FSH/LH secretion. In the female, however, it is not enough to maintain an adequate level of FSH/LH secretion; normal ovarian function demands a cyclic or episodic variation in gonadotrophin secretion, and particularly in LH secretion, to bring about rupture of the ovarian follicle and delivery of a fertilizable egg or eggs to the genital tract at the time when ovarian steroid hormones have acted upon the central nervous system to induce the appropriate changes that will lead to mating and subsequent fertilization of the ovum. Techniques for determining the occurrence of ovulation have been available for many years (except, unfortunately, in primates!) and this together perhaps with the dramatic nature of the event has made the study of ovulation a favourite subject for neuroendocrine investigations. A recent symposium (Villee, 1961) and an outstanding review by Everett (1964) provide many additional references to the specific topics discussed in this section. It is convenient to distinguish two types of ovulatory LH release (though as Everett emphasizes the distinction is not an absolute one), animals that normally ovulate only after coitus (the rabbit is the best known example but others are the cat and the ferret) and animals that ovulate spontaneously by which in this context is meant without the stimulus of coitus. This group includes rats, mice, dogs, the common farm animals and primates. The rat has been most intensively studied.

In the rabbit, release of sufficient gonadotrophin to cause ovulation occurs within an hour of a single brief mating. In contrast, prolonged association with the male and repeated intercourse may be necessary in the ferret and the cat. Drugs such as atropine or sympatheticolytic agents will block ovulation in the mated rabbit if administered within minutes of intercourse but not otherwise. The release of neuro-humoral transmitter substance from the hypothalamus occupies some time and probably proceeds over the whole period of gonadotrophin release after the initial triggering. The electrical changes in the hypothalamus after coitus or vaginal stimulation (the 'EEG after reaction') reported by Sawyer and his collaborators (Sawyer, 1959) were thought at first to be associated with the neural activation of the pituitary but they in fact appear to be related to the presence of a high level of gonadotrophin in the blood, and it has been suggested that they could represent a feedback of the protein hormones on the hypothalamus to regulate its own discharge. Ovulation can be induced in the rabbit by electrical stimulation of the tuber cinereum or mamillary bodies but not by stimulation of the pituitary itself as was demonstrated in the work described by Harris (1955a), and ovulation following coitus can be blocked by lesions in this area (Sawyer, 1959). Ovulation can be produced by the slow intra-pituitary infusion of extracts of rabbit or bovine median eminence which contain the polypeptide LH releasing factor that is secreted into the portal vessels following coitus (Harris, 1963; Guillemin and Schally, 1963; Harris and colleagues, 1966). The afferent pathways concerned in the transmission of the stimuli arising during coitus to the hypothalamus are not known in detail nor is the precise origin of these impulses. Local anaesthesia of the vulva and vagina does not prevent either coitus or ovulation, though the importance of these areas is indicated by the fact that consistently in the cat and in some cases in the rabbit mechanical stimulation of the vagina and cervix of the oestrus animal with a glass rod will provoke ovulation. It seems likely that a constellation of impulses via many sensory systems combine to initiate the hypothalamic activity that leads to ovulation.

Spontaneous ovulation in the rat is not a haphazard process; under standard lighting conditions (about 14 hours light and 10 hours dark) ovulation occurs every 4 or more rarely every 5 days in young adult females. The timed injection of atropine, barbiturates and other drugs (for a complete list of effective agents see Munson, 1963) enabled the time of LH discharge to be established as a 2 hour critical period during the afternoon of the day of prooestrus. A single dose of 'Nembutal' at this time blocks ovulation which would normally occur 10–12 hours later. Ovulation occurs 24 hours late if a single injection is given; if the treatment is repeated on the next afternoon, ovulation may be postponed for a further 24 hours. The time of the critical period is determined by the lighting schedule; by reversing the light and dark periods, LH release can be induced to occur in the early hours of the morning and ovulation in early afternoon (our time). The regular, daily discharge of LH could be a regular feature of the rat cycle and the 24 or 48 hour delay of ovulation by Nembutal might suggest this. However, it appears from the work of Schwartz and Bartosik (1962) that in fact it occurs only on the afternoon of prooestrus during the normal cycle and not at all in male rats. The stimulus for LH discharge in the female at a time determined by the environmental lighting conditions must be something else again; it appears likely that variations in circulating hormone levels are involved (see below). The Nembutal-blocked rat resembles the rabbit in that ovulation can be brought about by electrical stimulation of the hypothalamus or by intrapituitary infusion of bovine or rat median eminence extract (Nikitovitch-Winer, 1962). The exact site for electrical stimulation is uncertain. In the experiments of Critchlow (1958) stimulation of the median eminence was effective. Reports of effective stimulation from the suprachiasmatic region and the medial preoptic area may be related to chemical stimulation from the lesions produced rather than to the electrical stimulation *per se* (Everett, 1964). Such lesions are effective almost anywhere in the preoptic area or in the anterior hypothalamus in front of the caudal margin of the optic chiasma but not in the arcuate nucleus or tuberal region or median eminence. In contrast, Barraclough (1963) reports ovulation produced by the electrical stimulation of the posterior median eminence-arcuate nucleus area.

Apart from acute pharmacological blockade of ovulation, the rat may be rendered anovulatory (without failure of follicular ripening) by exposure to continuous illumination (such animals may ovulate after coitus), by lesions of the anterior basal hypothalamus, by isolated lesions of the suprachiasmatic nucleus or by exposure to androgenic hormones during the first few days of post-natal life. The proper functioning of the hypothalamic centres responsible for cyclic LH discharge can thus be upset by (i) an abnormal afferent pattern—the results of exposure to continuous darkness or blinding are contradictory (Critchlow, 1963); (ii) destruction of the neural elements involved without failure of gonadotrophin secretion; (iii) exposure to an abnormal hormonal environment during development. This last effect will be discussed in more detail later.

The role of circulating steroid hormone levels in the control of the ovulatory LH discharge is a complex one; there is not the appealing simplicity that exists in the analogous situation with regard to TSH and ACTH secretion, or, by comparison, in the relation of oestrogen and androgen levels to the basal secretion of FSH/LH discussed earlier. The effects of oestrogen will be discussed first. Large doses of oestrogen suppress FSH/LH secretion and lead to ovarian atrophy. Even quite small doses block LH secretion in the ovariectomized rat (Ramirez and McCann, 1963a, b, Parlow, 1964). Yet the prooestrous release of LH occurs at a time when oestrogen levels are high (Ramirez and McCann, 1964a) and oestrogen administration on the second day of dioestrus in a 5 day cyclic rat results in LH discharge 24 hours later (Everett, 1948), and oestrogen on day 4 or 5 of pregnancy or pseudopregnancy in the rat is reported to cause a discharge of LH 24 hours later that leads to ovulation (Everett, 1947). This effect can be blocked by appropriately timed injections of barbiturates or atropine (Sawyer, Everett and Markee, 1949). It is less certain whether oestrogen has a direct stimulatory effect on LH release in the rabbit; small doses at any rate appear to facilitate the release of LH following vaginal stimulation with a glass rod although large doses depress LH release and gonadotrophin secretion in general (Sawyer and Markee, 1959).

The effects of progesterone are equally perplexing; there seems little doubt that high levels of progesterone during the luteal phase of the cycle or during pregnancy suppress ovulation by inhibiting LH release (Rothchild, 1965) and the same effect is the basis of contraceptive methods involving the regular oral administration of synthetic steroids with progestational (progesterone-like) properties that are active when taken by mouth (see Pincus, 1956, for an account of the scientific background to this work). Yet progesterone also appears to be able to facilitate the release of LH and to induce premature laying in hens or, if given in the third day of dioestrous in 5 day cyclic rats, to advance ovulation by 24 hours. Both effects may involve the nervous system, as shown by the effectiveness of progesterone injected directly into the preoptic area in hens and the fact that the effect in rats can be blocked by atropine or Nembutal administration (Everett, 1964) but this is not certain (Brown-Grant, 1966b). The situation in the rat is rendered more complex by the fact that the same dose of progesterone given on the first day of dioestrous will delay ovulation for 24 hours or for 48 hours if the injection is repeated. A biphasic effect of progesterone is reported in the rabbit (Sawyer, 1959) the release of LH being first facilitated and later (24 hours) markedly depressed. The paradox of both stimulation and depression of trophic hormone release by target organ hormones is discussed in some detail by Jones and Ball (1962). Apart from its effects in normal animals, progesterone will restore regular cycles (i.e. periodic release of sufficient LH to cause ovulation) in rats in which constant oestrus has been induced by exposure to constant light or, in some cases, by hypothalamic lesions, but not in rats rendered anovulatory by androgen administration shortly after birth. In these animals it appears to depress LH release or stimulate its synthesis; the pituitary LH content rises and it becomes possible to produce ovulation by electrical stimulation of the median eminence-arcuate nucleus area of the hypothalamus (Barraclough, 1963).

The importance of the hypothalamus in regulating the cyclic discharge of LH that characterizes the pattern of gonadotrophin secretion of the non-pregnant female of spontaneously ovulating species is abundantly clear. The effects of the feedback of ovarian steroids is also of great importance but not simple to understand. The experimental evidence suggests that the same hormone can either stimulate or depress LH release when injected at different stages of the cycle, and it is not easy to formulate any useful hypotheses in terms of the relative levels of oestrogen and progesterone that will explain the observations. It is depressing that so much of the physiological evidence should be derived from two species, the rabbit and the rat, about which very little direct information concerning steroid hormone levels in the blood is available. Perhaps when measurements are made the picture will become a little clearer.

The characteristic cyclicity of gonadotrophin secretion in the female is not a property of the pituitary. Tissue from male rats transplanted beneath the median eminence of the female will sustain ovarian cycles. Nor is the presence of ovarian tissue sufficient to maintain cycles; ovarian grafts in intact or castrate male rats do not show cyclic changes. It seems that the capacity to initiate cyclic gonadotrophin secretion is a property of the female hypothalamus. Reference has been made already to the fact that injection of male hormone (or the grafting of a testis) in female rats or mice within a few days of birth results in an anovulatory state as adults. It appears that testosterone at this stage of development can permanently block the normal rhythmicity of the female hypothalamus (Brown-Grant, 1966b). Ovaries from such animals undergo normal cycles when transplanted to a normal female and their pituitaries can sustain normal cycles when transplanted to a normal female. The significance of these findings has been greatly increased by the more recent discovery that if the male rat is castrated within 24 hours of birth and then receives an ovarian transplant when adult, this graft, unlike those in male rats castrated several days after birth or as adults, shows cyclic ripening and rupture of follicles and corpus luteum formation. It appears that at least in the rat, endogenous testosterone from the testes of the newborn quickly induces a male pattern of hypothalamic activity or suppresses the female pattern of rhythmicity. A full discussion of recent work in this rapidly developing field of study has been given by Harris (1964a) who has also described the alterations in mating behaviour that have been observed in these changelings.

Control of the corpus luteum

This endocrine gland is formed from the tissues of the ovarian follicle after the egg has been expelled at ovulation. The life of the corpus luteum varies widely in different species. It may be a day or so as in the rat and mouse or two weeks as in the human and the duration of activity may be prolonged for weeks or months if the animal becomes pregnant. The mechanisms controlling the corpus luteum are quite obscure; the older teaching that prolactin (luteotrophin) was the hormone responsible was based on observations on rats and mice. LH, but not Prolactin, maintains the corpus luteum in the rabbit but in the guinea-pig, ewe and pig the corpus luteum, once formed, appears to be quite independent of the pituitary (Everett, 1964). The role of nervous stimuli from the uterus in 'informing' the brain that a fertilized ovum is present and that continuance of corpus luteum function is necessary for a successful pregnancy until the placenta (either by progesterone or gonadotrophin secretion) can take over has been investigated. No clear findings have yet emerged (the review of Anderson, Bowerman and Melampy (1963) summarizes the recent work on this topic). The control of prolactin as a luteotrophic gonadotrophin in rats and mice and as a pituitary hormone acting on the mammary gland in these and other species will be considered here.

The first point about prolactin secretion was made in an earlier section, namely that the normal influence of the hypothalamus over its secretion appears to be inhibitory. The most effective physiological stimulus to a removal of this inhibition in the rat or mouse is coitus. If the male partner is sterile a condition of pseudopregnancy results, which is probably the equivalent of the luteal phase of the more usual mammallian cycle, e.g. in primates. Afferent fibres from the uterus are involved in the pseudopregnancy response to mechanical stimulation of the cervix as it is prevented by spinal anaesthesia. In this connection the electrophysiological studies of Barraclough and Cross (1963), Cross (1964) and of Cross and Silver (1965) are of interest. They recorded from individual neurones in the lateral hypothalamus of rats and obtained changes in firing rate following a variety of stimuli including mechanical stimulation of the uterine cervix. The greatest proportion of neurones showing a *decrease* in firing rate following this stimulus was observed in prooestrus and oestrus rats. These are the stages of the cycle at which this stimulus is most effective in eliciting the pseudopregnancy response.

The second physiological stimulus is that of suckling. The role of afferent stimuli from the nipples in the release of oxytocin from the neurohypophysis was discussed in an earlier section. It appears to be a less effective means of inducing pseudopregnancy than stimulation of the cervix but is of considerable importance in the maintenance of prolactin secretion during lactation in most species. Afferents from this general area in man may initiate prolactin secretion, as for instance after thoracic surgery involving rib resection. The precise pathway of either set of nerve fibres to the hypothalamus or their termination there is not known and the site of the inhibitory centre in the hypothalamus is not firmly established. One of the difficulties here is that large basal lesions of the hypothalamus which destroy the median eminence may simply be acting to block all control of gonadotrophin secretion by destroying the pituitary stalk. This objection does not apply to the recent studies of Flament-Durand and Desclin (1964) who consistently induced pseudopregnancy by unilateral mechanical lesions of the anterior hypothalamus above

8*

and lateral to the paraventricular nucleus. The possibility of non-specific factors in these experiments has to be considered. A variety of traumatic procedures ('stresses') have been reported to cause pseudopregnancy in rats (Everett, 1964). Certain psychotropic drugs (tranquillizers) may also induce prolactin secretion; chlorpromazine for instance, regularly induces pseudopregnancy in rats (Barraclough and Sawyer, 1959) and has induced lactation in rabbits and in psychiatric patients undergoing prolonged treatments. This may be a form of pharmacological stalk-section as these drugs depress many hypothalamic functions.

A further type of nervous stimulus effecting prolactin secretion is olfactory. Recent work on the mouse suggests that the high incidence of 'spontaneous' pseudopregnancies in female mice housed together under crowded conditions may be due to olfactory stimuli from other females. The opposite effect, a failure of prolactin secretion (perhaps a restoration of the normal inhibition) appears to be involved in the failure of pregnancy in newly mated female mice when they are exposed to the smell of male mice of a strain other than that of the male they were mated with or even to the smell of the excreta of male mice of a different strain (see review by Parkes and Bruce, 1961).

From the type of stimuli affecting prolactin secretion, it is clear that some element of neural control is involved. There is also evidence that the feedback of steroid hormones may also be involved. Everett (1964) discusses the older work (based on cytological criteria and assays of pituitary prolactin content) that high doses of oestrogen increase the prolactin content of normal and of transplanted pituitaries in rats. More recently Kanematsu and Sawyer (1963a) have shown that in rabbits implants of oestradiol in the posterior median eminence resulted in ovarian atrophy, an increased pituitary *content* of prolactin and no evidence of mammary gland activation. Implants in the pituitary did not produce ovarian atrophy but pituitary prolactin content was slightly reduced and the mammary glands were stimulated and contained milk. These results are considered to show an effect of oestrogen in promoting prolactin *synthesis* by an action on the hypothalamus and in promoting prolactin *secretion* by a direct action on the pituitary. Different thresholds are suggested to explain this effect and the relationship of their findings to the inhibition of milk production during pregnancy is discussed. In contrast, Ramirez and McCann (1964b) obtained evidence of an increased secretion of prolactin following the implantation of oestradiol in either the median eminence or the pituitary of rats. A single large dose of oestrogen at oestrus in the rat will induce pseudo-pregnancy (Everett, 1964) as will androgens in large doses. Kane-matsu and Sawyer also noted mammary gland development following intrahypophysial testosterone implants in rabbits. The effects of pro-gesterone on prolactin secretion are possibly one of the few examples of 'positive feedback' in the endocrine system. Large doses of pro-gesterone daily result in the maintenance of large functional corpora lutea in the rat and single injection on the day of oestrus may induce a self-sustaining pseudopregnancy. Conceivably endogenous pro-gesterone also acts in this way to sustain the gland producing it (Rothchild, 1965). Whereas unilateral ovariectomy in the pig is compensated for in that a larger number of follicles ripen and ovulate in the remaining ovary at the next oestrus, a recent finding (Brinkley, and coworkers, 1964) shows that removal of an ovary containing cor-pora lutea does not induce any increase in size or functional activity. In fact some regression of the corpora lutea in the remaining ovary appears to occur, possibly due to loss of the positive feedback stim-ulus to prolactin production. What, in this case, terminates the activity of an established corpus luteum? It may be relevant that Zeilmaker (1963, 1964) has observed that when functional corpora lutea are induced in ovarian grafts in castrate male rats or in androgen sterilized rats their activity is *not* self-terminating as it is in normal females. Perhaps the pattern of LH secretion is involved (Rothchild, 1965).

An Assessment of the Present Position

As discussed elsewhere in this volume, an important and perhaps the most characteristic feature of biological systems is their capacity for self-regulation. Deviations from the optimal condition set in action restorative processes, their intensity being related to the degree of divergence from normal. This is the typical behaviour of a negative feedback control system and there are many examples of this type of mechanism in the endocrine system. The external closed feedback loop may be humoral, possibly some change in the composition of the blood. The simplest is shown in Figure 4A. The system consists of a gland secreting a hormone producing a rise or a decrease in some chemical constituent of the blood. Changes in the blood content of this sub-stance act on the gland to regulate the secretion of the hormone. Examples are: (i) the control of parathyroid hormone secretion which is inhibited by a rise and stimulated by a fall in the ionizable calcium level of the blood and acts to raise the blood calcium level, and (ii) the control of the secretion of glucagon by the blood sugar level. If

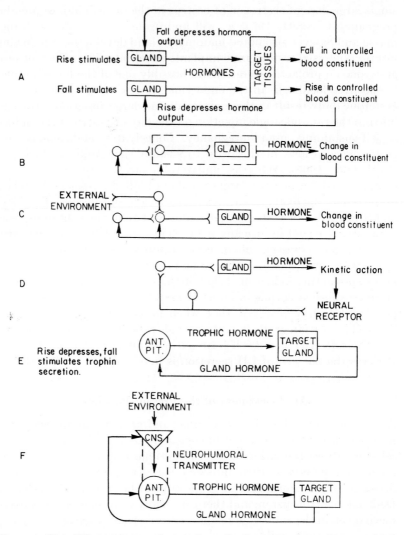

Figure 4. The different types of control mechanism encountered in the endocrine system.

there were no other regulatory mechanisms for the blood constituent controlled, then such a system would be adequate to prevent decreases in the controlled blood element but would have no greater response than to suppress hormone output to zero to control a rise in blood constituent. In the case of the blood sugar and blood calcium levels there also exist feedback systems involving increased secretion of insulin and calcitonin respectively in response to a rise in blood level. These hormones act to reduce the blood level.

These systems involve endocrine glands which are not now believed to have a functionally important nerve supply. When the endocrine gland is under neural control (or, as in the case of the posterior pituitary, nerve cell and hormone secreting cell are one and the same) a further complexity in the regulating mechanism is possible (Figure 4B). Changes in blood composition can still act directly upon the controlling neurone or upon other nerve cells connected with the controlling neurone to regulate hormone output. The introduction of a neural link introduces two further complications into the regulatory system. One is that changes in the external environment may enter the picture (Figure 4C). In this way one can picture the occurrence of emotional antidiuresis as described by Verney (1947) in water-loaded dogs where various extraneous disturbances caused a quite inappropriate inhibition of water diuresis through a release of vasopressin. The second is the possibility of a non-humoral feedback loop. There is now good evidence that via neural receptors (whose exact location is still disputed) changes in the volume of blood within the intrathoracic low pressure system (great veins plus right side of heart and pulmonary vessels and left atrium) influence the secretion of vasopressin, a decrease in volume resulting in increased secretion (Gauer and Henry, 1963). Changes elicited in this way are no longer dependent upon the immediate chemical changes produced in the blood by the actions of the hormone but can detect some change several steps removed from this. The existence of a neural link (Figure 4D) also provides for the monitoring of the effects of 'kinetic' as well as metabolic hormones (in the terminology of Jenkin, 1962). A major action of noradrenaline is upon the smooth muscle of the walls of arteries; if the cardiac output does not decrease, the vasoconstriction produced results in an increase in the systemic arterial blood pressure which is monitored by the baro-receptors whose impulses can in turn moderate the activity of the post-ganglionic sympathetic fibres and the secretion of noradrenaline from the adrenal medulla.

The endocrine glands which are the target organs for the trophic hormones of the anterior pituitary form a unique group. The activity of the target organ is almost abolished after removal of the pituitary and there is no good evidence that any physiologically significant variation in activity can occur. The pituitary-target organ system, through a double humoral link is able to act as a control mechanism to stabilize target organ hormone levels in the blood (Figure 4E).

Why the organism should want to do so and what is gained by such an arrangement over and above one depending on regulation by the intermediation of the effects produced by the hormones, is not immediately obvious. Perhaps this is not a legitimate question to ask, but is it possible that this arrangement was initially evolved to control the activity of glands that are not essential to the individual organism, the gonads? The survival of the species depends upon successful reproduction and, as pointed out earlier, the pattern of seasonal reproduction seen in most lower animals in the natural state depends for its organization on the pituitary control of the gonads, linked in turn to stimuli or signals obtained from the external environment. In the absence of effects of the gonadal hormones on the parent organism, that could be used as controlling signals to modulate pituitary activity, perhaps a system of monitoring the blood levels of hormones themselves was evolved instead. The regulation of thyrotrophin and adrenocorticotrophin secretion by thyroxine and corticosteroid levels has been discussed at length above; conceivably the pituitary control of the adrenal cortex and the thyroid is of later evolutionary development than its control of the gonads.

Much evidence was cited in earlier sections to indicate that the nervous system also influences the activity of the pituitary-target organ glands; Figure 4E must therefore be modified. There is no evidence for a functionally important innervation of the thyroid, adrenal cortex or gonads, so that neural influences must act via the pituitary. Considerable controversy at one time surrounded the question of the existence of a direct secretory nerve supply to the anterior pituitary (Harris, 1955a; Fraps, 1962) but at present the alternative idea of neural control exercised through the action of local hormones (neurohumoral transmitters) reaching the gland via the hypophysial portal vessels is widely accepted (Nalbandov, 1963). This is illustrated in Figure 4F. There is much evidence to suggest that the nervous system exerts a tonic stimulatory effect on trophic hormone secretion possibly via the continuous release of transmitter substances. There is also evidence that secretion of one pituitary hormone, prolactin, may be inhibited by the hypothalamus in some manner. One of the most intriguing problems at the moment for experimental endocrinologists is the isolation and analysis of the transmitter substances (Harris, 1964b; Harris, Reed and Fawcett, 1966) and the possibility that their half-lives in the systemic circulation may be long enough to permit their acting at a distance on transplanted pituitary tissue.

The morphological evidence (Green, 1951; Harris, 1962) suggests that a 'short route' to the anterior pituitary may be available in all vertebrates (but see discussion of the situation in teleosts by Barrington, 1960) and it is almost an article of faith that normally the transmitter substances pass this way. Recent work on certain elasmobranchs does, however, appear to reopen the question of a systemic route under physiological conditions. The elasmobranchs are an ancient group that have been separated from the rest of the vertebrates for at least 3×10^8 years and the type of fish studied by Dodd (*Scyliorhinus caniculus*) is known from as early as the Cretaceous era. Dodd (1960) described the ventral lobe of the pituitary in selachoid elasmobranchs (cartilaginous fishes of the dogfish and shark type) as being embedded in the cartilage of the base of the skull and as being connected to the rostral lobe (which is in vascular contact with the median eminence) only by a tenuous cord of cells. This ventral lobe, apparently devoid of portal vessel connections, increases in size during the breeding season and is believed to contain the gonadotrophin secreting cells. Removal of the ventral, but not the rostral, lobe arrests the regular egg laying of the spotted dogfish and leads to gonadal atrophy in both sexes. The existence of a breeding season and the regularity of egg laying suggest a neural control of gonadotrophin secretion but a 'short route' connection of the ventral lobe with the hypothalamus does not appear to exist. A study of the systemic blood for the presence of neurohumoral transmitter substance during the breeding season in this species might be rewarding.

In Figure 4F the level at which the feedback action of target organ hormones occurs is shown as being both the hypothalamus and the pituitary. The question of the exact site cannot be regarded as settled for any of the three groups of hormones, thyroidal, adrenocortical or gonadal, at present. Thyroid hormones can act at the pituitary level but, although no incontrovertible evidence is available at present, they may also act on the hypothalamus. Corticosteroids appear to act at the hypothalamic level though an action on the pituitary cannot be excluded and gonadal hormones appear to be capable of acting at both levels, perhaps producing different effects at each site.

The role of hormonal feedback in the control of pituitary trophic hormone secretion appears to be of more importance in the control of TSH secretion than in the case of ACTH where neural influences can override the feedback mechanism. The part played by ovarian hormones in the control of gonadotrophin secretion in the female is more difficult to assess. It is clear that after gonadectomy there is a relatively simple and relatively crude relationship between gonadotrophin (FSH and LH) secretion and the level of oestrogen or androgen and that a negative feedback relationship exists under these circumstances. Under more normal circumstances, however, the relationship is more complex; oestrogen, which depresses LH secretion in the castrate, can induce LH release and ovulation under certain circumstances in the rat; is this a positive feedback? Similarly progesterone can suppress LH release in the rat but under different circumstances may stimulate or at least facilitate it in rats and hens. In rabbits it may first stimulate and then depress secretion. What sort of a model can one envisage to explain observations of this sort? Is it possible that these are not, in fact, feedback mechanisms in the sense that the level of hormone is directly influencing the release of gonadotrophin, but rather that the hormone is in some way altering the sensitivity of the anterior pituitary to the neurohumoral transmitter substances or the pattern of activity of nerve cells in the hypothalamus or elsewhere in the central nervous system, so that afferent stimuli or intrinsic neuronal activity can now exert different effects on gonadotrophin secretion? Perhaps this is only begging the question; and perhaps, too, it is unreasonable to expect that electrophysiological investigations should be able to answer such questions at a time when the complexities of stimulation and inhibition at the anterior horn cell have not yet been resolved. Attempts to investigate this aspect of the problem have already met with some success, however (Cross and Silver, 1965).

A further point that should be made as regards the 'feedback' effects of ovarian hormones on gonadotrophin secretion is that many of the ideas in this field are based on the results of administration of arbitrarily chosen doses of exogenous hormone and the results of such experiments are then extended to explain effects postulated to be caused by the animals' own hormones. There is a reasonable body of information available about thyroid hormone and corticosteroid levels in the blood. The complexities introduced by the need to distinguish between free hormone and hormone temporarily attached to specific binding proteins are now appreciated. In contrast, knowledge of the blood levels of ovarian hormones in the common laboratory animals during the sexual cycle are as yet almost non-existent and a systematic attempt to fill this gap in our knowledge should be undertaken when methods are available. Such data might obviate the need to devise elaborate theories to explain the results of experiments that may have been grossly unphysiological.

A few of the concepts and much of the jargon of cybernetics and control system theory has entered into endocrinology, as may be obvious from this review. The general principles involved and the mathematical theory of simple servo or regulator systems involving linear equations are not difficult to grasp. Unfortunately, biological systems are not simple and the mathematical expression of models of the interesting biological systems appear inevitably to contain non-linear terms and equations of a high order necessitating recourse to computer methods to obtain solutions (Grodins, 1963). It is not, I think, that endocrinologists are in general reluctant to employ mathematical methods as Grodins appears to suggest; their record in this respect is quite good when the systems under consideration can be adequately defined and the various rate constants involved can be measured satisfactorily. Mathematical analyses of iodine metabolism are numerous (Riggs, 1952; Berson and Yalow, 1954) and thyroid hormone metabolism has been extensively analysed (Robbins and Rall, 1960; Ingbar and Freinkel, 1960). The theoretical basis of steroid hormone secretion-rate measurements by isotopic tracers has been examined critically and analysed mathematically (Gurpide, Mann and Lieberman, 1964; Tait, 1963). Tait and Burstein (1964) have made an extensive theoretical study of the consequences of the protein binding of steroid hormones.

No assumptions as to the method of working of regulators or error detectors are involved in this type of analysis. When an attempt is made to formalize and apply mathematical methods to a regulating system however, some decision as to how the system is to work must be made. In essence one is devising a model to have certain properties with regard to the regulation of input and output and the effect of applied variations in input. A model that has the known properties of a biological system may achieve these effects in an entirely different way and need not tell us anything as to the way the biological system achieves the same overall effect. Where the biological system is reasonably well known, analysis in terms of control theory does not really seem to have added very much to our understanding (see for instance the analyses by Bragdon, Nalbandov and Osborne (1953) and Bolie (1961) of the blood sugar regulating mechanisms). Where the biological system is not well understood it is possible to mistake the model for reality as in the attempts to explain the ACTH-adrenal cortex relationship in terms of a proportional controller with a variable set point component. 'The very act of formulating the problem in terms of block diagrams and mathematical relationships, requiring as it does the precise identification and rigorous definition of previously vague concepts, provides insight and clarification obtainable in no other way' (Grodins, 1963); it may also tend to lead to the construction of elaborate theoretical models based on hypothetical mechanisms and the elevation of generalizations to the status of precisely determined relationships at a time when the information available is inadequate to evaluate them.

Whether the individual research worker chooses to accumulate the data necessary for this type of analysis will be determined by the value he attaches to the type of approach exemplified by the quotation from Grodins (1963). If such an analysis is to be worthwhile then quantitative data on a massive scale in relation to a single experimental manoeuvre is essential. Although the changes in plasma and pituitary TSH or ACTH levels after thyroidectomy or adrenalectomy have been investigated in some detail by, for instance, Salaman (1964) and Fortier (1966), these studies represent but a fraction of the data that appears necessary in order to give validity to a quantitative mathematical analysis of the control system involved. Until a detailed knowledge of control system theory is more common among experienced investigators, it is an open question whether such work, except in highly specialized centres, represents the best use that can be made at the moment of the available scientific manpower, in view of the gaps in our present knowledge of the working of the endocrine system.

Two topics that have not been discussed as yet will serve to illustrate this opinion. The control of the secretion of growth hormone, a topic of very great fundamental and clinical importance, was until recently largely an unexplored field because of the lack of a precise and convenient short term method for assessing its action on the body and the absence of any satisfactory method for its assay in body fluids. Knobil and Hotchkiss (1964) concluded that there was no good evidence of neural control. The whole field, however, is clearly open to reexamination following the development of an immunological method for the assay of growth hormone, the demonstration of a marked rise in blood level induced by hypoglycaemia (Roth, Glick, Yalow and Berson, 1963) and the involvement of the hypothalamus in this response (Abrams, and colleagues, 1966). The control of aldosterone (the potent natural mineralocorticoid secreted by the adrenal cortex and regulating Na^+ and K^+ excretion by the kidney) is at present the most confusing topic in endocrinology, with a multitude of conflicting experimental results and numerous theories (Farrell and Taylor, 1962; Fortier, 1963; Ganong, 1963; Gauer and Henry, 1963; Blair-West, Coghlin, Denton, Goding, Wintour and Wright, 1963). Among mechanisms suggested are: a direct effect of plasma Na^+ and K^+ concentration on the adrenal, an effect of ACTH at high plasma levels, a humoral agent originating in the pineal gland or in the midbrain periaqueductal grey matter, angiotensin formed from plasma protein by the action of renin secreted in renal blood or lymph, neural afferents from vascular system receptors in the intrathoracic low pressure system, the arterial baroreceptors in the aorta, carotid arteries or in the thyrocevical trunk or a regulation by variations in the rate of hepatic removal of aldosterone. Clearly further experimental studies are badly needed in both these fields. The absence of information about sex hormone levels in the blood at different stages of the ovarian cycle in the common laboratory animals has been mentioned previously in connection with the difficulties involved in understanding the part played by steroid hormone feedback in regulating gonadotrophin secretion.

But whether one believes that either mathematical analysis of the feedback control of thyroid or adrenocortical function, an exploration of the almost untouched field of growth hormone regulation or the jungle of aldosterone control, or simply an attempt to determine the basic reason for the difference in gonadotrophin secretion between male and female, is the area in which most progress is likely to be made in the next few years, there can be no doubt that the study of control and regulation in the endocrine system, even at the currently unfashionable 'whole animal' level, still offers a wide variety of challenging problems of basic importance for the experimental biologist to investigate.

References

Aboul-Khair, S. A., Crooks, J., Turnbull, A. C. and Hytten, F. E. (1964). 'The physiological changes in thyroid function during pregnancy.' *Clin. Sci.*, **27**, 195–207.

Abrams, R. L., Parker, M. L., Blanco, S., Reichlin, S. and Daughaday, W. H. (1966). 'Hypothalamic regulation of growth hormone secretion.' *Endocrinology*, **78**, 605–613.

Anand, B. K. (1961). 'Nervous regulation of food intake.' *Physiol. Rev.*, **41**, 677–708.

Anderson, L. L., Bowerman, A. M. and Melampy, R. M. (1963). 'Neuro–utero–ovarian relationships.' In Nalbandov, A. V. (Ed.), *Advances in Neuroendocrinology*, University of Illinois Press, Urbana, chap. 11, pp. 345–373.

Andersson, B., Ekman, L., Gale, C. G. and Sundsten, J. W. (1962a). 'Thyroidal response to local cooling of the pre-optic "heat loss centre".' *Life Sci., Oxford*, **1**, 1–11.

Andersson, B., Ekman, L., Gale, C. C. and Sundsten, J. W. (1962b). 'Blocking of the thyroid response to cold by local warming of the pre-optic region.' *Acta Physiol. Scand.*, **56**, 94–96.

Andersson, B., Ekman, L., Gale, C. C. and Sundsten, J. W. (1963). 'Control of thyrotrophic hormone (TSH) secretion by the "heat loss centre".' *Acta Physiol. Scand.*, **59**, 12–33.

Andersson, B., Gale, C. C. and Hokfelt, B. (1964). 'Studies of the inter-action between neural and hormonal mechanisms in the regulation of body temperature.' In Bajusz, E. and Jasmin, G. (Eds.), *Major Problems in Neuroendocrinology*, Karger, New York, pp. 42–61.

Andersson, B., Gale, C. C. and Ohga, A. (1963). 'Suppression by thyroxine of the thyroidal response to local cooling of the "heat loss centre".' *Acta Physiol. Scand.*, **59**, 67–73.

Andik, I., Balogh, L. and Donhoffer, S. (1949). 'The effect of thyroxin in thyroidectomized rats treated with methylthiouracil.' *Experientia*, **5**, 249–250.

Aron, M., Van Caulaert, C. and Stahl, J. (1931). 'L'équilibre entre l'hor-mone préhypophysaire et l'hormone thyroïdienne dans le milieu intérieur à l'état normal et à l'état pathologique.' *Compt. Rend. Soc. Biol.*, **107**, 64–66.

Averill, R. L. W., Purves, H. D. and Sirett, N. E. (1961). 'Relation of the hypothalamus to anterior pituitary thyrotropin secretion.' *Endocrinology*, **69**, 735–745.

Baghdiantz, A., Foster, G. V., Edwards, A., Kumar, M. A., Slack, E., Soliman, H. A. and MacIntyre, I. (1964). 'Extraction and purification of calcitonin.' *Nature*, **203**, 1027–1028.

Bakke, J. L. and Lawrence, N. (1964). 'Influence of propylthiouracil and thyroxine on synthesis and secretion of thyroid stimulating hormone in the hypothyroid rat.' *Acta Endocrinol.*, **46**, 111–123.

Barrnett, R. S. and Greep, R. O. (1951). 'Regulation of secretion of adreno-tropic and thyrotropic hormones after stalk section.' *Am. J. Physiol.*, **167**, 569–575.

Barraclough, C. A. (1963). 'Discussion of paper by Flerko.' In Nalbandov, A. V. (Ed.), *Advances in Neuroendocrinology*, University of Illinois Press, Urbana, pp. 224–233.

Barraclough, C. A. and Cross, B. A. (1963). 'Unit activity in the hypo-thalamus of the cyclic female rat: effect of genital stimuli and pro-gesterone.' *J. Endocrinol.*, **26**, 339–359.

Barraclough, C. A. and Sawyer, C. H. (1959). 'Induction of pseudo-pregnancy in the rat by reserpine and chlorpromazine.' *Endocrinology*, **65**, 563–571.

Barrington, E. J. W. (1960). 'Some problems of adenohypophysial relation-ships in cyclostomes and fish.' *Symp. Zool. Soc. London*, **2**, 69–85.

Bayliss, W. M. (1915). '*Principles of General Physiology*,' 1st ed., Long-mans, London.

Berson, S. A. and Yalow, R. S. (1954). 'Quantitative aspects of iodine metabolism. The exchangeable organic iodine pool, and the rates of thyroidal secretion, peripheral degradation and faecal excretion of endogenously-synthesized organically-bound iodine.' *J. Clin. Invest.*, **33**, 1533–1552.

Berthet, J. (1963). 'Pancreatic hormones: glucagon.' In von Euler, U. S. and Heller, H. (Eds.), *Comparative Endocrinology*, Vol. 1, Academic Press, London, chap. 11, pp. 410–427.

Blair-West, J. R., Coghlan, J. P., Denton, D. A., Goding, J. R., Wintour, M. and Wright, R. D. (1963). 'The control of aldosterone secretion.' *Recent Progr. Hormone Res.*, **19**, 311–383.

Blanquet, P. (1962). 'Hypothalamus and thyroid.' *Advan. Biol. Med. Phys.*, **8**, 225–314.

Bogdanove, E. M. (1962). 'Regulation of TSH secretion.' *Federation Proc.*, **21**, 623–627.

Bogdanove, E. M. (1963a). 'Failure of anterior hypothalamic lesions to prevent either pituitary reactions to castration or the inhibition of such reactions by estrogen treatment.' *Endocrinology*, **72**, 638–642.

Bogdanove, E. M. (1963b). 'Local actions of target gland hormones on the rat adenohypophysis.' *Colloques internationaux du Centre National de la Recherche scientifique*, **128**, 163–181, C.N.R.S., Paris.

Bogdanove, E. M. (1963c). 'Direct gonad–pituitary feedback; an analysis of effects of intracranial estrogenic depots on gonadotrophin secretion.' *Endocrinology*, **73**, 696–712.

Bogdanove, E. M. (1964). 'The role of the brain in the regulation of pituitary gonadotropin secretion.' *Vitamins Hormones*, **22**, 205–260.

Bogdanove, E. M. and Crabill, E. V. (1961). 'Thyroid–pituitary feedback: direct or indirect? A comparison of the effects of intrahypothalamic and intrapituitary thyroid autotransplants on pituitary thyroidectomy reactions in the rat.' *Endocrinology*, **69**, 581–595.

Bolie, V. W. (1961). 'Coefficients of normal blood glucose regulation.' *J. Appl. Physiol.*, **16**, 783–788.

Bragdon, D. E., Nalbandov, O. and Osborne, J. W. (1953). 'The control of the blood sugar level.' In Quastler, H. (Ed.), *Essays on the Use of Information Theory in Biology*, University of Illinois Press, Urbana, pp. 191–207.

Brinkley, H. J., Wickersham, E. W., First, N. and Casida, L. E. (1964). 'Effect of unilateral ovariectomy on the structure and function of corpora lutea of the pig.' *Endocrinology*, **74**, 462–467.

Brodish, A. (1963). 'Diffuse hypothalamic system for the regulation of ACTH secretion.' *Endocrinology*, **73**, 727–735.

Brooks, R. V. (1962). 'Disorders of biosynthesis in man.' *Brit. Med. Bull.*, **18**, 148–153.

Brown, H., Willardson, D. G., Samuels, L. T. and Tyler, F. H. (1954). '17-Hydroxy-corticosteroid metabolism in liver disease.' *J. Clin. Invest.*, **33**, 1524–1532.

Brown-Grant, K. (1957). 'The "feedback" hypothesis of the control of thyroid function.' *Ciba Found. Colloq. Endocrinol.*, **10**, 97–116.

Brown-Grant, K. (1960). 'The hypothalamus and the thyroid gland.' *Brit. Med. Bull.*, **16**, 165–169.

Brown-Grant, K. (1963). 'Thyroid hormone metabolism in guinea-pigs, mice and rats.' *J. Physiol. (London)*, **168**, 599–612.

Brown-Grant, K. (1966a). 'The control of TSH secretion.' In Harris, G. W. and Donovan, B. T. (Eds.), *The Pituitary Gland*, Butterworths, London, chap. 7, pp. 235–269.

Brown-Grant, K. (1966b). 'The action of hormones on the hypothalamus.' *Brit. Med. Bull.*, **22**, 273–277.

Brown-Grant, K., Forchielli, E. and Dorfman, R. I. (1962). 'Reduction of 11-oxy and 11-deoxy steroids by mammalian liver Δ^4-hydrogenases.' *J. Endocrinol.*, **24**, 517–524.

Brown-Grant, K. and Gibson, J. G. (1955). 'The metabolism of exogenous and endogenous thyroid hormone in the rabbit.' *J. Physiol. (London)*, **127**, 341–351.

Brown-Grant, K., Harris, G. W. and Reichlin, S. (1954). 'The effect of emotional and physical stress on thyroid activity in the rabbit.' *J. Physiol. (London)*, **126**, 29–40.

Brown-Grant, K., Harris, G. W. and Reichlin, S. (1957). 'The effect of pituitary stalk section on thyroid function in the rabbit.' *J. Physiol. (London)*, **136**, 364–379.

Brown-Grant, K., Quinn, D. L. and Zarrow, M. X. (1964). 'Superovulation in the androgen-treated immature rat.' *Endocrinology*, **74**, 811–813.

Brown-Grant, K. and Tata, J. R. (1961). 'The distribution and metabolism of thyroxine and 3:5:3'-triiodothyronine in the rabbit.' *J. Physiol. (London)*, **157**, 157–176.

Brown-Grant, K., von Euler, C., Harris, G. W. and Reichlin, S. (1954). 'The measurement and experimental modification of thyroid activity in the rabbit.' *J. Physiol. (London)*, **126**, 1–28.

Brownie, A. C., Van der Molen, H. J., Nishizawa, E. E. and Eik-Nes, K. B. (1964). 'Determination of testosterone in human peripheral blood using gas–liquid chromatography with electron capture detection.' *J. Clin. Endocrinol. Metab.*, **24**, 1091–1102.

Burger, M. G., Kent, J. R. and Kellie, A. E. (1964). 'Determination of testosterone in human peripheral and adrenal venous plasma.' *J. Clin. Endocrinol. Metab.*, **24**, 432–441.

Cameron, A. T. (1945). *Recent Advances in Endocrinology*, 5th ed., Churchill, London.

Carlisle, D. B. and Jenkins, P. M. (1959). 'Terminology of hormones.' *Nature*, **183**, 336–337.

Chowers, I., Feldman, S. and Davidson, J. M. (1963). 'Effects of intra-hypothalamic crystalline steroids on acute ACTH secretion.' *Am. J. Physiol.*, **205**, 671–673.

Christensen, L. K. (1960a). 'Free non-protein bound serum thyroxine.' *Acta Med. Scand.*, **166**, 133–140.

Christensen, L. K. (1960b). 'Pituitary regulation of thyroid activity.' *Acta Endocrinol.*, **33**, 111–116.

Clements-Merlini, M. (1962). 'Metabolism of I^{131} by the endostyle and notochord of ammocoetes under different conditions of temperature.' *Gen. Comp. Endocrinol.*, **2**, 240–248.

Copp, D., Moghadam, H., Mensen, E. D. and McPherson, G. D. (1961). 'The parathyroids and calcium homeostasis.' In Greep, R. O. and Talmage, R. V. (Eds.), *The Parathyroids*, Thomas, Springfield, Illinois, pp. 203–219.

Critchlow, V. (1958). 'Ovulation induced by hypothalamic stimulation in the anaesthetized rat.' *Am. J. Physiol.*, **195**, 171–173.

Critchlow, V. (1963). 'The role of light in the neuroendocrine system.' In Nalbandov, A. V. (Ed.), *Advances in Neuroendocrinology*, University of Illinois Press, Urbana, chap. 12, pp. 377–402.

Cross, B. A. (1964). 'The hypothalamus in mammalian homeostasis.' *Symp. Soc. Exp. Biol.*, **18**, 157–193.

Cross, B. A. and Green, J. D. (1959). 'Activity of single neurones in the hypothalamus: effect of osmotic and other stimuli.' *J. Physiol. (London)*, **148**, 554–569.

Cross, B. A. and Silver, I. A. (1965). 'Effect of luteal hormone on the behaviour of hypothalamic neurones in pseudo-pregnant rats.' *J. Endocrinol.*, **31**, 251–263.

D'Angelo, S. A. (1955). 'Pituitary regulation of thyroid gland function.' *Brookhaven Symp. Biol.*, **7**, 9–29.

D'Angelo, S. A. (1963). 'Central nervous regulation of the secretion and release of thyroid stimulating hormone.' In Nalbandov, A. V. (Ed.), *Advances in Neuroendocrinology*, University of Illinois Press, Urbana, chap. 6, pp. 158–205.

Daughaday, W. H. (1960). 'The binding of corticosteroids by serum proteins *in vitro*.' In Antoniades, H. N. (Ed.), *Hormones in Human Plasma*, Little Brown and Co., Boston, chap. 14, pp. 495–512.

Davidson, J. M. and Feldman, S. (1962). 'Adrenocorticotrophin secretion inhibited by implantation of hydrocortisone in the hypothalamus.' *Science*, **137**, 125–126.

Davidson, J. M. and Feldman, S. (1963). 'Cerebral involvement in inhibition of ACTH secretion by hydrocortisone.' *Endocrinology*, **72**, 936–946.

Davidson, J. M. and Sawyer, C. H. (1961a). 'Evidence for an hypothalamic focus of inhibition of gonadotropin by androgen in the male.' *Proc. Soc. Exp. Biol. Med.*, **101**, 4–7.

Davidson, J. M. and Sawyer, C. H. (1961b). 'Effects of localized intracerebral implantation of oestrogen on reproductive function in the female rabbit.' *Acta Endocrinol.*, **37**, 385–393.

Davies, B. M. A. (1964). 'Blood corticotrophin in normal adults and in patients with Cushing's syndrome.' *Acta Endrocrinol.*, **45**, 55–67.

Desclin, L., Flament-Durand, J. and Gepts, W. (1962). 'Transplantation of the ovary to the spleen in rats with persistent oestrus resulting from hypothalamic lesions.' *Endocrinology*, **70**, 429–436.

Dodd, J. M. (1960). 'Gonadal and gonadotrophic hormones in lower vertebrates.' In A. S. Parkes (Ed.), *Marshall's Physiology of Reproduction*, 3rd ed., Vol. 1, Part 2, Longmans, London, chap. 11, pp. 417–583.

Donovan, B. T. and Van Der Werff ten Bosch, J. J. (1959a). 'The hypothalamus and sexual maturation in the rat.' *J. Physiol. (London).* **147**, 78–92.

Donovan, B. T. and Van Der Werff ten Bosch, J. J. (1959b). 'The relationship of the hypothalamus to oestrus in the ferret.' *J. Physiol. (London)*, **147**, 93–108.

Dowling, J. T., Razevska, D. and Goodner, C. J. (1964). 'Metabolism of thyroid hormones in frogs and toads.' *Endocrinology*, **75**, 157–166.

Engell, H. C., Winkler, K., Tygstrup, N. and Buus, O. (1961). 'Hepatic uptake of cortisol during surgical operations.' *Ann. Surg.*, **154**, 269–274.

Englert, E., Brown, H., Wallach, S. and Simons, E. L. (1957). 'Metabolism of free and conjugated 17-hydroxycorticosteroids in subjects with liver disease.' *J. Clin. Endocrinol. Metab.*, **17**, 1395–1406.

Escobar del Rey, F. and Morreale de Escobar, G. (1958). 'Studies on the peripheral disappearance of thyroid hormone.' *Acta Endocrinol.*, **29**, 176–190.

Escobar del Rey, F. and Morreale de Escobar, G. (1961). 'The effect of propylthiouracil, methylthiouracil and thiouracil on the peripheral metabolism of 1-thyroxine in thyroidectomized, 1-thyroxine maintained rats.' *Endocrinology*, **69**, 456–465.

Escobar del Rey, F., Morreale de Escobar, G., Garcia Garcia, M. D. and Mouriz Garcia, J. (1962). 'Increased secretion of thyrotrophic hormone in rats with a depressed peripheral deiodination of thyroid hormone and a normal or high plasma PBI.' *Endocrinology*, **71**, 859–869.

Estep, H. L., Island, D. P., Ney, R. L. and Liddle, G. W. (1963). 'Pituitary–adrenal dynamics during surgical stress.' *J. Clin. Endocrinol. Metab.*, **23**, 419–425.

Evans, H. M. (1924). 'The function of the anterior hypophysis.' *Harvey Lectures*, **19**, 212–235.

Everett, J. W. (1947). 'Hormonal factors responsible for deposition of cholesterol in the corpus luteum of the rat.' *Endocrinology*, **41**, 364–377.

Everett, J. W. (1948). 'Progesterone and estrogen in the experimental control of ovulation time and other features of the oestrous cycle in the rat.' *Endocrinology*, **43**, 389–405.

Everett, J. W. (1964). 'Central neural control of reproductive functions of the adenohypophysis.' *Physiol. Rev.*, **44**, 373–431.

Farer, L. S., Robbins, J., Blumberg, B. S. and Rall, J. E. (1962). 'Thyroxine–serum protein complexes in various animals.' *Endocrinology*, **70**, 686–696.

Farrell, G. and Taylor, A. N. (1962). 'Neuroendocrine aspects of blood volume regulation.' *Ann. Rev. Physiol.*, **24**, 471–490.

Flament-Durand, J. and Desclin, L. (1964). 'Observations concerning the hypothalamic control of pituitary luteotrophin secretion in the rat.' *Endocrinology*, **75**, 22–26.

Flerko, B. (1963). 'The central nervous system and the secretion and release of luteinizing hormone and follicle stimulating hormone.' In Nalbandov, A. V. (Ed.), *Advances in Neuroendocrinology*, University of Illinois Press, Urbana, chap. 7, pp. 211–224.

Florsheim, W. H. (1959). 'Influence of hypothalamus on pituitary–thyroid axis in the rat.' *Proc. Soc. Exp. Biol. Med.*, **100**, 73–75.

Florsheim, W. H., Austin, N. S. and Velcoff, S. M. (1963). 'Neuroendocrine control of thyrotropin secretion in the rat.' *Endocrinology*, **72**, 817–823.

Florsheim, W. H. and Knigge, K. M. (1961). 'Hypothalamic influences on heterotropic pituitary in the rat.' *Am. J. Physiol.*, **200**, 498–500.

Foà, P. P. (1964). 'Glucagon.' In Pincus, G., Thimann, K. V. and Astwood, E. B. (Eds.), *The Hormones*, Vol. 4, Academic Press, London, chap. 10, pp. 531–556.

Foà, P. P. and Galansino, G. (1962). *Glucagon: Chemistry and Function in Health and Disease*, Thomas, Springfield, Illinois.

Forchielli, E., Brown-Grant, K. and Dorfman, R. I. (1958). 'Steroid Δ^4-hydrogenases of rat liver.' *Proc. Soc. Exp. Biol. Med.*, **99**, 594–596.

Forchielli, E. and Dorfman, R. I. (1956). 'Separation of Δ^4-5α- and Δ^4-5β-hydrogenases from rat liver homogenates.' *J. Biol. Chem.*, **223**, 443–448.

Fortier, C. (1951). 'Dual control of adrenocorticotrophin release.' *Endocrinology*, **49**, 782–788.

Fortier, C. (1959a). 'Adenophypophysial corticotrophin, plasma free corticosteroids and adrenal weight following surgical trauma in the rat.' *Arch. Intern. Physiol.*, **67**, 333–340.

Fortier, C. (1959b). 'Pituitary ACTH and plasma free corticosteroids following bilateral adrenalectomy in the rat.' *Proc. Soc. Exp. Biol. Med.*, **100**, 13–16.

Fortier, C. (1963). 'Hypothalamic control of anterior pituitary.' In von Euler, U. S. and Heller, H. (Eds.), *Comparative Endocrinology*, Vol. 1, Academic Press, London, chap. 1, pp. 1–24.

Fortier, C. (1966). 'Nervous control of ACTH secretion.' In Harris, G. W. and Donovan, B. T. (Eds.), *The Pituitary Gland*, Butterworths, London, chap. 6, pp. 195–234.

Fortier, C., de Groot, J. and Hartfield, J. E. (1959). 'Plasma free corticosteroid response to faradic stimulation in the rat.' *Acta Endocrinol.*, **30**, 219–221.

Fortune, P. Y. (1955). 'Comparative studies of the thyroid function in teleosts of tropical and temperate habitats.' *J. Exp. Biol.*, **32**, 504–513.

Fotherby, K. (1964). 'The biochemistry of progesterone.' *Vitamins Hormones*, **22**, 153–204.

Fraps, R. M. (1962). 'Effects of external factors on the activity of the ovary.' In Zuckerman, S. (Ed.), *The Ovary*, Vol. 2, Academic Press, London, chap. 19, pp. 317–379.

Galton, V. A. and Ingbar, S. H. (1963). 'Role of peroxidase and catalase in the physiological deiodination of thyroxine.' *Endocrinology*, **73**, 596–605.

Ganong, W. F. (1963). 'The central nervous system and the synthesis and release of adrenocorticotrophic hormone.' In Nalbandov, A. V. (Ed.), *Advances in Neuroendocrinology*, University of Illinois Press, Urbana, chap. 5, pp. 92–144.

Ganong, W. F. and Forsham, P. H. (1960). 'Adenohypophysis and adrenal cortex.' *Ann. Rev. Physiol.*, **22**, 579–614.

Garren, L. D. and Lipsett, M. B. (1961). 'The effect of euthyroidal hypermetabolism on cortisol removal rates.' *J. Clin. Endocrinol. Metab.*, **21**, 1248–1253.

Gauer, O. H. and Henry, J. P. (1963). 'Circulatory basis of fluid volume control.' *Physiol. Rev.*, **43**, 423–481.

Glenister, D. W. and Yates, F. E. (1961). 'Sex difference in the rate of disappearance of corticosterone-4-C^{14} from plasma of intact rats: further evidence for the influence of hepatic $\varDelta^4$-steroid hydrogenase activity on adrenal cortical function.' *Endocrinology*, **68**, 747–758.

Goldberg, R. C. and Chaikoff, I. L. (1951). 'Failure of the dinitrophenol induced fall in plasma protein-bound iodine to stimulate augmented TSH production.' *Endocrinology*, **49**, 613–616.

Goldberg, R. C., Wolff, J. and Greep, R. O. (1957). 'Studies on the nature of the thyroid–pituitary interrelationship.' *Endocrinology*, **60**, 38–52.

Good, B. F., Hetzel, B. S. and Hogg, B. M. (1965). 'Studies of the control of thyroid function in rats: effects of salicylate and related drugs.' *Endocrinology*, **77**, 674–682.

Gorbman, A. and Bern, H. A. (1962). *A Textbook of Comparative Endocrinology*, Wiley, London.

Green, J. D. (1951). 'The comparative anatomy of the hypophysis, with special reference to its blood supply and innervation.' *Am. J. Anat.*, **88**, 225–312.

Greer, M. A., Matsuda, K. and Stott, A. K. (1966). 'Maintenance of the ability of rat pituitary homotransplants to secrete TSH by transplantation under the hypothalamic median eminence.' *Endocrinology*, **78**, 389–395.

Greer, M. A., Yamada, T. and Iino, S. (1960). 'The participation of the nervous system in the control of thyroid function.' *Ann. N.Y. Acad. Sci.*, **86**, 667–675.

Grodins, F. S. (1963). *Control Theory and Biological systems*, Columbia University Press, London.

Grossie, J. and Turner, C. W. (1962). 'Thyroxine secretion rates during food restriction in rats.' *Proc. Soc. Exp. Biol. Med.*, **110**, 631–633.

Grosvenor, C. E. (1962a). 'Thyroxine excretion in lactating rats.' *Endocrinology*, **70**, 75–78.

Grosvenor, C. E. (1962b). 'Methimazole: a goitrogen without apparent extrathyroidal effects in the rat.' *Endocrinology*, **70**, 934–936.

Grosvenor, C. E. (1963). 'Effect of propylthiouracil upon thyroidal I^{131} release in the methimazole–thyroxine treated rat.' *Endocrinology*, **73**, 122–124.

Grosvenor, C. E. and Turner, C. W. (1958). 'Effect of lactation upon thyroid secretion rate in the rat.' *Proc. Soc. Exp. Biol. Med.*, **99**, 517–519.

Guillemin, R. and Schally, A. V. (1963). 'Recent advances in the chemistry of neuroendocrine mediators originating in the central nervous system.' In Nalbandov, A. V. (Ed.), *Advances in Neuroendocrinology*, University of Illinois Press, Urbana, chap. 10, pp. 314–328.

Gurpide, E., Mann, J. and Lieberman, S. (1963). 'Analysis of open systems of multiple pools by administration of tracers at a constant rate or as a single dose as illustrated by problems involving steroid hormones.' *J. Clin. Endrocrinol. Metab.*, **23**, 1155–1176.

Hagen, A. A. and Troop, R. C. (1960). 'Influence of age, sex and adrenocortical status on hepatic reduction of cortisone *in vitro*.' *Endocrinology*, **67**, 194–203.

Harris, G. W. (1937). 'The induction of ovulation in the rabbit, by electrical stimulation of the hypothalamo-hypophysial system.' *Proc. Roy. Soc. (London), Ser. B*, **122**, 374–394.

Harris, G. W. (1948). 'Neural control of the pituitary gland.' *Physiol. Rev.*, **28**, 139–179.

Harris, G. W. (1950a). 'Oestrous rhythm, pseudopregnancy and the pituitary stalk in the rat.' *J. Physiol. (London)*, **111**, 347–360.

Harris, G. W. (1950b). 'The hypothalamus and endocrine glands.' *Brit. Med. Bull.*, **6**, 345–350.

Harris, G. W. (1955a). 'Neural control of the pituitary gland.' *Monographs of the Physiological Society*, **3**, Arnold, London.

Harris, G. W. (1955b). 'The function of the pituitary stalk.' *Johns Hopk. Hosp. Bull.*, **97**, 358–375.

Harris, G. W. (1955c). 'The reciprocal relationship between the thyroid and adreno-cortical responses to stress.' *Ciba Found. Colloq. Endocrinol.*, **8**, 531–550.

Harris, G. W. (1958). 'The central nervous system, neurohypophysis and milk ejection.' *Proc. Roy. Soc. (London), Ser. B*, **149**, 336–353.

Harris, G. W. (1959a). 'Neuroendocrine control of TSH regulation.' In Gorbman, A. (Ed.), *Comparative Endocrinology*, Chapman and Hall, London, pp. 202–222.

Harris, G. W. (1959b). 'Central control of pituitary secretion.' In *Handbook of Physiology*, Vol. 11, *Neurophysiology*, American Physiological Society, Washington, D.C., chap. 39, pp. 1007–1038.

Harris, G. W. (1962). 'The development of neuroendocrinology.' *Frontiers in Brain Research*, Columbia University Press, New York, pp. 191–241.

Harris, G. W. (1963). 'Evidence for a luteinizing-releasing factor (LRF).' In Nalbandov, A. V. (Ed.), *Advances in Neuroendocrinology*, University of Illinois Press, Urbana, chap. 7, pp. 233–236.

Harris, G. W. (1964a). 'Sex hormones, brain development and brain function.' *Endocrinology*, **75**, 627–648.

Harris, G. W. (1964b). 'The development of ideas regarding hypothalamic-releasing factors.' *Metab. Clin. Exp.*, **13**, 1171–1176.

Harris, G. W. and Jacobsohn, D. (1952). 'Functional grafts of the anterior pituitary gland.' *Proc. Roy. Soc. (London)*, *Ser. B*, **139**, 263–276.

Harris, G. W. and Michael, R. P. (1964). 'The activation of sexual behaviour by hypothalamic implants of oestrogen.' *J. Physiol. (London)*, **171**, 275–301.

Harris, G. W., Reed, M. and Fawcett, C. P. (1966). 'Hypothalamic releasing factors and the control of anterior pituitary function.' *Brit. Med. Bull.*, **22**, 266–272.

Harris, G. W. and Woods, J. W. (1958). 'The effect of electrical stimulation of the hypothalamus or pituitary gland on thyroid activity.' *J. Physiol. (London)*, **143**, 246–274.

Harrison, T. S. (1961). 'Some factors influencing thyrotropin release in the rabbit.' *Endocrinology*, **68**, 466–478.

Heller, H. (1966). 'The hormone content of the vertebrate hypothalamo-neurohypophysial system.' *Brit. Med. Bull.*, **22**, 227–231.

Heller, H. and Bentley, P. J. (1963). 'Comparative aspects of the actions of neurohypophysial hormones on water and sodium metabolism.' *Mem. Soc. Endocrinol.*, **13**, 59–65.

Herbst, A. L., Yates, F. E., Glenister, D. W. and Urquhart, J. (1960). 'Variations in hepatic inactivation of corticosterone with changes in food intake: an explanation of impaired corticosteroid metabolism following noxious stimuli.' *Endocrinology*, **67**, 222–238.

Hershman, J. M. (1964). 'Effect of 5- and 6-propylthiouracil on the metabolism of L-thyroxine in man.' *J. Clin. Endocrinol. Metab.*, **24**, 173–179.

Hershman, J. M. and Van Middlesworth, L. (1962). 'Effect of antithyroid compounds on the deiodination of thyroxine in the rat.' *Endocrinology*, **71**, 94–100.

Hodges, J. R. and Jones, M. T. (1963). 'The effect of injected corticosterone on the release of adrenocorticotrophic hormone in rats exposed to acute stress.' *J. Physiol. (London)*, **167**, 30–37.

Hodges, J. R. and Jones, M. T. (1964). 'Changes in pituitary cortico-trophic function in the adrenalectomized rat.' *J. Physiol. (London)*, **173**, 190–200.

Hogness, R., Wong, T. and Williams, R. H. (1954). 'I^{131} excretion after injection of radiothyroxine into hyperthyroid, hypothyroid or normal rats.' *Metab. Clin. Exp.*, **3**, 510–517.

Holland, R. C., Cross, B. A. and Sawyer, C. H. (1959a). 'Effects of intra-carotid injections of hypertonic solutions on the neurohypophyseal milk-ejection mechanism.' *Am. J. Physiol.*, **196**, 791–795.

Holland, R. C., Cross, B. A. and Sawyer, C. H. (1959b). 'EEG correlates of osmotic activation of the neurohypophyseal milk-ejection mechanism.' *Am. J. Physiol.*, **196**, 796–802.

Horowitz, S. and Van der Werff ten Bosch, J. J. (1962). 'Hypothalamic sexual precocity in female rats operated shortly after birth.' *Acta Endocrinol.*, **41**, 301–313.

Hoskins, R. G. (1949). 'The thyroid–pituitary apparatus as a servo (feed-back) mechanism.' *J. Clin. Endocrinol. Metab.*, **9**, 1429–1431.

Ichii, S., Forchielli, E., Perloff, W. H. and Dorfman, R. I. (1963). 'Deter-mination of plasma estrone and estradiol-17β.' *Analyt. Biochem.*, **5**, 422–425.

Ingbar, S. H. (1961). 'Clinical and physiological observations in a patient with an idiopathic decrease in the thyroxine-binding globulin of plasma.' *J. Clin. Invest.*, **40**, 2053–2063.

Ingbar, S. H. and Freinkel, N. (1960). 'Thyroid hormones.' In Antoniades, H. N. (Ed.), *Hormones in Human Plasma*, Little, Brown and Co., Boston, chap. 15, pp. 515–579.

Intoccia, A. and Van Middlesworth, L. (1959). 'Thyroxine excretion in-crease by cold exposure.' *Endocrinology*, **64**, 462–464.

Jenkin, P. M. (1962). *Animal Hormones*, Part I, Pergamon, London.

Jones, I. C. and Ball, J. N. (1962). 'Ovarian–pituitary relationships.' In Zuckerman, S. (Ed.), *The Ovary*, Vol. 1, Academic Press, London, chap. 7, pp. 361–438.

Jones, I. C., Phillips, J. G. and Bellamy, D. (1962). 'The adrenal cortex throughout the vertebrates.' *Brit. Med. Bull.*, **18**, 110–114.

Jones, S. L. and Van Middlesworth, L. (1960). 'Normal I^{131} L-thyroxine metabolism in the presence of potassium perchlorate and interrupted by propylthiouracil.' *Endocrinology*, **67**, 855–861.

Kanematsu, S. and Sawyer, C. H. (1963a). 'Effects of intrahypothalamic and intrahypophysial estrogen implants on pituitary prolactin and lactation in the rabbit.' *Endocrinology*, **72**, 243–252.

Kanematsu, S. and Sawyer, C. H. (1963b). 'Effects of hypothalamic and hypophysial estrogen implants on pituitary gonadotrophic cells in ovariectomized rabbits.' *Endocrinology*, **73**, 687–695.

Kanematsu, S. and Sawyer, C. H. (1964). 'Effects of hypothalamic and hypophysial estrogen implants on pituitary and plasma LH in ovari-ectomized rabbits.' *Endocrinology*, **75**, 579–585.

Kassenaar, A. A. H., Lameyer, L. D. F. and Querido, A. (1956). 'The effect of environmental temperature on the blood protein bound iodine content of thyroxine maintained thyroidectomized rats.' *Acta Endocrinol.*, **21**, 37–40.

Kendall, J. W. (1962). 'Studies on inhibition of corticotropin and thyrotropin release utilizing microinjections into the pituitary.' *Endocrinology*, **71**, 452–455.

Kendall, J. W., Matsuda, K., Duyck, C. and Greer, M. A. (1964). 'Studies of the receptor site for negative feedback control of ACTH release.' *Endocrinology*, **74**, 279–283.

Khazin, A. and Reichlin, S. (1961a). 'Thyroid regulatory function of intraocular pituitary transplants.' *Endocrinology*, **68**, 914–923.

Khazin, A. and Reichlin, S. (1961b). 'Thyroid regulating activity of intraocular pituitary grafts in hypophysectomized rats with massive hypothalamic lesions.' *Federation Proc.*, **20**, 186 (abstract).

Knigge, K. M. (1960a). 'Neuroendocrine mechanisms influencing ACTH and TSH secretion and their role in cold acclimation.' *Federation Proc.*, **19**, 45–51.

Knigge, K. M. (1960b). 'Time study of acute cold-induced acceleration of thyroidal I^{131} release in the hamster.' *Proc. Soc. Exp. Biol. Med.*, **104**, 368–371.

Knigge, K. M. and Bierman, S. M. (1958). 'Evidence of central nervous system influence upon cold-induced acceleration of thyroidal I^{131} release.' *Am. J. Physiol.*, **192**, 625–630.

Knigge, K. M. and Florsheim, W. M. (1960). Unpublished, quoted by Knigge, K. M. (see 1960a).

Knigge, K. M., Goodman, R. S. and Solomon, D. H. (1957). 'Role of pituitary, adrenal and kidney in several thyroid responses of cold-exposed hamsters.' *Am. J. Physiol.*, **189**, 415–419.

Kollross, J. J. (1942). 'Localized maturation of lid-closure reflex mechanism by thyroid implants into tadpole hindbrain.' *Proc. Soc. Exp. Biol. Med.*, **49**, 204–206.

Knobil,- E. and Hotchkiss, J. (1964). 'Growth hormone.' *Ann. Rev. Physiol.*, **26**, 47–74.

Lang, S. and Reichlin, S. (1961). 'Time course of thyroid response to rising and to falling blood levels of thyroxine.' *Proc. Soc. Exp. Biol. Med.*, **108**, 789–791.

Lauson, M. D. (1960). 'Vasopressin and oxytocin in the plasma of man and other mammals.' In Antoniades, H. N. (Ed.), *Hormones in Human Plasma*, Little, Brown and Co., Boston, chap. 9, pp. 225–293.

Lee, N. D., Henry, R. J. and Golub, O. J. (1964). 'Determination of the free thyroxine content of serum.' *J. Clin. Endocrinol. Metab.*, **24**, 486–495.

Leeman, S. E., Glenister, D. W. and Yates, F. E. (1962). 'Characterization of a calf hypothalamic extract with adrenocorticotropin-releasing properties: evidence for a central nervous system site for corticosteroid inhibition of adrenocorticotropin release.' *Endocrinology*, **70**, 249–262.

Leloup, J. and Fontaine, M. (1960). 'Iodine metabolism in lower vertebrates.' *Ann. N.Y. Acad. Sci.*, **86**, 316–353.

Lindner, H. R. (1961). 'Androgens and related compounds in the spermatic vein blood of domestic animals. 1. Neutral steroids secreted by the bull testis.' *J. Endocrinol.*, **23**, 139–159.

Lindner, H. R. (1964). 'Comparative aspects of cortisol transport: lack of firm binding to plasma proteins in domestic ruminants.' *J. Endocrinol.*, **28**, 301–320.

Lisk, R. D. (1960). 'Estrogen-sensitive centers in the hypothalamus of the rat.' *J. Exp. Zool.*, **145**, 197–208.

Lisk, R. D. (1962). 'Testosterone-sensitive centers in the hypothalamus of the rat.' *Acta Endocrinol.*, **41**, 195–204.

Lisk, R. D. (1963). 'Maintenance of normal pituitary weight and cytology in the spayed rat following estradiol implants in the arcuate nucleus.' *Anat. Record*, **146**, 281–291.

Littlejohn, B. M. and De Groot, J. (1963). 'Estrogen-sensitive areas in the rat brain.' *Federation Proc.*, **22**, 571 (abstract).

Maclagan, N. F. and Wilkinson, J. H. (1954). 'Some differences in the metabolism of thyroxine and triiodothyronine in the rat.' *J. Physiol. (London)*, **125**, 405–415.

Marshall, F. H. A. (1910). *The Physiology of Reproduction*, 1st ed., Longmans, London.

Marshall, F. H. A. (1922). *The Physiology of Reproduction*, 2nd ed., Longmans, London.

Marshall, F. H. A. (1936). 'Sexual periodicity and the causes which determine it.' *Phil. Trans. Roy. Soc. London, Ser. B*, **226**, 423–456.

Marshall, F. H. A. (1942). 'Exteroceptive factors in sexual periodicity.' *Biol. Rev. Cambridge Phil. Soc.*, **17**, 68–89.

McCormack, C. E. and Meyer, R. K. (1962). 'Ovulating hormone release in gonadotrophin treated immature rats.' *Proc. Soc. Exp. Biol. Med.*, **110**, 343–346.

McCormak, C. E. and Meyer, R. K. (1964). 'Minimal age for induction of ovulation with progesterone in rats: evidence for neural control.' *Endocrinology*, **74**, 793–799.

McGuire, J. S. and Tomkins, G. M. (1959). 'The effects of thyroxin administration on the enzymic reduction of Δ^4-3-ketosteroids.' *J. Biol. Chem.*, **234**, 791–794.

McKenzie, J. M. (1960). 'Bioassay of thyrotropin in man.' *Physiol. Rev.*, **40**, 398–414.

McKenzie, J. M. (1961). 'Studies on the thyroid activator of hyperthyroidism.' *J. Clin. Endocrinol. Metab.*, **21**, 635–647.

McKenzie, J. M. (1965). 'Review: pathogenesis of Graves' disease: role of the long-acting thyroid stimulator.' *J. Clin. Endocrin.*, **25**, 424–431.

Meites, J. Nicoll, C. S. and Talwalker, P. K. (1963). 'The central nervous system and the secretion and release of prolactin.' In Nalbandov, A. (Ed.), *Advances in Neuroendocrinology*, University of Illinois Press, Urbana, chap. 8, pp. 238–277.

Migeon, C. J. (1960). 'Androgens in human plasma.' In Antoniades, H. N. (Ed.), *Hormones in Human Plasma*, Little, Brown and Co., Boston, chap. 10, pp. 297–332.

Mills, I. H. (1962). 'Transport and metabolism of steroids.' *Brit. Med. Bull.*, **18**, 127–133.

Morreale de Escobar, G. and Escobar del Rey, F. (1961a). 'The effect of 2, 4-dinitrophenol on the tissue concentration of iodine-containing compounds in isotopically equilibrated intact rats.' *J. Physiol. (London)*, **159**, 1–14.

Morreale de Escobar, G. and Escobar del Rey, F. (1961b). 'The effect of 2,4-dinitrophenol on the 'uptake' of labelled thyroid hormones by red blood cells and rat diaphragms.' *J. Physiol. (London)*, **159**, 15–25.

Morreale de Escobar, G. and Escobar del Rey, F. (1962). 'Influence of thiourea, potassium perchlorate and thiocyanate and of graded doses of propylthiouracil on thyroid hormone metabolism in thyroidecto-mized rats, isotopically equilibrated with varying doses of exogenous hormone.' *Endocrinology*, **71**, 906–913.

Morreale de Escobar, G., Rodriguez, P. L., Jolin, T. and Escobar del Rey, F. (1963). 'Activation of the flavin photodeiodination of thyroxine by "thyroxine deiodinase" and other proteins.' *J. Biol. Chem.*, **238**, 3508–3515.

Munson, P. L. (1963). 'Pharmacology of neuroendocrine blocking agents.' In Nalbandov, A. V. (Ed.), *Advances in Neuroendocrinology*, University of Illinois Press, Urbana, chap. 13, pp. 427–444.

Munson, P. L., Hirsch, P. F. and Tashjian, A. H. (1963). 'Parathyroid gland.' *Ann. Rev. Physiol.*, **25**, 325–360.

Myant, N. B. (1956). 'Biliary excretion of thyroxine in humans.' *Clin. Sci.*, **15**, 227–237.

Myant, N. B. (1957). 'Relation between the biliary clearance rate of thyroxine and the binding of thyroxine by the serum proteins.' *J. Physiol.* (London), **135**, 426–441.

Naets, J. P. (1963). 'The role of the kidney in erythropoiesis in the dog.' *Mem. Soc. Endocrinol.*, **13**, 175–186.

Nalbandov, A. V. (1963), Ed., *Advances in Neuroendocrinology*, University of Illinois Press, Urbana.

Nelson, D. H., Meakin, J. W. and Thorn, G. W. (1960). 'ACTH-producing pituitary tumours following adrenalectomy for Cushing's syndrome.' *Ann. Internal. Med.*, **52**, 560–569.

Ney, R. L., Shimizu, N., Nicholson, W. E., Island, D. P. and Liddle, G. W. (1963). 'Correlation of plasma ACTH concentration with andreno-cortical response in normal human subjects, surgical patients, and patients with Cushing's disease.' *J. Clin. Invest.*, **42**, 1669–1677.

Nicoloff, J. T., Dowling, J. T. and Patton, D. D. (1964). 'Inheritance of decreased thyroxine binding by the thyroxine-binding globulin.' *J. Clin. Endocrin. Metab.*, **24**, 294–298.

Nikitovitch-Winer, M. B. (1962). 'Induction of ovulation in rats by direct intrapituitary infusion of median eminence extracts.' *Endocrinology*, **70**, 350–358.

Oppenheimer, J. H., Fisher, L. V., Nelson, K. M. and Jailer, J. W. (1961). 'Depression of the serum protein bound iodine level by diphenyl-hydantoin.' *J. Clin. Endocrinol. Metab.*, **21**, 252–262.

Oppenheimer, J. H. and Surks, M. I. (1964). 'Determination of free thyroxine in human serum: a theoretical and experimental analysis.' *J. Clin. Endocrinol. Metab.*, **24**, 785–793.

Oppenheimer, J. H., Squef, R., Surks, M. I. and Hauer, H. (1963). 'Binding of thyroxine by serum proteins evaluated by equilibrium dialysis and electrophoretic techniques. Alterations in non-thyroidal illness.' *J. Clin. Invest.*, **42**, 1769–1782.

Oppenheimer, J. H. and Tavernetti, R. R. (1962). 'Studies on the thyroxine–diphenylhydantoin interaction: effect of 5,5'-diphenylhydantoin on the displacement of L-thyroxine from thyroxine-binding globulin (TBG).' *Endocrinology*, **71**, 496–504.

Osorio, C. (1962). 'Effect of salicylate and dinitrophenol on the binding of thyroid hormones by human and rat serum proteins at pH 7·4.' *J. Physiol. (London)*, **163**, 151–159.

Osorio, C. and Myant, N. B. (1963). 'Effect of salicylate on the biliary excretion of thyroxine in rats.' *Endocrinology*, **72**, 253–258.

Parkes, A. S. and Bruce, H. M. (1961). 'Olfactory stimuli in mammalian reproduction.' *Science*, **134**, 1049–1054.

Parlow, A. F. (1964). 'Differential action of small doses of estradiol on gonadotrophins in the rat.' *Endocrinology*, **75**, 1–8.

Pincus, G. (1956). 'Some effects of progesterone and related compounds upon reproduction and early development in mammals.' *Acta Endocrinol. Suppl. XXVIII*, 18–36.

Pincus, G., Thimann, K. V. and Astwood, E. B. (Eds.) (1964). *The Hormones*, Vol. 4, Academic Press, London.

Pitt-Rivers, R. (1950). 'Mode of action of antithyroid compounds.' *Physiol. Rev.*, **30**, 194–205.

Pitt-Rivers, R. and Tata, J. R. (1959). *The Thyroid Hormones*, Pergamon, London.

Plager, J. E., Knopp, R., Slaunwhite, W. R. and Sandberg, A. A. (1963). 'Cortisol binding by dog plasma.' *Endocrinology*, **73**, 353–358.

Plager, J. E., Schmidt, K. G. and Staubitz, W. J. (1964). 'Increased unbound cortisol in the plasma of estrogen-treated subjects.' *J. Clin. Invest.*, **43**, 1066–1072.

Purves, M. D. (1960). 'On the mechanism of hypothalamic control of thyrotrophin secretion.' *Proceedings of 1st International Congress of Endocrinology*, pp. 21–24.

Raisman, G. (1966). 'Neural connexions of the hypothalamus.' *Brit. Med. Bull.*, **22**, 197–201.

Ramirez, V. D., Abrams, R. M. and McCann, S. M. (1963). 'Effect of estrogen implants in the hypothalamo-hypophysial region on the secretion of LH in the rat.' *Federation Proc.*, **22**, 506 (abstract).

Ramirez, V. D., Abrams, R. M. and McCann, S. M. (1964). 'Effect of estradiol implants in the hypothalamo-hypophysial region of the rat on the secretion of luteinizing hormone.' *Endocrinology*, **75**, 243–248.

Ramirez, D. V. and McCann, S. M. (1963a). 'Comparison of the regulation of luteinizing hormone (LH) secretion in immature and adult rats.' *Endocrinology*, **72**, 452–464.

Ramirez, V. D. and McCann, S. M. (1963b). 'A highly sensitive test for LH-releasing activity: the ovariectomized, estrogen progesterone-blocked rat.' *Endocrinology*, **73**, 193–198.

Ramirez, V. D. and McCann, S. M. (1964a). 'Fluctuations in plasma luteinizing hormone concentrations during the estrous cycle of the rat.' *Endocrinology*, **74**, 814–816.

Ramirez, V. D. and McCann, S. M. (1964b). 'Induction of prolactin secretion by implants of estrogen into the hypothalamo–hypophysial region of female rats.' *Endocrinology*, **75**, 206–214.

Reichlin, S. (1960). 'Thyroid response to partial thyroidectary, thyroxine and 2,4-dinitrophenol in rats with hypothalamic lesions.' *Endocrinology*, **66**, 327–339.

Reichlin, S. (1964). 'Function of the hypothalamus in regulation of pituitary–thyroid activity.' In Cameron, M. P. and O'Connor, M. (Eds.), *Brain–Thyroid Relationships, Ciba Found. Study Group*, No. 18, Churchill, London, pp. 17–32.

Riggs, D. S. (1952). 'Quantitative aspects of iodine metabolism in man.' *Pharmacol. Rev.*, **4**, 284–370.

Robbins, J. and Rall, J. E. (1960). 'Proteins associated with the thyroid hormones.' *Physiol. Rev.*, **40**, 415–489.

Rolleston, H. D. (1936). *The Endocrine Organs in Health and Disease, with an Historical Review*, 1st ed., Oxford University Press, London.

Rose, S. and Nelson, J. (1956). 'Hydrocortisone and ACTH release. *Australian J. Exp. Biol. Med. Sci.*, **34**, 77–80.

Roth, J., Glick, S. M., Yalow, R. S. and Berson, S. A. (1963). 'Hypoglycemia; a potent stimulus to secretion of growth hormone.' *Science*, **140**, 987–988.

Rothchild, I. (1965). 'Interrelations between progesterone and the ovary, pituitary, and central nervous system in the control of ovulation and the regulation of progesterone secretion.' *Vitamins Hormones*, **23**, 209–327.

Salaman, D. F. (1964). 'Thyrotrophic hormone in the plasma and anterior pituitary of the thyroidectomized rat.' *J. Endocrinol.*, **29**, 283–291.

Sandberg, A. A. (1960). 'The influence of extra-adrenal factors on blood corticosteroids.' In Antoniades, H. N. (Ed.), *Hormones in Human plasma*, Little, Brown and Co., Boston, pp. 363–398.

Sandberg, A. A., Eik-Nes, K., Samuels, L. T. and Tyler, F. H. (1954). 'The effects of surgery on the blood levels and metabolism of 17-hydroxy-corticosteroids in man.' *J. Clin. Invest.*, **33**, 1509–1516.

Sawyer, C. H. (1959). 'Nervous control of ovulation.' In Lloyd, C. W. (Ed.), *Recent Advances in the Endocrinology of Reproduction*, Academic Press, London, pp. 1–20.

Sawyer, C. H., Everett, J. W. and Markee, J. E. (1949). 'A neural factor in the mechanism by which estrogen induces the release of luteinizing hormone in the rat.' *Endocrinology*, **44**, 218–233.

Sawyer, C. H. and Markee, J. E. (1959). 'Estrogen facilitation of release of pituitary ovulating hormone in the rabbit in response to vaginal stimulation.' *Endocrinology*, **65**, 614–621.

Sawyer, W. H. (1963). 'Neurohypophyseal secretions and their origin.' In Nalbandov, A. C. (Ed.), *Advances in Neuroendocrinology*, University of Illinois Press, Urbana, chap. 4, pp. 68–80.

Sayers, G. (1950). 'The adrenal cortex and homeostasis.' *Physiol. Rev.*, **30**, 241–320.

Sayers, G., Redgate, E. S. and Royce, P. C. (1958). 'Hypothalamus, adenohypophysis and adrenal cortex.' *Ann. Rev. Physiol.*, **20**, 243–274.

Schreiber, V. (1963). *The Hypothalamo-hypophysial System*, Czechoslovak Academy of Sciences, Prague.

Schuetz, A. W. and Meyer, R. K. (1963). 'Effect of early postnatal steroid treatment on ovarian function in prepuberal rats.' *Proc. Soc. Exp. Biol. Med.*, **112**, 875–880.

Schwartz, N. B. and Bartosik, D. (1962). 'Changes in pituitary LH content during the rat estrous cycle.' *Endocrinology*, **71**, 756–762.

Selye, H. (1950). *The Physiology and Pathology of Exposure to Stress*, 1st ed., Acta Inc., Montreal.

Selye, H. and Rosch, P. J. (1954). 'Integration of endocrinology.' In *Glandular Physiology and Therapy*, 5th ed., Lippincott, London, chap. 1, pp. 1–10.

Shellabarger, C. J., Gorbman, A., Schatzlein, F. C. and McGill, D. (1956). 'Some quantitative and qualitative aspects of I^{131} metabolism in turtles.' *Endocrinology*, **59**, 331–339.

Shipley, E. G. (1962). 'Anti-gonadotrophic steroids, inhibition of ovulation and mating.' In Dorfman, R. I. (Ed.), *Methods in Hormone Research*, Vol. 2, *Bioassay*, Academic Press, London, chap. 5, pp. 179–274.

Smelik, P. G. (1963). 'Relation between blood level of corticoids and their inhibiting effect on the hypophyseal stress response.' *Proc. Soc. Exp. Biol. Med.*, **113**, 616–619.

Smelik, P. G. and Sawyer, C. H. (1962). 'Effects of implantation of cortisol into the brain stem or pituitary gland on the adrenal response to stress in the rabbit.' *Acta Endocrinol.*, **41**, 561–570.

Smith, P. E. (1916). 'Experimental ablation of the hypophysis in the frog embryo.' *Science*, **44**, 280–282.

Smith, P. E. (1927). 'The disabilities caused by hypophysectomy and their repair. The tuberal (hypothalamic) syndrome in the rat.' *J. Am. Med. Assoc.*, **88**, 158–161.

Smith, P. E. (1963). 'Postponed pituitary homotransplants into the region of the hypophysial portal circulation in hypophysectomized female rats.' *Endocrinology*, **73**, 793–806.

Stanbury, J. B. (1963). 'The metabolic errors in certain types of familial goiter.' *Recent Progr. Hormone Res.*, **19**, 547–572.

Stanbury, J. B., Brownell, G. L., Riggs, D. S., Perinetti, H., Itoiz, J. and del Castillo, E. B. (1954). *Endemic Goiter. The Adaption of Man to Iodine Deficiency*, Harvard University Press, Cambridge (Mass.).

Sterling, K. and Chodos, R. B. (1956). 'Radiothyroxine turnover studies in myxedema, thyrotoxicosis, and hypermetabolism without endocrine disease.' *J. Clin. Invest.*, **35**, 806–813.

Stockham, M. A. (1964). 'Changes of plasma and adrenal corticosterone levels in the rat after repeated stimuli.' *J. Physiol. (London)*, **173**, 149–159.

Strauss, W. F. and Meyer, R. K. (1962a). 'Neural timing of ovulation in immature rats treated with gonadotrophin: effect of light.' *Am. Zool.*, **2**, 563 (abstract).

Strauss, W. F. and Meyer, R. K. (1962b). 'Neural timing of ovulation in immature rats treated with gonadotrophin.' *Science*, **137**, 860–861.

Stuart-Mason, A. (1960). *Health and Hormones*, Penguin Books, Middlesex.

Suda, I., Koizumi, K. and Brooks, C. McC. (1963). 'Study of unitary activity in the supraoptic nucleus of the hypothalamus.' *Japan. J. Physiol.*, **13**, 374–385.

Swanson, H. E. and Van der Werff ten Bosch, J. J. (1964). 'The "early-androgen" syndrome; its development and the response to hemi-spaying.' *Acta Endocrinol.*, **45**, 1–12.

Szentágothai, J., Flerkó, B., Mess, B. and Halász, B. (1962). *Hypothalamic Control of the Anterior Pituitary*. Akadámiai Kiadó, Budapest, p. 218.

Tait, J. F. (1963). 'The use of isotopic steroids for the measurement of production rates *in vivo.' J. Clin. Endocrinol. Metab.*, **23**, 1285–1297.

Tait, J. F. and Burstein, S. (1964). '*In vivo* studies of steroid dynamics in Man.' In Pincus, G., Thimann, K. V. and Astwood, E. B. (Eds.), *The Hormones*, Vol. 5, Academic Press, London, chap. 4, pp. 441–557.

Taleisnik, S. and McCann, S. M. (1961). 'Effects of hypothalamic lesions on the secretion and storage of hypophysial luteinizing hormone.' *Endocrinology*, **68**, 263–272.

Talmage, R. V., Doty, S. B. and Yates, C. W. (1962). 'The effect of temperature on the uptake of radioiodine by the thyroid gland of the frog, *Rana pipiens.' Gen. Comp. Endocrinol.*, **2**, 266–272.

Tata, J. R. (1960). 'Transport of thyroid hormones.' *Brit. Med. Bull.*, **16**, 142–147.

Tata, J. R. (1961). 'The physiological significance of thyroxine dehalogenase.' *Acta Endocrinol.*, **37**, 125–134.

Tata, J. R. (1964). 'Distribution and metabolism of thyroid hormones.' In Pitt-Rivers, R. and Trotter, W. R. (Eds.), *The Thyroid Gland*, Vol. 1, Butterworths, London, chap. 8, pp. 163–186.

Trotter, W. R. (1962). *Diseases of the Thyroid*, Blackwell, Oxford, chap. 3, pp. 27–38.

Urquhart, J., Yates, F. E. and Herbst, A. L. (1959). 'Hepatic regulation of adrenal cortical function.' *Endocrinology*, **64**, 816–830.

Van Arsdel, P. P. and Williams, R. H. (1956). 'Effect of propylthiouracil on degradation of I^{131}-labeled thyroxine and triiodothyronine.' *Am. J. Physiol.*, **186**, 440–444.

Van Beugen, L. and Van der Werff ten Bosch, J. J. (1961). 'Effects of hypothalamic lesions and of cold on thyroid activity in the rat.' *Acta Endocrinol.*, **38**, 585–597.

Van der Werff ten Bosch, J. J. and Swanson, H. E. (1963). 'The hypothalamus and propylthiouracil-induced goitre in the rat.' *Acta Endocrinol.*, **42**, 254–262.

Van der Werff ten Bosch, J. J., Van Rees, G. P. and Wolthuis, O. L. (1962). 'Prolonged vaginal oestrus and the normal oestrous cycle in the rat. 2. ICSH in serum and pituitary gland.' *Acta Endocrinol.*, **40**, 103–110.

Van Middlesworth, L. (1960). 'Reevaluation of certain aspects of iodine metabolism.' *Recent Progr. Hormone Res.*, **16**, 405–438.

Van Middlesworth, L., Jagiello, G. and Vanderlaan, W. P. (1959). 'Observations on the production of goiter in rats with propylthiouracil and on goiter prevention.' *Endocrinology*, **64**, 186–190.

Van Middlesworth, L. and Jones, S. L. (1961). 'Interference with deiodination of some thyroxine analogues in the rat.' *Endocrinology*, **69**, 1085–1087.

Van Rees, G. P. and Wolthuis, O. L. (1962). 'Influences of testosterone, progesterone and oestradiol on the FSH-release of hypophyses grafted under the kidney capsule.' *Acta Endocrinol.*, **39**, 103–109.

Verney, E. B. (1947). 'The antidiuretic hormone and the factors which determine its release.' *Proc. Roy. Soc. (London)*, Ser. B, **135**, 25–105.

Villee, C. A. (1961). *Control of Ovulation*, Pergamon, London.

Vincent, S. (1922). *Internal Secretion and the Ductless Glands*, 2nd ed., Arnold, London.

von Euler, C. and Holmgren, B. (1956a). 'The thyroxine "receptor" of the thyroid–pituitary system.' *J. Physiol. (London)*, **131**, 125–136.

von Euler, C. and Holmgren, B. (1956b). 'The role of hypothalamo–hypophysial connexions in thyroid secretion.' *J. Physiol. (London)*, **131**, 137–146.

von Euler, U. S. (1956). *Noradrenaline*, Thomas, Springfield, Illinois.

Wagner, J. W. and Brown-Grant, K. (1965). 'Studies on the time of luteinizing hormone release in gonadotrophin-treated immature rats.' *Endocrinology*, **76**, 958–965.

Werner, S. C., Hamilton, H. and Nemeth, M. (1952). 'Graves' disease: hyperthyroidism or hyperpituitarism?' *J. Clin. Endocrinol. Metab.*, **12**, 1561–1571.

Williams, W. C., Island, D., Oldfield, R. A. A. and Liddle, G. W. (1961). 'Blood corticotropin (ACTH) levels in Cushing's disease.' *J. Clin. Endocrinol. Metab.*, **21**, 426–423.

Woeber, K. A. and Ingbar, S. H. (1963). 'Effect of salicylate on the binding of thyroxine by human serum proteins at pH 7·4.' *Endocrinology*, **73**, 118–119.

Woeber, K. A. and Ingbar, S. H. (1964). 'The effects of non-calorigenic congeners of salicylate on the peripheral metabolism of thyroxine.' *J. Clin. Invest.*, **43**, 931–942.

Wolff, J., Rubin, L. and Chaikoff, I. L. (1950). 'The influence of 2,4-dinitrophenol on plasma protein-bound iodine.' *J. Pharmacol. Exp. Therap.*, **98**, 45–48.

Wolff, J., Standaert, M. E. and Rall, J. E. (1961). 'Thyroxine displacement from serum proteins and depression of serum protein-bound iodine by certain drugs.' *J. Clin. Invest.*, **40**, 1373–1379.

Woods, J. and Bard, P. (1959). 'Thyroid activity in the chronic decerebrate cat with an isolated "island" of hypothalamus and pituitary.' *Federation Proc.*, **18**, 173 (abstract).

Yamada, T. (1959). 'Studies on the mechanism of hypothalamic control of thyrotropin secretion: comparison of the sensitivity of the hypothalamus and of the pituitary to local changes of thyroid hormone concentration.' *Endocrinology*, **65**, 920–925.

Yamada, T. and Greer, M. A. (1959). 'Studies on the mechanism of hypothalamic control of thyrotropin secretion: effect of thyroxine injection into the hypothalamus or the pituitary on thyroid hormone release.' *Endocrinology*, **64**, 559–566.

Yates, F. E., Herbst, A. L. and Urquhart, J. (1958). 'Sex difference in rate of ring A reduction of Δ^4-3-keto-steroids *in vitro* by rat liver.' *Endocrinology*, **63**, 887–902.

Yates, F. E., Leeman, S. E., Glenister, D. W. and Dallman, M. F. (1961). 'Interaction between plasma corticosterone concentration and adrenocorticotropin-releasing stimuli in the rat: evidence for the reset of an endocrine feedback control.' *Endocrinology*, **69**, 67–80.

Yates, F. E. and Urquhart, J. (1962). 'Control of plasma concentrations of adrenocortical hormones.' *Physiol. Rev.*, **42**, 359–443.

Yates, F. E., Urquhart, J. and Herbst, A. L. (1958). 'Effects of thyroid hormones on ring A reduction of cortisone by liver.' *Am. J. Physiol.*, **195**, 373–380.

Young, F. G. (1963). 'Pancreatic hormones: insulin.' In von Euler, U. S. and Heller, H. (Eds.), *Comparative Endocrinology*, Vol. 1, Academic Press, London, chap. 10, pp. 371–411.

Zander, J. (1959). 'Gestagens in human pregnancy.' In Lloyd, C. W. (Ed.), *Recent Progress in the Endocrinology of Reproduction*, Academic Press, New York, pp. 255–277.

Zarrow, M. and Brown-Grant, K. (1964). 'Inhibition of ovulation in the gonadotrophin-treated immature rat by chlorpromazine.' *J. Endocrinol.*, **30**, 87–95.

Zarrow, M. X. and Quinn, D. L. (1963). 'Superovulation in the immature rat following treatment with PMS alone and inhibition of PMS-induced ovulation.' *J. Endocrinol.*, **26**, 181–188.

Zeilmaker, G. H. (1963). 'Experimental studies on the regulation of corpus luteum function in castrated male rats bearing a transplanted ovary.' *Acta Endocrinol.*, **43**, 246–254.

Zeilmaker, G. H. (1964). 'Aspects of the regulation of corpus luteum function in androgen-sterilized female rats.' *Acta endocrinol.*, **46**, 571–579.

Zuckerman, S. (1962) (Ed.), *The Ovary*, Vols. 1 and 2, Academic Press, London.

III
DEVELOPMENT and GENETICS

9

Early Studies of Biological Regulation: An Historical Survey

J. S. WILKIE

Department of the History and Philosophy of Science,
University College, London

For the sake of convenience it is possible to classify the regulative processes of organisms thus:

1. Those which adjust the behaviour of the *whole organism* to particular environmental circumstances or to objects in the environment. Such processes are brought into play in tracking and in avoiding obstacles, to give two elementary examples.

2. Those which adjust the organism to more general environmental conditions; for example, processes controlling the temperature of the organism or its posture. These differ from the first class principally in the relative simplicity of their evoking stimulus, and in the relatively stereotyped relation of the nervous and endocrine systems to the other systems principally concerned in the response.

It is possible to think of processes of this second type as occurring at the level of organs or organ-systems, because the external stimuli being simple, the often complex mechanisms are concerned principally with the relations of organs or of organ-systems to one another.

3. Those which concern almost exclusively the relations of parts of the organism to one another, and which have little direct dependence upon environmental circumstances. To this type belong regeneration of parts in the adult and readjustment of the developing embryo after accidental or experimentally inflicted traumata.

Although in this case whole organs, such as limbs, may be affected, it is tempting to consider such regulative processes as taking place at the cellular level, and as primarily directed by cells or populations of cells.

This Chapter aims at presenting a brief outline of the history of studies of regulative processes of the third class.

What makes regulative processes of this class so interesting is that they seem to belong to organized materials as such. It is possible to think of them as abilities of living matter that may be lost or much diminished in the course of evolution (as seems to be the case with regeneration of limbs), but which it is difficult to think of as ever having been acquired.

It is easy, or at least possible, to think of the first two classes of regulative processes as having been acquired phylogenetically by the natural selection of chance variations; but it is extremely difficult to believe that the ability of snails to regenerate their heads, however convenient this may be to animals having their structure and habits, was in the first instance acquired as a consequence of natural selection of small variations, though it doubtless may have been favoured by this evolutionary process. It is even more difficult to believe that natural selection has somehow provided embryos with abilities designed to compensate for the interventions of experimental embryologists, many of which interventions are excessively unlikely to be paralleled in any natural environment.

If it is true that regulative processes of the third kind are accidental concomitants of the organization of matter into cells, just as the ability of the two halves of a spherical droplet to round themselves off into two new spheres is an accidental concomitant of the particular system of molecules in the initial drop, then the study of such processes in organisms is of peculiar interest. For any theory of the fundamental organization of living materials must be able to account for them, and their existence may, in turn, throw some light upon the nature of this fundamental organization.

The studies which I intend to describe do no more than ascertain the extent of the problem, but I hope that some interest may attach to the stages by which our knowledge of the principal possibilities of cellular regulation has been acquired.

Regulative phenomena of the kind which concerns us here have been known since it was realized that plants can be propagated by cuttings. This fact was known to Aristotle, and both he and Theophrastus refer to it as to something long and widely known. It seems that the practice must have been common in Greece long before it excited the interest of the philosophers who recorded it. Probably it was introduced from the older civilizations of the East.

Aristotle writes:

> Some plants are formed from seed, some from slips planted out, others by sideshoots (e.g. the onion tribe). [*De Gen. Anim.* 761 b, transl. A. L. Peck.]

Theophrastus, as we should expect, gives a much fuller and more interesting account:

> However, in all the trees which have several methods of originating the quickest method and that which promotes the most vigorous growth is from a piece torn off, or still better from a sucker, if this is taken from the root. And, while all the trees which are propagated thus or by some kind of slip seem to be alike in their fruits to the original tree, those raised from fruit, where this method of growing is also possible, are nearly all inferior, while some quite lose the character of their kind; as, vine, apple, fig, pomegranate, pear. [*Hist. Plant.*, ii, 2, transl. Sir Arthur Hort.]

Aristotle well understood the theoretical importance of such phenomena. He refers to them in an interesting passage of the *De Anima* (411 b, 19–22), but the analysis of this passage would require a space which could hardly be justified in the present context. However, a briefer passage from another of his works expresses very neatly his view of what cuttings demonstrate:

> The parts of a divided plant live on, for in every part of a plant are potentially roots and shoots. [*De Vita longa et brevi*, 467 a.]

It is interesting to compare this formulation with one of similar import in the writings of Hugo de Vries (De Vries, 1889):

> Botanists have been of opinion that all, or at least the greater majority of the cells of the plant-body are equivalent in respect of their latent characters.

Hugo de Vries draws attention particularly to a paper by Beijerinck on galls produced by insects: *Cecidomyia poae*, for example, provoked well-formed roots on *Poa nemoralis*, on parts which neither in this nor in any other grass could normally produce roots.

> Thus the larvae exploit, in this case, a potentiality (*ein Vermögen*) the existence of which, without their intervention, we

should probably never have suspected, and should still less have been able to prove (*op. cit.*).

The first experimental study of regeneration of which I have been able to find any trace is that made by Thevenot and Perault in the seventeenth century.

In 1686 Thevenot exhibited before the *Academie Royale des Sciences* a green lizard the tail of which he cut off. Within a month a considerable regeneration had taken place, but when the regenerated portion was dissected it was found that, though the skin appeared to be normal, no muscles had been formed, and in the place of the vertebrae there was a hollow cartilage 'of the size of a large pin'.

A more systematic and more successful study of a case of regeneration was made by Réaumur, an account of which he published in 1714. Réaumur's experiments on the regeneration of the limbs of the crayfish (Figure 1) were extremely careful and systematic, and his account of them well deserves to be read in full.

The paper begins with a brief account of observations made on a coast where there was a very large population of crabs. Not only were there many crabs having the two claws of different sizes, but the smaller of the two claws could be found in all stages of growth, in crabs apparently full-grown and having the other claw of normal size. This fact persuaded Réaumur that the popular belief in the crab's power of regenerating its claws was probably well-founded, and he proceeded to experiment. He selected the freshwater crayfish for study, and easily demonstrated regeneration of the claw. Not content, however, with merely proving the existence of the phenomenon, he studied also the circumstances of its production.

He found that the claws regenerate more rapidly in hot than in cold weather. He also found that the claw regenerates more quickly if it has been broken off 'near the fourth joint'. His drawing shows that he means 'broken off within the segment of the limb nearest to the body', and by this is probably meant what is now called the basipodite. This, he says, is also the place at which the limb is most often broken off accidentally. If the limb is broken off by the experimenter at any point distal to this region of natural weakness, the missing part is regenerated more slowly; but this experiment is difficult to perform, for usually a limb thus broken is broken again before it can regenerate; the second break occurring at the point of weakness and being produced intentionally, so Réaumur believed, by the animal itself.

Réaumur also studied the regeneration of the other limbs. The walking legs regenerated more slowly than the claws. Both the antennae and the third maxillipedes regenerated.

The 'tail' of the animal, as he calls it, meaning the hinder part of the body, lacked this ability.

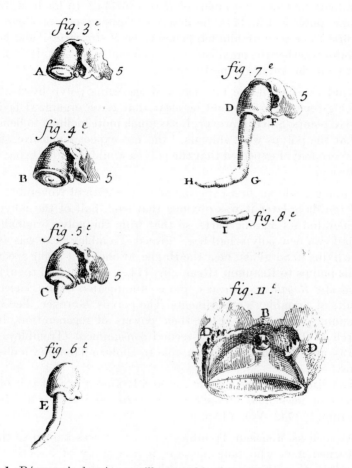

Figure 1. Réaumur's drawings to illustrate his observations on the regeneration of the limbs of the crayfish. (Réaumur, 1714.)

In spite of the excellence of Réaumur's paper, it seems to have excited little interest in regeneration, and the study of this subject made no major advances till Trembley rediscovered the organism *Hydra* in 1740, and noticed its remarkable regenerative powers. Réaumur and Bernard de Jussieu (probably in the same year) named the organism '*Polype*'.

Trembley pointed out that animals of this kind had been seen by Leeuwenhoek and mentioned by him in 1703, in an article in the *Transactions of the Royal Society* (No. 283); but it is clear that Trembley knew nothing of Leeuwenhoek's observations when he, Trembley, first saw specimens of *Hydra* in 1740. In his book on the *Polype*, published in 1744, he describes three species of *Hydra*, but the first to be seen by him happened to be *Hydra viridis*, and he was in doubt whether the organism was an animal or a plant. He was thus led to cut the animal in half:

> I judged that, if the two parts of the same polyp lived after being separated, it would be clear that these organized bodies were plants. Since, however, I was much more inclined to believe that the polyps were animals, I did not expect much from this experiment: I expected that the polyps would die after being cut through.

Trembley made his first experiment on the 25th of November 1740, and ten days later it was obvious that each half of the polyp had regenerated its missing parts, so that from the polyp originally cut in half two new polyps had been formed. Trembley, who was at that time living in Sorgvliet, near the Hague, at once sent living specimens of his polyps to Réaumur (Réaumur, 1742), who showed them in the *Academie Royale des Sciences*, and communicated to the Society the results of Trembley's experiments. The polyps, however, died before Réaumur was able to study their powers of regeneration; but in March 1741 Trembley sent a second consignment (Trembley, 1744, p. 19), and Réaumur was then able to confirm Trembley's observations.

The discovery made a great stir, and Réaumur says that it became a piece of news '*dont on s'est beaucoup entretenu à la Cour & à la ville*' (Réaumur, 1742, Vol. VI, p. li).

> As soon as Monsieur Trembley's discovery was known to those learned men who take pleasure in the study of insects, they judged that the power of multiplying in this curious manner could not have been accorded to polyps alone . . . A large number of aquatic insects was soon exposed to the danger of a cruel death under the instruments used with the intention of multiplying them. However, Monsieur Bonnet did not take long to discover a species of worm, about 15 or 16 lines in length, but extremely thin, which lives in the mud at the bottom of the water, and which can in fact be multiplied when it is cut in pieces (Réaumur, 1742, Vol. VI, p. lv).

By 1742, when this was printed, a considerable body of research had been carried out. Réaumur failed to find any power of regeneration in leeches, but was successful with animals which he calls *Sangsuës-limaces*, and which his description shows to have been free-living flatworms. He also repeated another of Trembley's observations: a small freshwater annelid, about half-an-inch in length, called by Trembley *Mille-pieds* (though clearly not related to millepedes) was cut in half, and each half produced an entire worm. Gerard de Villars, a physician of La Rochelle, cut sea anemones in two longitudinally, and observed that both halves survived, presumably with some regeneration. He also saw the first stages in the regeneration of an arm of a starfish. Both Réaumur and Bonnet experimented with earthworms. The experiments were successful, each end of the worm regenerating the other end. Réaumur knew of the position of the reproductive organs of the worm, and asserted that they were regenerated by the hinder part of the worm, when the cut had passed behind these organs in the operated animal.

Trembley sent polyps to England, and the regeneration experiment was repeated on them by Folkes (*Trans. Roy. Soc.*, 1743, No. 467).

Trembley did not publish his results till 1744, but he then produced a detailed monograph on the three species of *Hydra* he had discovered. He described the external form of the polyps, and as much as could be seen of their internal structure with the microscopes of the time; also their movements, including locomotion. He described their feeding habits and the variation of their numbers with the seasons. He saw that they reproduced by budding; but, though he probably saw their sexual organs, he did not realize their nature. The number and variety of the experiments he made is extremely impressive. As well as cutting the animals in two, both transversely and longitudinally, he cut them transversely into several parts and observed the regeneration of these. He also cut into the animal longitudinally, but without dividing it, and saw the production of two heads as a result of this operation. He further divided the heads thus produced, and, continuing the series of operations, produced the many-headed animal shown in Trembley's Fig. 11 (Figure 2). In describing these many-headed polyps as '*des Hydres*' he applied to them the name which was later to be given, less aptly, to the genus.

His most remarkable achievement was to turn polyps inside-out, and to keep them in this condition till they recovered from the operation and started to feed. (Recovery is due, I believe, to a migration of cells, effectively reversing the consequences of the operation.) In performing the operation, the hydra is first given an insect larva to eat. This greatly distends its mouth (Figure 2, Trembley's Fig. 12). The hydra is then rested on the operator's left hand, and a pig's bristle is pressed against the end furthest from the mouth (Figure 2, Trembley's Figs. 12 and 13).

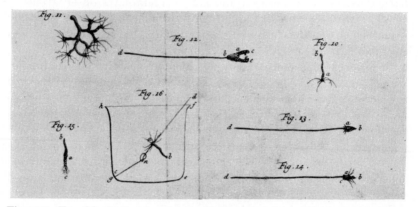

Figure 2. Trembley's drawings to illustrate various operations on his 'polype.' (Trembley, 1745.)

Trembley writes that often after the operation has been completed:

The lips turn outwards, as though the polyp wished to turn itself back again. This is, in fact, what it is trying to do, and what it often succeeds in doing...I therefore attempted to keep it turned inside-out. In some cases it was enough, as the lips turned back over the body, to push them down with the point of a fine paintbrush. However, success is achieved too rarely by this means for it to suffice, if the experiment is to be often repeated. I had recourse to a more reliable expedient. This was to *impale* the polyps which had been turned inside-out, using for this purpose a bristle from a pig or from a wild boar, and pushing it through near to the anterior end of the polyp...The knot which is near the lower end of the bristle I put there so that, if the polyp slips down by its own weight, it is stopped by the knot and does not free itself from the bristle [Figure 2, Trembley's Fig. 16]. To be impaled is no great matter for a polyp. I have tried it in various ways with polyps which were not turned inside-out, and they were not thereby prevented from eating or from reproducing themselves.

The expression 'reproducing themselves' refers, of course, to budding, since Trembley did not know of sexual reproduction in his polyps.

Trembley's monograph, it will be remembered, was published in 1744. In the next year, 1745, Bonnet published a work entitled *Traité d'Insectologie*. This consists of two small volumes, of which the first is devoted to studies of aphids and the second to records of observations on the regeneration of some species of freshwater annelids. In his studies of aphids Bonnet was able to show, by a series of most carefully controlled observations, that during any one year several generations of aphids are produced asexually. The importance of this discovery for the theory of reproduction in general has resulted in a concentration of interest upon the first volume of the *Traité d'Insectologie*, but the second volume, though the facts it records were less surprising (for they seemed only to enlarge the scope of a principle established by Trembley), is at least as worthy of attention.

Bonnet is usually judged to have been rather a theorist than a practising scientist, and this judgement is certainly correct if the whole body of his publications be taken into account; but anyone who approaches the *Traité d'Insectologie* with this in mind will be astonished by the extremely high quality of Bonnet's empirical work. The most interesting of Bonnet's results are the following.

1. In the case of the first species examined by him, the individual worm can be cut into as many as twelve parts, each of which will regenerate a perfect worm.

2. In another species, if the worm is cut across at or behind the middle of its length, the hinder part regenerates, not a front part, but another hind part; so that a worm with a tail at each end is produced. This happens regularly in this species. If, however, such worms are divided nearer the head, the hinder part regularly regenerates a front end, thus producing a perfect worm.

3. A regenerated head is nearly always in line with the body which produces it, but sometimes oblique to the line of the body. Bonnet did not, however, observe, what was probably the case, that an oblique regenerate followed an oblique division of the operated worm.

4. The speed of regeneration is in all cases very sensitive to temperature. If the initial operation of division is performed in the Summer, and the regeneration is not completed before the Winter, then regenerative growth ceases in the Winter and is resumed in the Spring.

5. Other things being equal, smaller fragments regenerate, if at all, more rapidly than larger. The reverse of this seems to have been found by more recent workers, but Bonnet's results are very circumstantial, and from his figures, given in the form of a table, interesting curves can be plotted (Figure 3).

6. In general, the nearer a fragment is to the head in the operated worm, the more rapidly it regenerates; but the most detailed of Bonnet's results, based it is true only on one worm, show that the region which grows most quickly is not the head end itself, but that which lies next to it (Figure 4).

From 1745 to 1768 there seems to have been a dirth of new work on regeneration. That this was indeed the case is implied in a letter

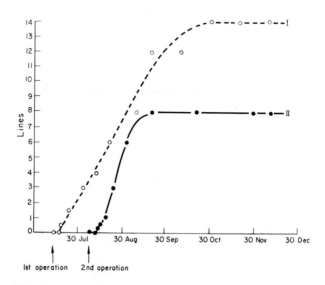

Figure 3. Curves constructed from data given by Bonnet, 1745, in tabular form.

The upper line shows the growth of the tail regenerated by the second (middle) portion of a worm cut into three pieces.

The lower line shows the growth of tail regenerated by the sixth portion (counting the head as the first) of a worm cut into twelve pieces.

The smaller portion grows absolutely more rapidly than the larger, and relatively very much more rapidly.

1 line = $\frac{1}{12}$ of an inch.

addressed by Bonnet to Spallanzani in 1765 (Rostand, 1951, p. 73). In this letter Bonnet expressed his astonishment that regeneration in the earthworm had been so long known, and that nevertheless so little work had been done on this animal. He suggested that Spallanzani should investigate regeneration in earthworms, and accordingly we find that their regeneration is the first subject treated in Spallanzani's tract of 1768.

This remarkable work was published in the form of a programme for a large book, which, however, never appeared; and we find in it both brief records of investigations actually carried out and suggestions for further experiments. Thus, for example, Spallanzani had realized the importance of ascertaining the result of dividing worms

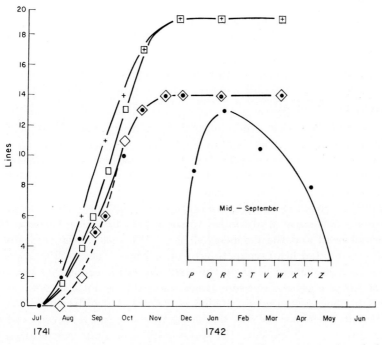

Figure 4. Curves constructed from data given by Bonnet, 1745, in tabular form. Speeds of regeneration of tail by parts of a worm cut into the ten pieces: $P, Q, R, S, T, V, W, X, Y, Z$.

Q, R and S all regenerated at the same speed.

T and V regenerated at the same speed.

W and X perished.

Taking the heights of the curves at mid-September as measures of the relative speeds of growth, we can construct the figure ('mid-September') showing a kind of axial gradient.

Symbols: P ● T, V □

        Q, R, S + Y ◇

The points are taken from Bonnet's tables. (Bonnet, 1745.)

obliquely as well as at right-angles, but he does not report any experiments actually performed to elucidate this problem. He experimented on the following animals: (i) several species of earthworm; (ii) a freshwater worm called by him *Lombrico a batello*; (iii) tadpoles; (iv) snails; (v) slugs; (vi) salamanders; (vii) frogs; (viii) toads.

Earthworms. The head is regenerated 'after a long time and with difficulty', and not in all species. The middle portions regenerate both head and tail, provided not too much of the head end has been removed. If too much head has been removed, a tail is grown in place of a head.

Lombrico a batello. This also seems to have been an annelid, for he says '*è composto d'anelli*'. Its powers of regeneration were found to be comparable with those of earthworms.

Tadpoles. The younger the tadpole the more rapid the regeneration. Though fasting tadpoles neither grow nor metamorphose, they do regenerate their tails.

Snails and slugs. Few experiments were made on slugs. Their 'horns' regenerated as easily as those of snails, but they were greatly inferior in their ability to regenerate their heads. Spallanzani lays stress on the complexity of structure of the head of the snail, referring to Swammerdam's studies, published in the *Biblia Naturae*; he says he confirmed Swammerdam's observations on the anatomy. The most careful anatomical study failed to show any difference between the normal head and the regenerated head. Whereas in the earthworm the regenerated head seems to grow as a whole from its first appearance as a miniature head, in the snail the head regenerates part by part (*pezzo a pezzo*), and sometimes some part is not reformed, even after as long as two years. Spallanzani notes that different results in different cases may be due to the amount of head removed, and to the direction of the cut. He states explicitly that regeneration is possible even if the brain (*cervello*) has been removed by the operation.

The work on snails aroused immense interest, and innumerable attempts were made to repeat it. Among many others, Voltaire and Lavoisier, independently, operated on snails to see whether Spallanzani's report was reliable (Rostand, 1951).

Salamanders. The experiments on salamanders were in one way less astonishing to Spallanzani's contemporaries, because the regeneration of the head of the snail might be supposed to concern the seat of consciousness, while in the salamanders only limbs were involved. In another respect, however, the experiments on salamanders were more impressive, because of their closer relation to man. Spallanzani writes:

> Now, when the legs and tail of salamanders are cut off, we have the production of new vertebrae and of other new bones. A phenomenon both astonishing and, as far as I am aware, unknown to natural philosophers.

The regeneration of bones impressed Spallanzani particularly, and he enumerated the bones which must be regenerated after all the legs have been cut off, making the number ninety-nine in all. (Surprisingly, for we must suppose the number to be even.)

The observations are extremely systematic. Spallanzani, like Bonnet, found that regeneration ceased in the Winter. Studies of regeneration of the tail showed that it occurs in all species of salamander, at all ages, both in water and in air, and whatever the length of the portion removed.

The speed of regeneration of the tail varies with the length removed, with the species providing that the age is constant and with the age within any one species.

These rules remain constant for the second, third and further regenerations of the same limb.

The legs also regenerate in all species. If all four are removed, the front legs regenerate more quickly than the hind legs. Fasting salamanders regenerate their legs as quickly as those which have been fed.

Certain irregularities, extremely rare in normal salamanders, are not uncommon among those with regenerated limbs : these being, too many or too few digits; excess or defect in number of joints of digits.

No limit was found to the number of times a leg or tail could regenerate, and the speed of regeneration seemed to be the same for all regenerates of the same part.

The jaws, both lower and upper, regenerate, forming 'new teeth, new bones, new cartilages, muscles, veins and arteries'.

Frogs and Toads. In frogs and in toads, of which two species seem to be distinguished (*botte e rospi*), legs regenerate in young animals but regeneration is slow and of inconstant occurrence. Spallanzani notes that the difference in facility of regeneration between salamanders and frogs and toads cannot be due to the difference of medium, for salamanders regenerate their limbs as easily in air as in water.

Much of Spallanzani's work was repeated by Bonnet, who gave excellent drawings (Bonnet, 1781, Tome V, pp. 246–358). By the end of the eighteenth century the most striking phenomena of regeneration were thus known, including the systematic regeneration, in some annelids, of a tail in place of a head. This last fact is clearly of cardinal importance, for it destroys the possibility of a crudely teleological or animistic explanation of the phenomena.

Apart from adding examples of the same kind there was little more that the next century could do to increase our knowledge of the facts of regeneration, and if the theory became less confused it can hardly be said to have become more illuminating. The nineteenth century, however, added to knowledge of the powers of regulation at the cellular level the knowledge of two sets of phenomena: the production of new adaptive structure in the system of trabeculae in bones, and regulation in embryonic development. These are in one respect more significant than the phenomena of regeneration; for in regeneration, as the name implies, nothing new is produced, whereas in the newly discovered sets of phenomena there is, in the one set, a production of new adaptive structure, while, in the other, though the normal structure is produced, it is produced by stages essentially novel. Perhaps it should be conceded that regeneration is also a production of normal structure by unusual means, but in the regulative development of embryos the fact that the process is essentially novel is a great deal more apparent than it is in the case of regeneration, if only because it is much easier to specify precisely what the normal course of development is in the earlier stages of embryonic growth.

Knowledge of the adaptive rearrangement of trabecular structure could not, of course, be attained until the normal structures had been observed and interpreted as adapted to support the stresses to which the normal bone is subjected. The first suggestion that the trabecular or cancellous tissue of bones has a structure amenable to analysis on mechanical principles seems to have been made by Bougery in 1832, and an approximately correct interpretation of the system of trabeculae in the head of the femur was given by F. O. Ward in 1838:

> The arrangement of the cancellous tissue in the ends of the femur is very remarkable; and as it illustrates the general mechanical principles which determine the structure of this tissue throughout the skeleton, it should engage our particular attention. In the lower extremity of the bone, it consists of numerous slender columns, which spring on all sides from the interior surface of the compact cylinder, and descend, converging towards each other, so as to form a series of inverted arches, adapted, by their pointed form, to sustain concussion or pressure transmitted from below. . . . The cancellous tissue in the upper extremity of the shaft presents a similar arrangement—the convexities of the arches being here, however, directed upwards.

The structure of the neck is shown at Fig. 1 in the subjoined diagram [Figure 5, 1]. It resembles in its mechanical principles a bracket of the kind represented in Fig. 2, in which *a* is the principal support, and *b* a cross piece tying *a* to the wall or column which sustains the whole. It is evident that the piece *a* contributes by its *rigidity*, and the piece *b* by its *tenacity*, to the support of the weight; in other words, that the weight tends to *bend* the former, and to *stretch* the latter. Referring to Fig. 1 (in which the direction of the principal fibres is shown with fictitious distinctness—the cross filaments, &c. being purposely omitted) we find the pieces *a* and *b* of the bracket respectively represented by the sets of bony fibres marked *aa* and *bb*. (Ward, 1838.)

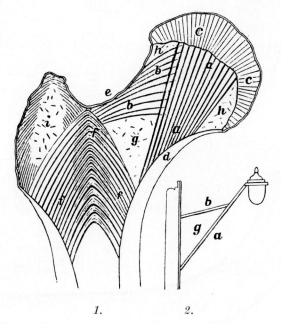

Figure 5. Ward's drawing to illustrate his mechanical interpretation of the trabecular structure of the head of the human femur. (Ward, 1838.)

Ward's suggestion was criticized in 1857 by Wyman, who improved upon it and gave interpretations of the 'cancellated' structure of some other bones. The subject was also discussed by J. Engel in 1851 and by G. M. Humphry in 1858.

A new stimulus to systematic study and interpretation of this kind was given by engineer and mathematician Culmann, at the time a professor of the *Polytechnische Schule* in Zürich. Culmann happened to be present at a meeting in Zürich at which Hermann von Meyer demonstrated preparations of bones showing the cancellous structure. He at once appreciated the mechanical significance of this structure and later collaborated with von Meyer and with Julius

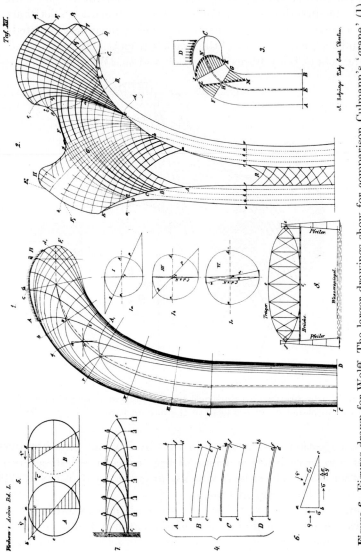

Figure 6. Figures drawn for Wolff. The larger drawings show for comparison Culmann's 'crane' (1) and Wolff's drawing of the internal structure of the head of the human femur (2). The section from which Wolff made his drawing is shown in Figure 7. (Wolff, 1870.)

Wolff in its analysis. Culmann had refined and developed the mathematical techniques required for the accurate representation of 'lines of tension and of pressure' in structures of given shapes subjected to known stresses. At the request of Julius Wolff he prepared an outline drawing of a shape schematically representing the upper part of the human femur, and he asked his students to draw in the lines of tension and of pressure, assuming the outline to be that of a 'crane' supporting a weight. The students were not told of the purpose of the drawing. The resulting figure is shown as 1 in my Figure 6, and on the right of it is seen Wolff's drawing of a slightly schematized section through a human femur (Figure 6, 2). The degree of schematization in Wolff's drawing may be judged by comparing it with the photograph (Figure 7, 1) of the actual section from which it was prepared.

In the paper of 1870 in which these drawings were published, Wolff predicted that bones, such as those of rachitic patients, which had been deformed by a disease process, and had then supported the patient's weight for many years, would show a new internal structure, and 'it will be possible to show that the static conditions have directly determined the architecture'.

During the next twenty years Wolff published a number of papers in which this prediction finds considerable confirmation, and in 1892 he published a book, from which my Figure 8 is taken. I have chosen this photograph of a section of an ankylosed joint (femur and pelvis) because it seems to me to be, among the figures given by Wolff, the most striking example of a functionally determined arrangement of trabecular structure. The great mass of 'bony fibres' sweeping down, in the lateral part of the section (Figure 8, on the right), in a curve from the pelvis to the shaft of the femur is clearly new, and one can intuitively understand its functional significance.

Wolff worked for some time in collaboration with Culmann, so that we may have every confidence in his analytic interpretations of such new structures, but I have not been able to find that he gives many such analyses, certainly none that can be quoted in a few lines. I have therefore chosen to give this photograph (Figure 8) as the most immediately persuasive.

It may perhaps be worth recording that Wolff's sections were cut with a saw designed for the cutting of very thin plates of ivory for ladies' fans. There is an example of such a fan, made entirely of ivory, in the Victoria and Albert Museum.

Wolff's book, *Das Gesetz der Transformation der Knochen*, which appeared in 1892, seems to have been projected at least as early as

1885. It was intended that it should contain a section written by
Wilhelm Roux, describing and analysing the trabecular structure of
an ankylosed knee-joint. Roux, however, found that he could not do

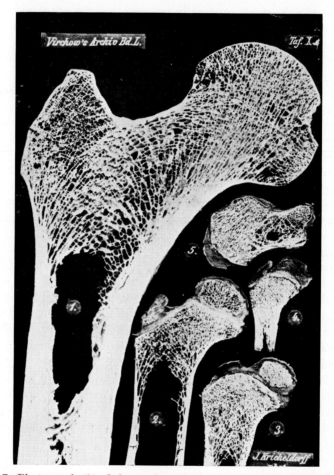

Figure 7. Photograph (1) of the section of a human femur from which Wolff
prepared the drawing of which Figure 6, 2 is a copy made by a lithographer.
The photographs 2, 3 and 4 are stages in the growth of the human femur.

justice to the subject within the limits which would have been imposed by the intended plan of the book. He therefore devoted to the
subject a long paper, which he was able to produce in 1885 (Roux,
1885a), though without photographs, these being given in Wolff's
book of 1892.

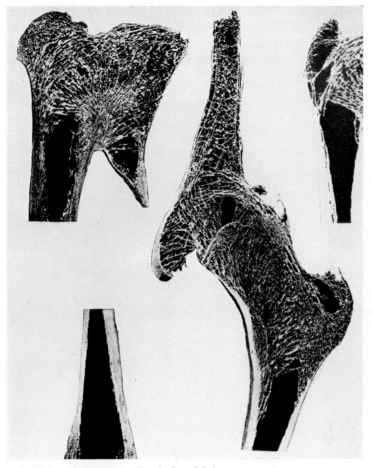

Figure 8. Internal structure of ankylosed joints.

On the right, a section through an ankylosed hip-joint, showing an internal structure reorganised so as to take the strains imposed by long use of the ankylosed joint, and which could no longer be taken by the muscles.

On the left, another ankylosed hip-joint showing a new internal structure, but one more difficult to interpret. (Wolff, 1892.)

The subject was a dried specimen from the collection of pathological anatomy at Würzburg, and Roux cut this specimen into ten sections parallel to the sagittal plane. I have given here the photograph of one of Roux's sections, together with his interpretative schematization (Figures 9 and 10). It may be thought that Roux believed himself able to see a great deal more than was actually present in the preparation, but anyone familiar with photographs of gross anatomical preparations will know how much of what is undubitably present is often omitted. We have to remember also that much more might be seen if all ten sections were examined together. Moreover, Roux explicitly states that some parts of the structure could only be made

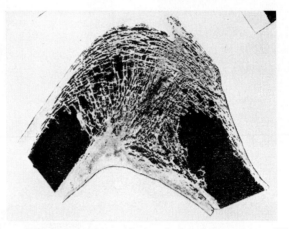

Figure 9. One of Roux's sections of an ankylosed knee-joint. (Wolff, 1892.)

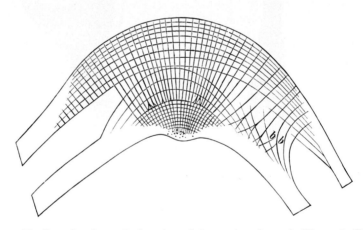

Figure 10. Roux's schematic drawing of the section shown in Figure 9. (Roux, 1885.)

out by studying several consecutive sections in transmitted light, and for this reason did not appear at all clearly in the photographs.

Figure 11 shows for comparison Roux's schematization of the structure of the normal knee-joint.

A feature of particular interest in Roux's treatment of the subject

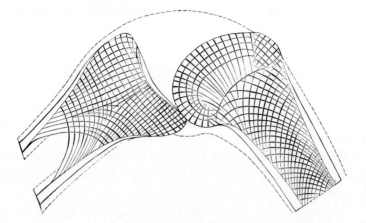

Figure 11. Roux's schematic drawing of a section of the normal human knee-joint. (Roux, 1885.)

was his use of a technique, devised by himself, for producing and rendering visible the 'lines of tension and of pressure' in models subjected to the same stresses as the ankylosed knee-joint: the stresses, that is, produced in the joint by the patient using the rigid leg in walking.

Under the heading *Functionelle Bedeutung der neuen Structur*, Roux writes as follows:

> The interpretation of these complicated arrangements is in fact very difficult and, as I learned from fruitless enquiries addressed to constructional engineers, not difficult only for dilettanti in graphical statics. I should therefore hardly be in a position to offer to the reader anything upon which reliance might be placed had I not hit upon an appropriate *method of self-realization of trajectories.* . . . On a rubber model of the object to be interpreted it is possible to render overtly perceptible the lines of strongest pressure and of tension that arise on pressing, pulling, bending or twisting the model. This is done by first coating the surface very evenly with fluid paraffin-wax, the paraffin being allowed to solidify, which may take half an hour. On deforming by bending, cracks appear in the wax, and these lie, in the regions of maximum tension at right-angles to the tension, and in the regions of maximum pressure at right-angles to the pressure.

These models were exhibited at the World's Fair at Saint Louis in 1904.

It will have been noticed that the method shows only one system
of lines, and this system may correspond to the less obvious groups
of trabeculae in the bone, though it happens that in the photographs
(particularly in Figure 9) some of the trabeculae that lie at right-
angles to the principle directions of force are clearly visible. The
second series of lines, that which is collinear with the lines of force,
is constructed in drawings of the models by drawing a second set of
lines such that each line of this set cuts each line of the already given
system at right-angles.

Figure 12 shows Roux's drawing of his model intended to reproduce
the circumstances of the section shown in Figure 9 and interpreted

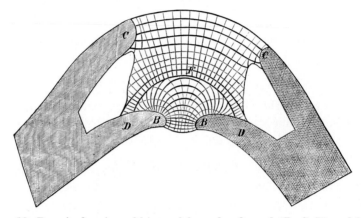

Figure 12. Roux's drawing of his model, made of wood (*B, C, D*) and hard
rubber (*F*) coated with paraffin wax. The model was exhibited at the World's
Fair in St. Louis, Mo., in 1904. (Roux, 1885.)

in Figure 10. The model consists of a central piece of hard rubber
coated with wax, and of two lateral pieces of wood intended to simu-
late the compact bone and the cavities above and below the region of
ankylosis. The pieces of wood were fastened to the rubber with glue.

The production of a new adaptive structure in the organism is
clearly of great interest, though it probably requires for its explanation only the assumption of a tendency in the osteoblasts to arrange
themselves along lines of force, such lines being, we may suppose,
first realized in an initially casual matting of collagen fibres.

If we now turn to the last group of regulative phenomena with
which this paper is concerned, regulation in embryonic development,
we find ourselves again confronted with the name of Wilhelm Roux,
for he is usually considered to have been the founder of that branch

of biology known as *experimental embryology*. Claims of this kind are difficult to establish with any certainty, but I have not found any reason to suppose that this particular claim should be rejected.

Experiments upon embryos had, of course, been made before Roux took up the subject, but no one before him seems to have worked upon it so systematically, with so clear a grasp of the problems and principles involved, or with such important consequences.

Roux worked upon a comparatively small number of problems, but all were of central importance, and the results he achieved were significant in the highest degree. His first paper dealing with a problem in embryogeny was published in 1883 (Roux, 1883a). His method in this investigation can hardly be called 'experimental', for the only manipulation it involved was the fixing of frogs' eggs to the substratum so as to observe the relation between the direction of the first cleavage-plane of the egg and the median plane of the embryo. He observed the coincidence of these planes, which was independently observed by Pflüger in the same year. In his paper of 1883 Roux also suggested that one direction of the first cleavage-plane might be determined by the point at which the sperm entered the egg. That this is in fact the case had already been observed by Newport (1851–54) some forty years earlier, but it is clear that Roux first learned of Newport's work in 1887. In his paper of 1883 Roux also reports a constant relation between the second cleavage-plane and the antero-posterior direction of the embryo. Thus 'all the principal directions of the embryo are fixed at the time of formation of the second cleavage'.

Roux's next investigation, of which the results were published in 1884, was in the full sense of the word experimental. This work was undertaken to refute Pflüger's supposition that the gravitational field is of decisive importance in development. For this work Roux borrowed from a colleague an apparatus which had been designed and used for botanical experiments. This was, in fact, a form of the apparatus first devised by Andrew Knight, at the beginning of the nineteenth century, and still familiar to botanists: a disc or wheel which can be rotated at constant speeds about either a vertical or a horizontal axis. Roux placed the axis horizontally and attached frogs' eggs to the wheel. The eggs were themselves free to rotate within their coat of jelly, and when the wheel was rotated at high speeds the eggs arranged themselves each with its principal axis along a radius of the wheel, they then developed normally in the artificial gravitational field. When the wheel rotated too slowly to produce any effective gravitational field of its own, the eggs, which rotate extremely slowly within their jelly, remained in the positions in which they had been attached so that, like seeds fixed to the wheel, they were effectively withdrawn from the influence of the earth's field. They developed normally, thus showing that no gravitational field is necessary for their development.

In the next year (1885) Roux began the regular publication of his series of papers entitled *Beiträge zur Entwickelungsmechanik des Embryo*, the paper of 1884 being republished as *Beitrag II*. The whole series consists of seven papers, of which the last was published in 1893. The first of these experimental *Beiträge* (Roux, 1885b) was supplied with a substantial theoretical introduction which, so Roux tells us (Roux, 1923), contains the solution of problems over which he had pondered for years. He writes, '*Diese Einleitung und Beitrag I sind wohl das Wichtigste meiner Lebensarbeit*'.

The conclusions announced by Roux in his *Einleitung* now seem almost platitudes, but they were clearly important at the time of their publication, a time when it was still possible for rival theorists to use the expressions 'preformationist' and 'epigeneticist' as significant terms of abuse. However, one of Roux's conclusions proved to be not more but less obvious than he himself supposed. In his *Einleitung* he wrote:

> No special argumentation seems required to show that, although the theory of evolution has thrown light upon the particular results realized by the processes of development (ontogeny) in each of its stages, the processes as such require a causal investigation peculiar to themselves.

He was later to write, in his autobiography (1923):

> How little I was prepared for the resistance to anything new, in this case arising from the exponents of descriptive and comparative embryology, which was then all-powerful! . . . Haeckel never forgave me for '*die überüssflige törichte Entwickelungsmechanik*'.

The following citations from the *Einleitung* of 1885 show that he was from the first free from any inclination to assert dogmatically that all development must be mosaic development, that is, he was in no sense a 'preformationist'.

> Concerning particular parts of the egg or embryo, it is therefore possible to enquire whether their development is self-differentiation or dependent development. On the decision of this question

in favour of one or of the other of these alternatives will depend the resolution of a number of fundamental problems.

In the first place, should it be found that many parts of the egg differentiate purely by their own proper powers lying within themselves, and that it is in this way that the rich variety of parts of the later stages arises, then it must be concluded that the egg must from the first have been constituted of many different parts, and that, consequently, development is essentially the change of one manifold into another, or *preformation* in the sense we choose to give to the word; and this must be true, notwithstanding the existence of that overt *epigenesis* which is associated with the name of C. F. Wolff. . . Thus the determining conditions of development would have to be referred principally to molecular processes, and would for the present be for the most part such as to evade further investigation. The whole egg which has undergone cleavage is thus, it may be, merely the sum of these independent parts, and there is during the time of this independent differentiation of parts no unified coordination in the whole, and thus the whole can exercise no regulating formative influence upon the parts. . . .

If, on the other hand, development is essentially dependent upon reactions between many or all of the parts, it must follow, conversely, that the fertilized egg needs to consist only of a few different parts which by reciprocal interaction produce little by little great complication. Development will then be a true production of complexity, an *epigenesis*, in our sense of the word. There will be a mutual cooperation of parts working towards a whole, in which case there can be exercised in the reverse sense an influence of the whole upon the parts; and there lies open to us, in the discovery of such conditions, a rich field of investigation already exploitable with the means now at our disposal. . . But, finally, in the organic process, independent and dependent differentiation of parts, and hence preformation and epigenesis, might be combined in complex interaction with one another; and it would then be necessary for us to display twice as much caution and twice as much ingenuity in order to disentangle the part played by the one factor from that played by the other.

From the first, then, Roux considered it at least possible that the fate of a cell in the embryo might depend not only on the set of 'determinants' which that cell had received from its parent-cell, but also upon the properties of the cells in contiguity with it. This is already at variance with Weismann's theory of a rigidly hierarchical relation of cells in the embryo (Weismann, 1892). However, Roux's concession of the possibility of interaction of cells in development also contains implicitly the supposition that, if the normal course of development were to be artificially disturbed, the cells might so adapt themselves to the novel circumstances as to produce a normal end-result. This possibility is not, I think, explicitly considered in the *Einleitung*, but in the paper which follows it (*Beitrag I*) Roux records the complete development of eggs which, as he believed, he had damaged significantly.

In fact, neither in his theorizing nor in his discussion of the results of experiments did Roux ever dismiss the possibility of regulation in embryogeny. Moreover, although in his first *Beitrag* he commonly uses the word *Regeneration* when speaking of regulative processes in the embryo, he does also speak of *Selbstregulation* and says that he believes it to be 'the most characteristic of the properties common to all organisms' (*die wesentlichste allgemeine Eigenschaft des Organischen*).

In his third *Beitrag*, published in the same year as his first (Roux, 1885c), Roux expounds a theory which seems extremely close to that of Weismann (to whom he refers); but a few pages further on in the same paper he makes it quite clear that the theory is intended to refer only to normal undisturbed development:

> If then the nuclear development is able accurately to separate the qualities [the various factors determining development] from one another, it is also necessary for the mode of development referred to above that the products of division be brought into the correct spatial relations to the various parts of the cell-body and to those neighbouring cells which have already been formed.
>
> We might be inclined to assume that this positioning was so rigidly connected with the chronological sequence of divisions that, through inner causes, each successive division would take up a particular definite position in relation to the direction of the previous division, and the typical pattern of cleavages of the animal egg and the arrangement of divisions of the apex-cell of plants appear to lend support to this supposition.
>
> But such a rigid arrangement would exclude any possibility of regulation (*Selbstregulation*), and a single faulty division would set wrong the whole following series of divisions. The facts show, however, that anachronistic cleavages are tolerated without ill effect.

Probably the best known of the phenomena first observed by Roux is the half-embryo produced from one of the first two blastomeres of the frog's egg. This was certainly claimed by Roux as a striking instance of 'self-differentiation', of the power possessed by parts of the embryo to develop independently of what may happen to other parts. But a very large part of his first paper (Roux, 1888) devoted exclusively to this subject (the phenomenon is recorded in the first *Beitrag*, 1885) is given to the discussion of the reorganization (*Reorganisation*) of the damaged blastomere so as to restore the missing half of the embryo. That, in this instance, Roux was almost certainly mistaken as to the facts (Morgan, 1927, pp. 383–385) is of secondary importance; the important point is that he not only allowed that regulation might take place, but believed he had observed the process. He was therefore justified in the bitter complaints to which he gave vent in a page-long footnote added to the reprint of this paper in his *Gesammelte Abhandlungen*, printed in 1895.

> I have been accused of erecting the principle of mosaic development (*das Princip der Mosaikarbeit*) into a general principle of development, and of denying the part played by differentiating correlation [the mutual influence of adjacent parts] in ontogenesis, with the result that a whole literature has grown up in refutation of this 'error'.

It seems that the mistaken accusation arose from Roux's theory of cell division (Roux, 1883b) which certainly suggested that all development must be mosaic, and which, I think, he never modified so as to bring it into harmony with his ideas on regulation. The conflation of the ideas of Roux with those of Weismann implied in the term 'The Roux–Weismann Theory' is therefore not quite as unjust as might appear from the quotations I have given from Roux's work.

However, whatever injustice may or may not have been done to Roux in the 'literature' to which he refers, there is no doubt that the work which owed its being, at least in part, to a desire to refute the Roux–Weismann Theory (the theory of exclusively mosaic development) resulted in a most important acquisition of new knowledge, and includes some extremely elegant experiments. The work of Driesch is well known, at least in outline, and to consider it here would lead us into a too prolonged discussion of the theoretical consequences which Driesch imagined his results to support. It is possible, however, to quote in isolation one of his results, from which an inference may be drawn as to the extent of possible regulation in some early embryos, an inference which in no way depends upon or entails Driesch's more

speculative conclusions. The passage quoted below is taken from
E. B. Wilson's book *The Cell in Development and Inheritance*, pub-
lished in 1896, and concerns chiefly an experiment conducted by
Wilson himself, but suggested by the work of Driesch.

The theory of qualitative nuclear division has been practically
disproved in another way by Driesch, through the pressure-
experiments already mentioned. Following the earlier experiments of Pflüger and Roux on the frog's egg, Driesch subjected
segmenting eggs of the sea-urchin to pressure, and thus obtained
flat plates of cells in which the arrangement of the nuclei differed
totally from the normal; yet such eggs when released from pressure continue to segment, *without rearrangement of the nuclei*, and
give rise to perfectly normal larvae. I have repeated these experiments not only with sea-urchin eggs, but also with those of an
annelid (*Nereis*), which yield a very convincing result, since in
this case the histological differentiation of the cells appears very
early. In the normal development of this animal the archenteron
arises from four large cells or macromeres (entomeres), which
remain after the successive formation of three quartets of micromeres (ectomeres) and the parent-cell of the mesoblast. After the
primary differentiation of the germ-layers the four entomeres do
not divide again until a very late period (free-swimming trocho-
phore), and their substance always retains a characteristic
appearance, differing from that of the other blastomeres in its
pale non-granular character and in the presence of large oil-drops.

If the unsegmented eggs be subjected to pressure, as in
Driesch's echinoderm experiments, they segment in a flat plate,
all the cleavages being vertical. In this way are formed eight-
celled plates in which all the cells contain oil-drops. If they are
now released from the pressure, each of the cells divides in a
plane approximately horizontal, a smaller granular micromere
being formed above, leaving below a large clear macromere in
which the oil-drops remain. The sixteen-cell stage, therefore,
consists of eight deuteroplasm-laden macromeres and eight
protoplasmic micromeres (instead of four macromeres and twelve
micromeres, as in the normal development). These embryos develop into free-swimming trochophores containing eight instead
of four macromeres, which have the typical clear protoplasm
containing oil-drops. In this case there can be no doubt whatever that four of the entoblast nuclei were normally destined for
the first quartet of micromeres, from which arise the apical

ganglia and the prototroch. Under the condition of the experiment, however, they have given rise to the nuclei of cells which differ in no wise from the other entoderm cells. Even in a highly-differentiated type of cleavage, therefore, the nuclei of the segmented egg are not specifically different, as the Roux-Weismann hypothesis demands, but contain the same materials even in cells that undergo the most diverse subsequent fate. But there is, furthermore, very strong reason for believing that this may be true in later stages as well...The strongest evidence in this direction is afforded by the facts of regeneration.

I quote this passage because it describes a particularly conspicuous and striking example of the powers of regulation possessed by some early embryos, and because in its last phrases it draws together two of the themes of this chapter; but it has the following further advantage. Even were it possible to believe that some other cases of regulation could be considered as consequences of a process of natural selection, it would clearly be absurd to suppose that the eggs of *Nereis* had been in the past frequently subjected to the treatment to which Wilson subjected them. In addition, even were that absurd supposition to be made, to say that those eggs which happened to be capable of regulation survived while those not so endowed perished, would admit that the regulation could not itself be a product of that natural selection for which it would be a necessary condition.

If then we are not to despair of any further analysis of the phenomena, as Driesch in effect did, we must assume that the embryo is capable of regulation because it is the kind of physical system it is, just as a spherical drop, if divided, produces not two half-drops but two droplets, because it is that sort of physical system.

References

Bonnet, C. (1745). *Traité d'Insectologie*, Durand, Paris.
Bonnet, C. (1781). *Oeuvres*, Fauche, Neuchatel.
Bourgery, J. M. (1832). *Traité Complet de l'Anatomie de l'Homme*, Delaunay, Paris.
Culmann, C. (1866). *Die graphische Statik*, Reimann, Zürich.
De Vries, H. (1889). *Intracellulare Pangenesis*, Fischer, Jena.
Engel, J. (1851). 'Ueber die Gesetze der Knochen-Entwickelung.' *S.B. Akad. Wiss. Wien.*, **7**, 591–684.
Folkes, M. (1743). 'Some account of the insect called the fresh-water polypus.' *Phil. Trans. Roy. Soc. London, Ser. B*, **42**, No. 467, ii–xvii.
Humphry, G. M. (1858). *A Treatise on the Human Skeleton*, Macmillan, Cambridge.

Leeuwenhoek, A. (1703). 'Concerning green weeds growing in water, and some animalcula found about them.' *Phil. Trans. Roy. Soc. London, Ser. B*, **23**, No. 283, 1306–1307.

Meyer, H. von (1867). 'Die Architectur der Spongiosa.' *Reichert und Du Bois-Reymond's Archiv*, (see Wolff, J., 1892, pp. 7–8).

Morgan, T. H. (1927). *Experimental Embryology*, Columbia University Press, New York.

Newport, G. (1851–54). 'On the impregnation of the ovum in the amphibia.' *Phil. Trans. Roy. Soc. London, Ser. B*, [1851], 169–242; **143**, 233–290; **144**, 229–244.

Pflüger, E. (1883–84). 'Ueber den Einfluss der Schwerkraft auf die Theilung der Zellen und auf die Entwickelung des Embryo.' *Pflüger's Archiv*, **31**, 311–318; **32**, 1–79; **34**, 607–616.

Réaumur, R. A. F. (1714). 'Sur les diverses reproductions qui se font dans les ecrevisses, les omars, les crabes, etc.' *Histoire de l'Academie Royale des Sciences*, 1714.

Réaumur, R. A. F. (1742). *Mémoires pour Servir à l'Histoire des Insectes*, De l'Imprimerie Royale, Paul Lechevalier, Paris.

Rostand, J. (1951). *Les Origines de la Biologie Experimentale et l'Abbé Spallanzani*, Fasquelle, Paris.

Roux, W. (1883a). *Ueber die Zeit der Bestimmung der Hauptrichtungen des Froschembryo*, Engelmann, Leipzig.

Roux, W. (1883b). *Ueber die Bedeutung der Kerntheilungsfiguren. Eine hypothetische Erörterung*, Engelmann, Leipzig.

Roux, W. (1884). 'Ueber die Entwickelung der Froscheier bei Aufhebung der richtenden Wirkung der Schwere.' *Breslauer Ärztliche Zeitschrift* [22 Marz], (see Roux, W. (1895). *Ges. Abhandl.*, **2**, 256–276).

Roux, W. (1885a). 'Beschreibung und Erläuterung einer knöchernen Kniegelenkankylose.' *Arch. Anat. Physiol. Lpz., Anat. Abth.*, [1885], 120–158.

Roux, W. (1885b). 'Einleitung zu den Beiträgen zur Entwickelungsmechanik des Embryo.' *Z. Biol.*, **21**, 411–428.
Beitrag I. 'Zur Orientirung über einige Probleme der embryonalen Entwickelung.' *Z. Biol.*, **21**, 429–524.

Roux, W. (1885c). 'Ueber die Bestimmung der Hauptrichtungen des Froschembryo im Ei und über die erste Theilung des Froscheies.' *Breslauer Ärztliche Zeitschrift* [1885], (see Roux, W. (1895). *Ges. Abhandl.*, **2**, 277–343).

Roux, W. (1887). 'Die Bestimmung der Medianebene des Froschembryo durch die Copulationsrichtung des Eikernes und des Spermakernes.' *Arch. Mikr. Anat.*, **29**, 157–211.

Roux, W. (1888). 'Ueber die künstliche Hervorbringung "halber" Embryonen durch Zerstörung einer der beiden ersten Furchungszellen, sowie über die Nachentwickelung (Postgeneration) der fehlenden Korperhälfte.' *Virchow's Archiv*, **114**, 113–291.

Roux, W. (1895). *Gesammelte Abhandlungen über Entwickelungsmechanik der Organismen*, Engelmann, Leipzig.
Roux, W. (1923). 'Autobiographie.' In Grote, L. R. (Ed.), *Die Medizin der Gegenwart in Selbstdarstellung*, Meiner, Leipzig.
Spallanzani, L. (1768). *Prodromo di un'Opera da imprimersi sopra le Riproduzioni animali*, Giovanni Montanari, Modena.
Swammerdam, J. (1737–38). *Bybel der Nature*, Severinus, Leyden.
Trembley, A. (1744). *Mémoires pour Servir à l'Histoire d'un Genre de Polypes d'Eau Donce, à Bras en forme de Cornes*, Durand, Paris.
Ward, F. O. (1938). *Outlines of Human Osteology*, Renshaw, London.
Weismann, A. (1883). *Ueber die Vererbung*, Fischer, Jena.
Weismann, A. (1892). *Das Keimplasma. Eine Theorie der Vererbung*, Fischer, Jena.
Wilson, E. B. (1896). *The Cell in Development and Inheritance*, The Macmillan Company, New York.
Wolff, J. (1870). 'Ueber die innere Architectur der Knochen und ihre Bedeutung für die Frage vom Knochenwachsthum.' *Virchow's Archiv*, **50**, 389–450.
Wolff, J. (1872). 'Beiträge zur Lehre von der Heilung der Fracturen.' *Arch. Klin. Chir.*, **14**, 270–312.
Wolff, J. (1891). 'Ueber die Theorie des Knochenschwundes durch vermehrten Druckentlastung.' *Arch. Klin. Chir.*, **42**, 303–324.
Wolff, J. (1892). *Das Gesetz der Transformation der Knochen*, Hirschwald, Berlin.
Wolff, J. (1896). 'Die Lehre von der functionellen Pathogenese der Deformitäten.' *Arch. Klin. Chir.*, **53**, 831–905.
Wyman, J. (1857). 'On the cancellated structure of the bones of the human body.' *Boston J. Nat. Hist.*, **6**, 125–140.

10

Regulation of Plant Growth

P. R. BELL

Department of Botany, University College, London

Living things are never stable. Some kind of change is always going on inside them, even though the rate of change, as in dormant seeds, may be extremely slow. One of the most conspicuous forms of change is growth, during which both size and mass of the organism usually increase. Sometimes, of course, one part of an organism may grow at the expense of another, so that there is no net increase in weight; there may in fact be a loss because of respiration. An example in plants is provided by potato tubers sprouting in the dark. The buds expand at the expense of the food stored in the tuber, and the system as a whole diminishes in mass. Nevertheless, the buds undoubtedly show the phenomenon of growth.

Growth does not often consist of mere increase in mass and volume. The growing organism almost always generates definite patterns and shapes, both internally and externally, repeated with remarkable fidelity from one generation to the next. Although we shall consider the quantitative and qualitative aspects of growth separately, this is for convenience of discussion. The amount of growth in an organism and its spatial distribution are in fact rarely independent of each other. A particular morphology, for example, may limit the amount of growth that can take place because an increase in size beyond a certain limit would lead to a situation impossible physiologically. Relationships of this kind may explain why the only wholly spherical plants are a few small aquatic algae. Since the surface of a sphere does not increase in proportion with the volume, increasing size of a spherical organism must be accompanied by an increasing rate of transport of metabolites through each unit of surface. This rate, and hence the size, will eventually reach a limit set by the physico–chemical properties of the bounding membranes.

Growth Considered as Quantitative Increase

The increase in mass of an organism may come about by an increase in the number of its cells, by an increase in the size of its cells, or by a combination of these two processes.

Growth principally by cell multiplication, with very little change in the sizes of the cells involved, is shown quite generally by bacteria

Figure 1. *Acetabularia.* This marine alga, despite its complex form, consists of a single uninucleate cell.

and other unicellular microorganisms. An example from higher plants is provided by the layer of cells on the outside of a leaf. During the final expansion of the leaf, this layer increases in area by sporadic divisions, and there is little variation in individual cell size (Lewis, 1930).

The marine alga *Acetabularia* (Figure 1) shows the converse kind of growth. Although this plant consists, throughout its vegetative life, of a single uninucleate cell, it enlarges nevertheless from a minute sphere into a mushroom-like shape, some 10 cm high, attached to the sea bottom by complicated rhizoids. There are other plants in which, although growth is wholly of one cell, the protoplast becomes multinucleate. This form of growth, termed *coenocytic*, shows that cell division is not an inevitable consequence of nuclear division.

In higher plants growth usually consists of a combination of cell multiplication and cell expansion, although the relative amounts of multiplication and expansion differ in different regions and in different tissues. This makes the task of investigating the control of these phenomena more difficult, and some investigators have consequently sought to stop one process by experimental means in order to facilitate the study of the other. In multicellular plants cell multiplication is, of course, basic to all normal growth, and it will be appropriate to begin by examining the factors which control it.

The control of cell division

Cell multiplication inevitably depends upon a supply of metabolites adequate to build up the materials of the new cells. Quite apart from this, however, it is a common experience that in many systems cell multiplication may stop or be slowed down, even though the supply of metabolites is still quite adequate. A culture of bacteria, yeast, or of a unicellular alga, for example, shows a more or less S-shaped growth curve, even though abundant nutrients may remain in the medium. An annual plant shows a similar growth curve (Figure 2), but there is no question of the soil on which it is growing being exhausted. Clearly some factors are inhibiting cell multiplication.

Before we can profitably consider what these factors are, or might be, we must first look at what is involved in cell division. First is the synthesis of deoxyribonucleic acid, and its associated proteins, making it quantitatively possible for the nucleus to divide. Second is the synthesis of ribonucleic acid, proteins and lipids, so that the daughter cells may be fully stocked with the enzymes and intracellular structures essential for life. Third is the synthesis of new wall material so that each cell comes to occupy its own compartment. Although these activities are somewhat consecutive in the order given, they also show quite wide overlapping and interrelation.

Interference with one or more of these synthetic processes will disturb cell division, either preventing it completely or disorganizing it. All will suffer if some basic feature of the metabolism, such as respiration, is disturbed. The factor which actually causes cessation of cell multiplication and ultimate death may be different in different instances. In an anaerobic colony of yeast, for example, cell division ceases as the concentration of ethyl alcohol in the medium increases and becomes toxic. Loss of meristematic activity in the growing points of an annual plant is probably a much more complex phenomenon, connected with the cytochemistry of flowering, and as yet unsolved.

Quite apart from the general death of a colony of cells or of a multicellular organism, cell multiplication in a multicellular organism is more often than not confined to one or more regions of the plant

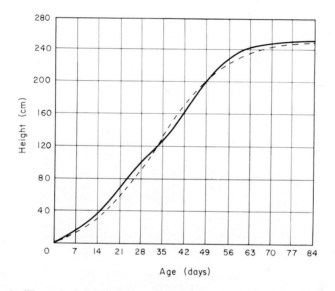

Figure 2. The relationship between height and age in the sunflower. The height (cm) is the mean value obtained from 58 similar plants of equal age about 20 cms apart in an experimental plot. The broken curve is that which would have been expected assuming growth to follow a mathematical relationship analogous to that of autocatalysis. (From Reed and Holland, 1919.)

body, and suspended elsewhere. This enables growth to be ordered. A familiar example of suspended growth is seen in the dormant buds present in almost every higher plant. This irregular distribution of growth is usually accompanied by apical dominance, i.e. so long as the uppermost terminal bud remains active, the lateral buds below it remain suppressed. This suppression clearly depends upon the activity of the terminal bud, since following its removal, one or more of the lateral buds will grow out. An analogous situation occurs in much simpler plants. The sexual generation (*gametophyte*) of most ferns, for example, is a small plant, flattened and heart-shaped. It grows from a meristem at the base of the apical notch. If this meristem ceases activity, or is removed, cells around the margin and on the surface of the gametophyte begin to proliferate, and the symmetry of the plant is destroyed (Figure 3). It appears therefore that normally the tendency for these cells to grow is effectively suppressed by some influence from the apical region, and so the cordate form remains stable (Albaum, 1938).

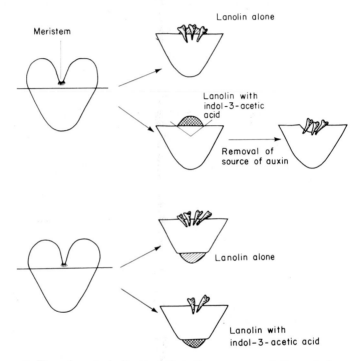

Figure 3. Experiments indicating that the presence of the apical meristem suppresses outgrowths elsewhere in the gametophyte of a fern. A source of indol-3-acetic acid in place of the meristem has a similar effect; the same auxin is much less effective when applied to the basal end.

The suppression of growth in one region of a plant while it continues in another is an example of *correlation*, a growth phenomenon that continues to excite interest and research. So far as apical dominance is concerned, there is much evidence that it is maintained by growth-regulating substances (*auxins*). One of these substances is indol-3-acetic acid, a substance related to the ubiquitous amino acid tryptophane. Indol-3-acetic acid is produced very actively in meristematic regions, and such regions consequently set up gradients of auxin within the plant. Gradients of this kind seem to be very import-ant in controlling the development of a wide range of plants. In fact, it is often possible to maintain correlative inhibitions in a plant from which the apex has been removed by putting in its stead a source of indol-3-acetic acid.

Little is known about how indol-3-acetic acid, and related growth-regulating substances, actually bring about the inhibition of the lateral buds. It is possible that the effect is a direct one upon the syn-thesis of deoxyribonucleic acid in the nucleus. The nuclei in the meristematic area of dormant buds are larger than those in the surrounding cells, but they contain only the diploid amount of de-oxyribonucleic acid. It seems as if they are in a condition to double the amount of deoxyribonucleic acid, as in the normal interphase of dividing nuclei, but for some reason are unable to do it. As soon as the apical bud is removed, the deoxyribonucleic acid in these nuclei begins to increase, reaches the tetraploid level, and then the nucleus divides (Naylor, 1958). When a lateral bud begins to be active, it will also begin to produce auxin, and consequently to have correlative effects throughout the plant.

Despite the extensive evidence in favour of the auxins being re-sponsible for the correlation of growth, there are many who believe other factors to be involved. With regard to apical dominance, for example, some students of tree growth have preferred to regard this phenomenon as a consequence of the unequal distribution of nutrients within the tree, believing the auxin effects to be complementary or even secondary. They maintain that an actively growing bud draws, in a manner as yet unexplained, so much of the available nutrition to itself that little if any growth is possible elsewhere in the shoot system. There is also evidence, mainly from herbaceous plants, that some buds grow out less readily than others, even when all competition is re-moved. This 'growth potential' of buds awaits further investigation, and the extent of its influence on the distribution of growth in a higher plant is not yet known.

The stimulation of cell division

Although earlier we thought of generalized cell division as something to be continually suppressed if ordered growth is to occur, it is also true that there are occasions in the life of a plant when renewed meristematic activity is required. A renewal of cell division, for ex-ample, occurs regularly in woody plants when the cambium appears in the stem, and adds new vascular tissue to that already present. The seasonal periodicity in cambial activity is probably related to the variations in the meristematic activity at the apices and the flow of growth-regulating substances from them, the substances in this instance have a stimulatory rather than inhibitory effect upon cell division. The cambium is also affected in other ways. There is evidence, for example, that pressures and tensions in a stem control cambial activity in such a way that the mechanical stability of the growing organ is maintained. An anatomical result of this property of the cambium is that trunks and branches which have been subjected to bending and torsion frequently show an eccentric structure.

As a cell ages and becomes vacuolate it divides less readily than formerly, but so long as it retains a protoplast it probably never wholly loses the power of division. Renewed meristematic activity in such cells, which would normally have remained quiescent indefinitely, is usually the result of adjacent cells being wounded, or of irritation by the products of gall insects or bacteria. These products are probably growth-regulating substances similar to those already present in the plant, but which, being out of place, cause disordered growth. The response to wounding may have a similar explanation, the growth-regulating substances being released as a result of a breakdown of the cytoplasmic substructure and permeability changes in the injured cells. Some botanists, however, have believed in the existence of special wound hormones liberated only in the event of injury.

The control of cellular growth

In a unicellular plant such as *Acetabularia*, which increases its volume very many times during its life cycle, it is not surprising to find that the growth is accompanied by considerable synthesis of ribonucleic acid, protein, and wall material. A similar, but less marked, situation is found in expanding cells in multicellular plants. Substances which interfere with these synthetic activities within cells will therefore affect cell expansion (Key, 1964).

With respect to ribonucleic acid and protein, the dependence of growth upon uninterrupted synthesis has been demonstrated by feeding to growing systems compounds which, although unnatural, are nevertheless incorporated into these substances. Examples of such growth inhibitors are certain unnatural purines, pyrimides and amino acids, now used extensively in the study of growth. The influences affecting the metabolism of expanding cells *in vivo* must, of course, be more subtle, and there is evidence that well-known growth-regulating substances are again involved. Certain concentrations of indol-3-acetic acid, for example, promote the elongation of newly-formed cells in stems and coleoptiles of grasses. Gibberellic acid behaves similarly. These effects are very striking, and similar mechanisms undoubtedly operate in normal growth, although the actual substances involved may be different in different instances. Despite considerable research, very little is yet known in detail of how the growth-regulating substances act upon individual cells. It seems very unlikely that it is upon only one feature of their metabolism.

Clearly, in both unicellular and multicellular plants, there are also factors limiting expansion of individual cells. *Acetabularia*, for example, does not go on increasing indefinitely in size, no matter how abundant the supply of nutrients. Similarly, in the axes of higher plants, cells do not elongate indefinitely, even in the continued presence of auxins. The amount of expansion in the cells of a plant must therefore be the resultant of promoting and inhibiting factors.

The level of this resultant can sometimes be altered by external influences, often in a striking way. The shading of an herbaceous plant, for example, results in the growing shoots becoming very much longer. In those plants tolerant of shade, the leaves produced in reduced illumination are larger in area, often by just that amount necessary to intercept more of the incident light, and maintain the relative growth rate at an almost constant level (Evans and Hughes, 1961). Investigations have shown that changes of this kind in length of axes and area of leaves are principally the result of greater cell expansion, rather than of an increase in the number of cells.

The way in which light brings about these growth effects is still unknown. It is, however, becoming clear that cells can be affected by light in many different ways. Quite apart from photosynthesis in green cells, and the photoperiodic and phototropic effects of light to be discussed later, light is able to influence both the amount and kind of protein synthesis in growing cells and, probably contingently, their shapes and sizes. A photoreceptor able to bring about effects of this kind, recently isolated from plants, is *phytochrome*. This is a protein, existing in two forms, active and inactive respectively in promoting growth. The proportion of the two forms depends upon the relative amounts of energy being absorbed by the plant in the red (660 μ) and far-red (730 μ) regions of the spectrum. Phytochrome is certainly responsible for some of the morphogenetic effects of light, but there are undoubtedly additional systems awaiting analysis (Mohr, 1964).

Quantitative correlation of growing regions

The correlative influence of one region of expanding cells upon another is difficult to demonstrate, since expanding cells are rarely sharply separated from dividing cells. There is, however, much evidence that growing regions in which both cell multiplication and expansion are occurring affect each other. This underlies the phenomenon of *relative growth*, demonstrated mathematically by a logarithmic relationship between the sizes of the various growing regions of one organism. An example is provided by the radish, where the mass of the shoot is logarithmically related to that of the bulb. It seems highly unlikely that simple competition for nutrients between the growing regions is responsible for these relationships, and a balance of auxins is probably involved as well.

In the growth of an organ or tissue, quantitative increase does not of course take place uniformly throughout its mass. Growth takes place more in some directions than in others, so that definite and sometimes complex shapes are generated. Where the shape during growth remains regular, increase in one dimension is often logarithmically related to that in another. The growth of gourds provides a good example of this phenomenon. Here studies of bottle gourds have shown that the logarithmic relationship between growth and width follows the same course in dwarf varieties as in giant. The difference lies in the fact that the dwarf varieties stop growing, i.e. reach maturity, earlier. Thus, there is not only a definite interrelationship between growth in different dimensions, but also a general control over how long the process is allowed to go on. Since the size of the gourd is governed by factors showing Mendelian inheritance, the time of onset of maturity in the growth of the fruit must be determined by the nucleus, but the mechanism is unknown. The spatial interrelationships of the growth of the gourd are presumably determined by differences in the amount of cell division and expansion in different directions, possibly influenced by the pressures and tensions generated in the enlarging fruit, but precise details are lacking.

A somewhat simpler, but comparable situation is encountered in the growth of grass leaves. During the growth of the first leaf of wheat, there is a logarithmic relationship between length and width. Both cell division and expansion are involved in this growth. By exposing the seeds to certain doses of γ-radiation, seedlings can be obtained in which cell division is almost wholly suppressed. The expanding leaf reaches nevertheless about half the normal size, and up to this point the logarithmic relationship between length and width is exactly as in the untreated plants (Figure 4). The relationship of the growth curves is thus similar to that in the gourds mentioned previously. Here, however, we know that in the smaller leaf there has been no cell division, so that, although the logarithmic relationships are the same in the two leaves, the behaviour of the individual cells must have been different. The shape of the first leaf, although characteristic of the race of wheat used, does not therefore follow from its being composed of a definite number of cells of

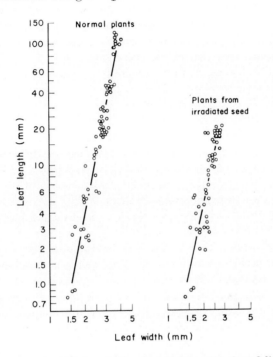

Figure 4. Changes of width in relation to length in the first foliage leaves of wheat seedlings. The plants from irradiated seed lacked cell division.

certain shapes and sizes. Instead it must be envisaged as the result of the interplay of gradients of metabolites and growth-regulating substances which change in an ordered manner during growth, and which to some extent are independent of the number of cells involved (Haber, 1962).

The Quality of Growth

The regular arrangement of the tissues in a plant stem, visible in a transverse section even with a hand lens, reveals at once that plants generate intricate internal patterns of differentiated cells, as well as the familiar external patterns of branching and foliage. The growth of a shoot of a higher plant provides in fact an excellent demonstration of the complex interrelationships involved in the development of plant form.

The growing point of a shoot is more or less cone-shaped, and it is pushed forward in consequence of the multiplication of cells just below the surface of the extreme tip. The surface cells divide to keep pace with the elongation, and in addition groups of cells against the flanks of the cone remain embryonic. These cells, although superficially similar, give rise below the extreme tip to either leaves or branches. The leaves, however, usually develop first, and the fact that they appear at more or less regular intervals suggests that they are a response to some rhythmical physiological process (Turing, 1952). Behind the apical growing point is a region in which the cells elongate and differentiate. The cells here, initially anatomically indistinguishable, may develop into a number of different kinds of mature cells. The way in which they develop is found to depend upon their position in the stem, and its relation to those of the developing leaves and branches. This suggests at once that the external morphological differentiation and the internal anatomical differentiation are intimately connected. This is supported by the fact that the experimental removal of a very young leaf or branch from a growing point leads to considerable modification of the pattern of differentiation at and below the point of removal.

Apart from the ordered cell multiplication and differentiation that takes place internally, the shoot apex has the remarkable property of being able to generate an ordered system of branching, orientated in a particular manner in space. This means that, in addition to mechanisms controlling the extent and kind of growth, the shoot must possess others, responsive to such factors as light and gravity, which control its general direction.

The analysis of so complex a system is clearly a formidable problem, and it first demands answers to questions which, although simple to pose, are difficult to answer. For example, although the direction of division of some cells in the extreme tip of the growing shoot may be random, most of the cells in the apex, if they divide at all, do so in a definite direction related to their position. This raises the problem of what determines the direction of cell division, and this in turn takes us to quite fundamental features of cell organization. We will consequently make this our starting point in considering the control of form in growth.

The direction of cell division

Starting with the cell, the very simple matter of the direction in which a cell divides may have profound consequences on the growth form generated. A fern spore (a single cell), for example, first grows into a filament of cells, i.e. the cell divisions all lie along one axis. If they were not orientated in this regular fashion, the spore would give rise to an irregular nodule of cells, a situation which can be brought about by certain experimental treatments. The regular orientation of the divisions occurring normally must reflect some definite alignment of the macromolecules in the cell. Often the first division of a spore or fertilized egg will polarize all subsequent development, i.e. the growth from one of the daughter cells will be different from that of the other. A beautiful example is provided by the fertilized egg of the seaweed, *Fucus*. When the egg divides, one of the daughter cells gives rise to the branching photosynthesizing part of the plant which is suspended in the water, and the other to a system of root-like branches which anchors the plant to rock or stones. It has been shown experimentally that if the germinating egg is subjected to gradients of light, temperature, salinity, or in fact of a whole range of environmental factors, division always occurs in the direction of the gradient, and the dividing wall forms transverse to it. Consequently, one daughter cell is always in an environment slightly different from that of the other. Whether it is this slight difference which causes the subsequent development of the daughter cells to be so different is not known. It may be that the difference in development follows inevitably from a macromolecular asymmetry in the ungerminated egg, and all that the external gradients do is to determine the orientation of this asymmetry.

Observations on *Tradescantia* indicate that a lack of uniformity in the cytoplasm, of the kind envisaged in the foregoing, may be operative in controlling the germination of pollen grains. The first division of the nucleus of the grain takes place in the anther, and the appearance and subsequent behaviour of the two daughter nuclei are strikingly different. The direction of this first division is normally aligned with what was a radius of the tetrad in which the grain was formed, but the spindle can be displaced by almost 90° by giving the anther heat shocks. When this is done, the division still completes itself, but the daughter nuclei remain alike and the grain is capable of no further development. This points to the differentiation of the two nuclei, and the associated capacity of the grain to grow and function, being dependent upon an orientated irregularity in the cytoplasm. Since this orientation is related to the symmetry of the tetrad, factors which ultimately control the development of the male gametophyte appear to arise during the division of the preceding pollen mother cell (Sax, 1935).

Claims have also been made that external pressure influences the direction of cell division, but conclusive proof that it is actually the mitotic spindle that is so orientated has yet to be provided. It appears as if a cell subjected to pressure will divide so that the plane of the dividing wall is parallel to the direction of the pressure, and a cell subjected to tension so that the plane of the wall is transverse to the direction of the tension. This, of course, may result just as well from movements of the daughter cells in the later stages of division, when the new cell wall is still plastic, as from a primary orientation of the spindle. Readjustments of this kind probably play a large part in the growth of meristematic regions, such as the apices of stems, for it is clear from the response to shallow cuts that the surface layers are often under considerable tension. These physical factors may indeed, by controlling the arrangements of the newly forming cells, play a large part in causing the regularity with which the primordia of leaves and side-shoots are formed at the shoot apices of higher plants.

Studies of the development of the leaves of grasses have revealed yet another kind of control over the direction of cell division. The grass leaf grows from its base, so by examining a leaf from its basal growing zone upwards we can see the steps by which specialized regions, such as stomata, develop. With stomata the situation is made particularly clear by their being produced in lines, each line separated from neighbouring lines by unspecialized cells. In the stomatal line, the cell which will give rise to the two guard cells is already made conspicuous close to the growing region by the size of its nucleus. Before this nucleus divides, and the guard cells are formed, it evidently has a profound influence over the nuclei in the cells lying at each side of it, even though they are by no means exactly adjacent to it. When these latter nuclei divide the spindle is always so orientated that one of the daughter nuclei in each instance comes to lie exactly by the side of the nucleus of the stomatal mother cell. A transverse rank of three cells is thus formed, and these ultimately give rise to the stoma and its two associated cells. This pattern of development must follow from the nucleus of the stomatal mother cell exerting some control, by a means as yet obscure, over the macromolecular arrangements in the adjacent cells (Stebbins and Shah, 1960). There seems no reason why internuclear influences of this kind should be confined to the leaves of grasses; similar influences probably occur regularly in the growth of all organs and tissues.

The differentiation of cells

Differentiation almost always accompanies growth, and occurs both within cells, and within organs and tissues.

The single cell of *Acetabularia* provides a striking example of a complex and changing morphology during growth, parts of the cell even being shed. In multicellular higher plants differentiation of certain classes of cells, such as latex ducts and idioblasts, again involves the generation of elaborate shapes. More usual in higher plants is intracellular differentiation resulting in intricate patterns of thickening on the walls, a process particularly well shown by the development of the conducting elements. The control of the differentiation of a cell is certainly complex. In *Acetabularia*, the nucleus is known to play a central role, and this probably results from its general control of protein synthesis (Hämmerling, 1963). Feeding abnormal amino acids to growing *Acetabularia* leads to the production of quite new forms, presumably a consequence of the unusual protein metabolism. In higher plants, the way in which the laying down of elaborate arrangements of thickening on the cell walls is controlled is largely unknown. There is some evidence that the appearance of these thickenings is preceded by some kind of differentiation in the cytoplasm, but little more can yet be said.

Differentiation within a multicellular organ implies that the development of the cells is not uniform, but may follow a number of different paths. This differentiation between cells almost always takes place in a regular pattern and, since physiological gradients are very likely to occur in growing regions, we can infer that this pattern reflects environmental differences within the mass of cells. Support for this view has come from the study of tissue cultures. It is now possible in certain conditions to isolate individual cells of higher plants, capable of growth and division in appropriate media. In liquid culture these cells first give rise to a nodule of similar cells, but when the nodule reaches a certain size, differentiation sets in (Steward, Mapes and Smith, 1958).

The requirement that must be met for differentiation is not that the nodule of cells should reach a critical mass, but that some cells should get cut off from direct contact with the surrounding medium. Once there are cells which are dependent upon other cells for the passage and supply of nutrients, it is possible for considerable environmental differences to arise between the internal and the external cells of the nodule. It is presumably these that cause the cells to behave differently. Even within a stem apex, similar conditions may apply. There are almost certainly radial gradients of various kinds in the differentiating cells at the tip of the stem, but they are difficult to demonstrate conclusively since any sectioning or dissection of the apex inevitably disturbs the system. Once a differentiated system of cells has arisen at an apex, this system probably tends to stabilize and develop itself, since the different kinds of cells will have different permeabilities. Nutrients and auxins, for example, moving longitudinally will be transported at different rates through the different groups of cells, and in consequence regional differences in the differentiating zone will be accentuated. The prime factor in differentiation within a tissue is thus probably a mechanism, possibly fairly simple, which will switch development into one path or another. These initial stages no doubt consist of macromolecular reorganization, invisible externally. In the stem they probably occur in the zone immediately above that in which differentiation is seen to begin. This upper zone is often referred to in consequence as the *zone of determination*.

If differentiation is indeed largely dependent upon local environments in the growing regions, it is not surprising that it is affected by changes in the external environment that would tend to disturb the internal equilibrium. Subjecting a growing stem to traction, for example, promotes the differentiation of fibrous tissues (Table 1), and

Table I. The effect of traction on differentiation in the stem of *Helianthus annuus*. The experiment lasted 29 days, and the plants were subjected to a traction increasing from 25 to 300 gm wt. All values are means of five. (From Bordner, 1910.)

	Breaking strength (kg)	Area of walls of hard phloem fibres in transverse section (arbitrary units)	Total area of xylem in transverse section (cm^2)
Experimental plants	5·75	6·287	0·670
Control	3·65	5·423	0·476
% difference	+56	+16	+41

the bending of branches of trees leads to the formation of an anatomically distinct reaction wood in the affected regions. The mechanism of these effects, sometimes producing conspicuous changes in the anatomy of a plant, must be a complex one, and involve fairly extensive modifications of the metabolism in the differentiating regions.

The differentiation of organs

In a higher plant, the plant body itself is usually strikingly differentiated. Two regions, the shoot and the root, differing in appearance and function, can usually be distinguished without difficulty. Each region develops from a primordium in the embryo, but the factors which establish this primary differentiation are difficult to investigate. Some indication of the factors involved can however be obtained from culture experiments. Isolated pieces of stem, for example, cultured *in vitro* will develop buds, the nature of which can be varied according to the composition of the medium. Indol-3-acetic acid proves to be particularly effective in promoting the formation of roots, and adenine (a purine, and a component of the nucleic acids) of shoots (Skoog and Miller, 1957). Whether meristematic cells give rise to a shoot or a root may therefore depend upon the particular balance available to them of auxins, on the one hand, and, on the other, the purines and pyrimidines of the nucleic acids. Following a fairly simple switch mechanism of this kind, inevitable metabolic sequences may then lead to the morphological and functional differences between roots and shoots.

Biochemical limitations may, of course, prevent an undifferentiated tissue giving rise to certain organs. A striking feature of callus cultures is that, if any organized growth occurs at all, it is first of roots, and only at a later stage, and then in more restricted conditions, of shoots. Moreover, shoots will usually yield roots readily, but shoots are much less commonly produced from roots. A high level of cytochemical complexity may therefore be essential before shoot buds can be formed.

Within the shoot system itself there are striking differences in symmetry of growth between axes and leaves which cannot yet be fully explained, although experiments are again revealing the kind of factors likely to be operative. Very young leaf primordia, only a fraction of a mm in height, can be cut from the apices of ferns and grown aseptically on artificial media (Sussex and Steeves, 1953). These primordia yield leaves, smaller than normal, but of precisely normal form. The 'leaf' nature of a primordium is thus apparently determined at a very early stage, when it consists of little more than embryonic cells. At the same time, surgical experiments upon apices have shown that, if minute cuts are made between the apex and subapical sites at which leaves would be expected, stem-like outgrowths are produced instead. It seems very likely that, because of the actively meristematic cells at the tip, the subapical region is one of steep concentration gradients of various kinds. The lateral primordia therefore begin to develop in an environment which, in the direction of their growth, shows a strong transverse inequality. The effect of cuts on the apical side of the lateral primordia will be to reduce or eliminate this inequality, and it is significant that this results in a radially symmetrical outgrowth. The transverse gradients normally present may therefore impose some macromolecular organization (see p. 301) upon the cytoplasm of the cells of the lateral primordia such that their subsequent growth is determinate and principally in one plane.

The lateral buds of higher plants usually become distinct lower down the apex than the leaf primordia, and consequently distant from the intense concentration gradients of the summit region. This may account for their predominantly radial symmetry. Bilateral symmetry can in fact often be still detected in the lateral buds of some species. Sometimes, as in species of *Gardenia*, this symmetry persists and the branching of a lateral is more or less confined to one plane. No erect axes are produced even if a lateral is struck as a cutting and becomes an independent plant. Nuclear change does not appear to account for the stability of these growth forms since seeds, produced in reproductive organs formed in the strongly bilateral regions, yield progeny with the normal erect primary axis. We are thus possibly again concerned with cytoplasmic changes which, persisting throughout the normal vegetative multiplication of cells, control the symmetry of growth.

Additional evidence in support of cytoplasmic factors of this kind has come from fungal genetics. A number of morphological variants arising spontaneously in cultures of *Aspergillus* were found to be transmitted indefinitely by spores produced asexually. If, however, a culture were allowed to go through a sexual cycle, the progeny were always of the original form. The most satisfactory explanation of this striking result was that the variations were cytoplasmic in origin, and that oogenesis provided an opportunity for the nucleus to 'restandardize' the cytoplasm (Mather and Jinks, 1958).

Flowering

A special problem of growth is that of flowering, an event which, in an annual plant, heralds the closing phase of its life. The onset of this striking change in the manner of growth can be influenced by external factors, particularly in many plants by the length of the daily period of illumination, a phenomenon known as photoperiodism. Unfortunately, despite extensive experimentation, it is still very difficult to identify the general principles governing the control of flowering. Some plants (e.g. chrysanthemum) respond to short days, others (e.g. henbane, and most winter rosette plants) to long days, while yet others appear to be quite indifferent to day-length. It has been found with many photoperiodic plants that, after exposure for a certain minimum period to the required day-length, flowering regularly follows, no matter what the duration of the succeeding daily illumination. This suggests that, as elsewhere in development, we are concerned with a switch mechanism, growth inevitably following a particular path after some initial determinative event. In certain species a purely vegetative shoot grafted on to a plant in which flowering is imminent will itself produce flowers, even though the whole plant may have been returned to the conditions of vegetative growth (Figure 5). This has been held as evidence of a specific flowering hormone, capable of diverting growth into reproductive structures, but it could also indicate that the metabolism peculiar to reproduction, once initiated, comes to dominate the whole structure irrespective of graft unions. The biochemical features peculiar to flowering are as yet little known, but evidence is accumulating that as an apex becomes reproductive, increasing amounts of ribonucleic acid, possibly qualitatively different from that associated with vegetative growth, are found in its cells. Partial interference with the ribonucleic acid metabolism has now been shown in a number of plants to cause the growth to remain vegetative, even though the conditions favour flowering (Hess, 1959). Why there should be these changes in the ribonucleic acid metabolism, and presumably also in the protein metabolism in response to certain periods of daily illumination is an unsolved problem.

Direction of growth

It is a familiar feature of higher plants that growth is directional. Main stems grow vertically upwards (they are said to be 'negatively geotropic') and main roots vertically downwards (hence 'positively geotropic'). Lateral branches frequently grow more or less horizontally, and this also appears to be a response to the gravitational field, not merely a result of indifference to gravity. If a root or stem is displaced from its normal orientation, its subsequent growth will be redistributed and remain so until the direction of growth is restored. Many lower plants are also susceptible to gravity, and sometimes simple filaments consisting of only one elongated cell will show geotropic curvatures. Gravity therefore plays an important part in determining the shape of plants, and botanists await with excitement the bizarre results which may follow when a plant is grown in a space ship removed from the earth's field.

Figure 5. Grafting experiments with *Bryophyllum*, a short day plant. The arrow points to the site of the graft. The stock has been brought into the flowering condition, but returned to long days at the time of the grafting. The scion has come from a purely vegetative plant kept in long days, and although it never experiences short days it changes over to reproductive growth.

Where geotropic curvature takes place in a single cell, as in some fungi, the 'perception' of the stimulus and the response are necessarily confined to one protoplast. In higher plants, however, numerous experiments have shown that, following a displacement, the movements which redirect growth take place at some distance from the regions most sensitive to gravity, usually close to the meristematic tip. The curvature thus results not from an influence on the orientation of cell division, but on the disposition of growth by cell expansion. The transmission of the stimulus to the elongating region, and the differential growth that leads to curvature, appear to depend upon changes in the distribution of the growth-regulating substances leaving the apex. These, of course, include the familiar indol-3-acetic acid, and growth curvatures can readily be produced in decapitated stems by applying a source of this auxin asymmetrically to the cut surface. The difficult problem of geotropism is to discover how the direction of gravitational field is initially 'perceived' by the plant. It seems likely that gravity affects the distribution of special particles or macromolecules in the cell. If these are caused to move by a permanent displacement of the organ, then the protein substructure of the cytoplasm may be disturbed. Since this substructure almost certainly depends upon a complex of electric charges between the various proteins it is plausible to suppose that the disturbance is transmitted as a wave to the bounding membrane of the cell. Once the proteins forming part of this membrane are altered, changes in permeability are almost certain to follow. This may then result in a diminished or increased secretion or transmission of growth-regulating substances, and thus ultimately in a visible change in direction of growth.

A root or shoot which is correcting its position as a consequence of displacement will always 'overshoot'. This will lead to a renewed geotropic stimulus which will tend to bring the organ back again. There is thus a certain amount of oscillation about the vertical before the system becomes restabilized. Shoots which normally grow horizontally behave similarly if they are permanently displaced. The response is thus similar to the 'actual' response of Figure 17 in Chapter 3. That the oscillatory effect is not merely the result of 'overshooting' and renewed stimulation is, however, shown by experiments in which, after the initial stimulation, the organ concerned is rotated horizontally. Horizontal rotation does not of course eliminate the gravitational field, but it does cause it to act transversely to the longitudinal axis in every direction in turn, so that a shoot, for example, taken from its normal vertical position and so rotated will remain horizontal. If the organ has been stimulated by gravity before the rotation, both the curvature and the final oscillation still occur. Some factor other than renewed stimulation must thus be responsible for, or contribute to, the final correcting movements. It appears as if the 'perceptive' cells have a 'preconception' (no doubt ultimately explicable in molecular terms) of what their orientation in space should be.

The orientation of some simple filaments, of growing stems, and to a much smaller extent of roots, is also affected by light. Unilateral illumination at low intensity will cause a stem to align its growth parallel to the incident light and towards its source. Here again there is often separation between the 'perceptive' region and that in which the differential growth producing the realignment takes place. As with geotropism, the growth effects can be satisfactorily explained by changes in the distribution of growth-regulating substances (Figure 6). The problem lies again in the mechanism of 'perception',

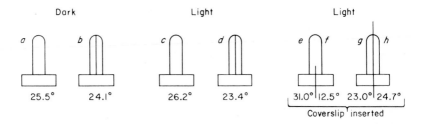

Figure 6. The effect of unilateral illumination of low intensity on the downward diffusion of auxin in the coleoptiles of maize. The coleoptiles are standing on blocks of agar gel, and the figures indicate the concentration of auxin in the blocks at the end of the experiment. *a*, intact coleoptiles kept in darkness; *b*, coleoptiles split longitudinally but the halves in contact, in darkness as *a*; *c*, intact coleoptiles illuminated unilaterally; *d*, coleoptiles split longitudinally as in *b*, illuminated as in *c*; *e* and *f*, coleoptiles partially split longitudinally, a thin glass coverslip inserted into the split, and the auxin from the light and dark sides collected separately; *g* and *h*, coleoptiles completely split longitudinally, the halves separated by a coverslip, and the auxin collected as in *e* and *f*. It is clear that the amount of auxin diffusing from the coleoptile is not seriously disturbed by either the experimental treatment or the illumination. The auxin does, however, migrate to the shaded side. (From Briggs, Tocher and Wilson, 1957.)

but the situation here is not so intractable. Several classes of macromolecules capable of taking up light energy are known to occur naturally in the cells of both unicellular and multicellular plants, although there is still some uncertainty about those actually concerned with phototropic effects (see p. 297). Once taken up, the energy would be capable of causing the same kind of wave-like disturbance as that envisaged in the cells sensitive to gravity, with similar physiological consequences. As with geotropism, 'overshooting' and correcting movements occur after phototropic stimulation, strengthening the view that these arise from some innate property of the sensitive cells.

Conclusion

The foregoing account, necessarily brief, reveals the complexity of plant growth. It becomes clear that the way in which a plant grows is not solely the consequence of inherited potentialities. We have to envisage a continual interaction between the specific kind of growth determined by the chromosomes, and possibly in part by the inherited cytoplasm, and the environment in which the plant is situated. This interaction is of course reciprocal. Many features of the environment, e.g. the atmospheric humidity and the properties of the soil adjacent to a plant, are continually changed by its growth, quite apart from fluctuations in the environment arising from cosmic causes.

In view of this continual and unstable interaction between plant and environment, it is not surprising that the mechanisms which regulate the total growth of a plant, and which channel, direct and determine its kind, are numerous and not sharply set off from each other. Moreover, most mechanisms that have been proposed, e.g. those relating to apical dominance and growth movements, are not wholly satisfactory. This probably results inevitably from the fact that physiology, to facilitate research, is forced to regard growth as consisting of a number of different aspects, and to study each more or less independently and frequently in highly artificial conditions. Nevertheless, much more research has to be done along these relatively simplified lines before we can begin very profitably to consider how these empirical relationships may be modified and integrated in natural vegetation.

References

General

Audus, L. J. (1959). *Plant Growth Substances*, Leonard Hill, London.
Bünning, E. (1953). *Entwicklungs- und Bewegungsphysiologie der Pflanze*, Springer-Verlag, Berlin.
Salisbury, F. B. (1963). *The Flowering Process*, Pergamon, Oxford.
Sinnott, E. W. (1960). *Plant Morphogenesis*, McGraw-Hill, New York.
Wardlaw, C. W. (1965). *Organization and Evolution in Plants*, Longmans, London.

Special

Albaum, H. G. (1938). 'Normal growth, regeneration and adventitious outgrowth formation in fern prothallia.' *Am. J. Botany*, **25**, 37–44.
Bordner, J. S. (1910). 'Influence of traction on the formation of mechanical tissue in stems.' *Botan. Gaz.*, **48**, 251–274.

Briggs, W. R., Tocher, R. D. and Wilson, J. F. (1957). 'Phototropic auxin redistribution in corn coleoptiles.' *Science*, **126**, 210–212.

Evans, G. C. and Hughes, A. P. (1961). 'Effect of artificial shading on *Impatiens parviflora*.' *New Phytologist*, **60**, 150–180.

Haber, A. H. (1962). 'Nonessentiality of concurrent cell divisions for degree of polarization of leaf growth.' *Am. J. Botany*, **49**, 583–589.

Hämmerling, J. (1963). 'The role of the nucleus in differentiation especially in *Acetabularia*.' *Symp. Soc. Exp. Biol.*, **17**, 127–137.

Hess, D. (1959). 'Die selektive Blockierung eines an der Blühinduktion beteiligten Ribosenucleinsäure—Eiweisssystems durch 2-Thiouracil.' *Planta*, **54**, 74–94.

Key, J. L. (1964). 'Ribonucleic acid and protein synthesis as essential processes for cell elongation.' *Plant Physiol.*, **39**, 365–370.

Lewis, F. T. (1930). 'A volumetric study of growth and cell division in two types of epithelium.' *Anat. Record*, **47**, 59–99.

Mather, K. and Jinks, J. L. (1958). 'Cytoplasm in sexual reproduction.' *Nature*, **182**, 1188–1190.

Mohr, H. (1964). 'The control of plant growth and development by light.' *Biol. Rev. Cambridge Phil. Soc.*, **39**, 87–112.

Naylor, J. M. (1958). 'Control of nuclear processes by auxin in axillary buds of *Tradescantia paludosa*.' *Can. J. Botany*, **36**, 221–232.

Reed, H. S. and Holland, R. H. (1919). 'The growth rate of an annual plant (*Helianthus*).' *Proc. Nat. Acad. Sci. U.S.*, **5**, 135–144.

Sax, K. (1935). 'The effect of temperature on nuclear differentiation in microspore development.' *J. Arn. Arb.*, **16**, 301–310.

Skoog, F. and Miller, C. O. (1957). 'The chemical regulation of growth.' *Symp. Soc. Exp. Biol.*, **11**, 118–130.

Stebbins, G. W. and Shah, S. S. (1960). 'Cytological features of stomatal development in the Graminae.' *Develop. Biol.*, **2**, 477–500.

Steward, F. C., Mapes, M. O. and Smith, J. (1958). 'Growth and organized development of cultured cells.' *Am. J. Botany*, **45**, 693–703.

Sussex, I. M. and Steeves, T. A. (1953). 'Growth of excised fern leaves in sterile conditions.' *Nature*, **172**, 624–627.

Turing, A. M. (1952). 'The chemical basis of morphogenesis.' *Phil. Trans. Roy. Soc., London, Ser. B*, **237**, 37–72.

11

Regulation in Animal Development

D. R. NEWTH

Department of Zoology, University of Glasgow

Embryonic Regulation

The importance of regulation in development was first acknowledged by Heider (1900) sixty-five years ago. He used the word to describe the reactions of the eggs and embryos of certain kinds of animal to gross experimental interference. Contrary to the perfectly rational expectation that these small, soft, and delicate-looking pieces of living matter would prove highly susceptible to the smallest distortions, wounds, or other injuries, they proved capable of withstanding the most extraordinary assaults, surviving, and completing their developmental programme successfully.

The expectation of delicacy was rational for several reasons. It was already understood that the price to be paid for sexual reproduction is that the individuals of each generation have to pass through a stage when they consist of a single cell. This throws a two-fold burden upon the species. It has to contrive a machinery for converting a single cell into an anatomically complex multicellular organism, but it is impossible to do this and at the same time ensure that the early stages of the process are as effective in exploring and mastering their environment, and as successful at protecting themselves from it, as will be the adults into which they turn. So the early stages of most animals—the eggs and embryos—are the passive beneficiaries of parentally contrived protection. They inhabit shells, jellies, capsules, or wombs. They are endowed with food reserves to see them through a period when they are unable to feed themselves, and since this makes them attractive to predators they are laid in obscure places, or camouflaged, or both. If these devices are, for any reason, impracticable the eggs are laid in vast numbers. All this suggests that embryos are both vulnerable to interference from without and heavily preoccupied with their own affairs.

These, as the 19th century embryologists had discovered, were of a complexity that was awe-inspiring. To turn one cell into a population of cells by repeated cell division is only the beginning. The population thus formed must become capable of entering upon a multiplicity of roles in the adult organism. Its members must be marshalled into groups which follow different paths (not only metaphorically, but literally as a result of morphogenetic movements) to different fates. Changes in form, displacements of material, the assumption of new functions, and the establishment of new interrelationships between cell groups—all would appear to involve minute control of timing, of spacing, and of metabolic change. To the early embryologists the intricacy of the processes they studied seemed as far removed from the working of the adult finished product as is that of the automatic machine producing matchboxes today from the simple box that it produces.

Clearly, if this impression were soundly based we could predict two things. Firstly, that an animal's vulnerability to death from outside interference would increase as one proceeded backwards in time towards its conception. Secondly, that the after-effects of sublethal injury would be more pronounced the earlier the injury was inflicted. After all, small mistakes in the early stages of any constructional work tend to be magnified as they are built upon. These expectations are realized in some cases. A crude example is that certain species of nemertine worm can be transected when adult, and each half will survive and proceed to form an anatomically perfect replica of a complete worm. Nemertine embryos, when transected in certain planes, will produce anatomically inadequate and inviable end-products of development. Nature, too, provided examples, for it had early (Chun, 1880) been noticed that severe storms in the Bay of Naples were followed, after an interval, by the appearance of ctenophores (comb-jellies) with 2 or 4 instead of the usual 8 rows of comb-plates. These were the viable but anatomically incomplete products of embryos broken into pieces by the turbulent water.

Hence the discovery by Driesch (1892) and others that a totally different reaction to injury was possible, caused, and deserved to cause, a profound impression in embryology. Driesch demonstrated that cells normally destined to give rise to one half or one quarter of a sea-urchin embryo—and hence it might fairly be assumed to half or a quarter of the adult—could, when isolated, give rise to a whole one. Whole, that is, in the sense of anatomically complete, viable, and differing only in size from the normal pattern of their kind. We now know that many kinds of animal (including higher mammals) react in this way. It was such behaviour that Heider described as regulative, in contrast to the 'mosaic' behaviour of eggs such as those of ctenophores whose various parts tended, when isolated, to give rise to no more and no less than they would in normal circumstances.

It is perhaps worth stressing the paradox that the mosaic, non-regulative, behaviour of part of a developing system is that which is apparently unaffected by violent change; that is not thrown off by unforseeable events; that persists calmly in a course which is doomed only because the rest of the system is not there or is not playing its part. Regulative behaviour may achieve a normal and viable end-result, but only by virtue of making each part almost hysterically sensitive to the condition of the whole.

We no longer regard the distinction between regulative and mosaic eggs as absolute or invariant, because many eggs and embryos have been shown to be capable of both kinds of behaviour though at different stages in development. Indeed, while an embryo as a whole may be 'mosaic', *within* the separate parts whose behaviour appears so rigidly circumscribed, considerable regulation may be possible. If we still, for convenience, describe eggs as regulative or mosaic, it is with the developmental period which lies between the first cleavage division and the beginning of gastrulation in mind.

Regulation for Quality

When a developing system is divided into two and one or both parts successfully regulate, our criterion for their success is primarily qualitative, although the effect of our operation is clearly quantitative as well. We are impressed by the normality of form and proportion, by the presence of all appropriate organs and by the correctness of their anatomical relations. In fact most of these things have a quantitative aspect, and many can be wholly or partly expressed in quantitative terms. Nevertheless, we are conscious of being in the presence of a phenomenon which sees so great a shifting of the presumptive fates of egg parts that only qualitative descriptions are relevant. If a cell which is normally destined to give rise to skin becomes a neurone instead, in obedience to the demands of a regulating system, then though we may believe the difference to be due to molecular changes capable of quantitative definition we can only, at the present time, describe the change in qualitative terms.

There are, indeed, situations in which regulative forces are brought into play in systems which are not subjected to quantitative change. For example, parts of an amphibian embryo may be caused to disintegrate into isolated cells by placing them into a calcium-free medium or one containing a chelating agent. On restoring available calcium, reaggregation and normal development may follow (Curtis, 1957). No change in the total mass of living material has occurred, and yet this is as striking a demonstration of regulative behaviour as one could wish. Somehow the cells *either* have got back into their correct normal places *or* have at least gone to places from which normal developmental behaviour can be provided.

It is this aspect of regulatory behaviour which is furthest removed from the homeostatic mechanism of the adult and is the most difficult to interpret in terms of electrical, mechanical or chemical models. It seemed, in fact, to imply that embryonic cells not only have some faculty for recognizing the whole design to which they are committed but can direct their activities towards its realization. Such words may carry a flavour of unacceptable mysticism and have in the past been associated with vitalistic conceptions in biology. But there is a case for saying that it would have been foolish to conceal our ignorance and that to make statements in unsatisfactory terms underlined the challenge that our ignorance presented.

Regulation for Quantity

Initial mass and final mass

Since eggs are necessarily smaller than the females which produce them, the reproductive cycle by which one female produces another must include growth processes. Strictly speaking this is not a necessary feature of the production of males, since the minute size of spermatozoa makes it theoretically possible for males to be no greater in mass than the egg from which they come. That this is not a wholly fanciful idea is shown by the worm *Bonellia* whose male nearly achieves such a self-effacing extreme.

However, growth need not be uniform in rate throughout post-fertilization life, or even during the shorter time between fertilization and sexual maturity. Indeed, it is very rare for eggs and young embryos to absorb nutrients from the environment, although they may absorb water, exchange gases, and discharge nitrogenous waste products. There are, it is true, a few known cases (for example, among sponges and mammals) in which very young embryos feed, but more often they are at first content to mobilize for metabolic use the food reserves (yolk) provided for them by the mother. Where this is so the early embryonic life of the individual sees little growth in mass, provided that the yolk is counted as part of the embryo and not regarded as outside it. Yet shortly before, or at the time of, or at the latest shortly after, the consumption of the last of the yolk, true feeding must begin, and very active growth usually starts at this time. One question naturally arises: does the rate of growth, or its duration, or any other factor influencing the definitive size towards which the individual grows, depend at all upon the initial size of the egg?

Egg size does vary within species, and within the output of a single female. But while females of races with small body size in some cases tend to produce eggs of a small mean size, it does not follow that within a given clutch the smaller eggs will necessarily grow into smaller adults. There are good reasons for believing that they need not.

One striking example of size-regulation in growth comes from our own species. Approximately 1 in 80 of all live births in European populations are twin births and over one-third of these are monozygotic. The mean weight of twins at birth is not, of course, really helpful, because the length of pregnancy is very variable and a woman might tend to go into labour earlier if she had twins—for example, if the timing of parturition was in part dependent upon the total weight of the womb and its contents. Indeed, we find that twins are, on average, much lighter at birth than single children, though much heavier than triplets. Thus the mean weight at birth of one sample of single-ton boys was 3260 grams. The boy twins weighed on average 2466 grams each, and even if only those thought to be born after a full 38 weeks of pregnancy were counted their mean weight was 2778 grams (Drillen, 1964).

But if final size were strictly related to initial size we might expect that the population of adult twins would be of two kinds: dizygotic (fraternal) twins falling within the normal range of adult size, and monozygotic ('identical') twins, being approximately half-sized. Since this does not happen we may assume that regulation for final size occurs. However, this is not a really decisive case because we assume, without knowing, that human identical twins are initially (i.e. at the time of twinning) half-sized. This may well be so, but the direct evidence of controlled experiment is more convincing. For example, rabbit half-eggs have been grown to maturity and their size, though low at birth, is within the normal range at maturity (Seidel, 1952).

These cases show that whatever the instructions to the egg about its ultimate size may be, they are not given in the form of a multiple of the initial size of the zygote. The implications of this are obvious. Total size control cannot be exercised by performing some simple operation, say doubling a volume, a preset number of times. It must be exercised by a feedback system which is capable of monitoring the instantaneous values of the size achieved.

Organ size and cell-number

The allocation of subpopulations of the cells formed during cleavage to their different fates involves granting to each future organ a standard volume and mass of cells or, which normally comes to the same thing, a standard number of cells. Hence the machinery controlling this segregation could act by measuring volume or mass, or could achieve the same result by counting cells. We can find out which by altering one variable but not the others.

This is most easily done by altering the ploidy of the newly fertilized egg while leaving its volume intact. Haploid cells are very roughly half the volume of normal diploid ones, triploid half as large again, and tetraploid cells are bigger still. Embryos consisting wholly of cells with any one of these ploidies can be persuaded to develop some of the way, and the higher ploidies are fairly viable. But since initial total volume is constant the change in ploidy n, $2n$, $3n$, $4n$ onwards means a decrease in embryonic cell numbers as cell size increases. The rudiments of organ systems involved in these cases retain approximately normal volume relations. Hence no organ in a haploid animal is 'satisfied' by its normal number of cells, and no organ in a polyploid 'demands' its normal number. The initial allocation must be done by volume or by mass.

In most animals, unlike most plants, total body size does not appear to vary as the ploidy of their cells. Mass or volume, rather than cell number, is the quantity to which the controlling system is sensitive. Nevertheless some invertebrates do show increase in body size with the ploidy of their cells.

Inductive relations, too, can be volume-dependent, for after making grafts between newt embryos with different ploidies it is found that the induced structure (i.e. lens) has the normal volume rather than the normal number of cells. This, of course, matches it to the size of the inductor (eye-cup) into which it has to fit. However, similar experiments in which donor and recipient embryos are of species which differ in lens size show that the volume of the induced lens rudiment is an inherent property of the lens forming cells (Twitty, 1955).

Integral Counting and Number Regulation

Yet there may be exceptions to these findings where small numbers of cells are involved. We have only to consider the instance where the dimensions of an organ are normally related to the linear dimensions of a single cell to realise that volume regulation may be incomplete. The thickness of cell epithelia is not constant through the series of different ploidies. Mauthner's fibres in the spinal cord of urodeles are normally present as a single pair. In tetraploids they are large and hence occupy an abnormally high proportion of the animal's total volume. In haploids they are small, but occasionally make good this deficiency by doubling in number. My colleague, Dr L. Hamilton, has suggested that one striking abnormality common in haploid amphibian larvae—lordosis, or a concave back—may result from a disproportionate shortening of each of the fixed number of myotomes running in pairs down the animal's back. Each is one cell long, and in haploids consequently rather shorter than in diploids.

If true, this is but one example of the ability of developing systems to count fairly accurately, an ability which has been discussed recently by Maynard Smith (1960).

Qualitative effects of quantitative change

The mass or volume of material participating in a developmental process can often be increased or decreased without qualitative change. This, for example, happened when Mangold and Seidel (1927) performed the complement to the newt twinning experiment by fusing two 2-cell stages together. The result was good regulation in this case too, for the combined eggs gave a single, large, but well proportioned embryo. Recently it has proved possible to fuse mouse eggs together and produce viable young (Tarkowski, 1964). Yet there are cases in which change in the initial quantity of material leads to incomplete regulation and an end result that is capable of qualitative distinction from the normal.

Within limits one can reduce the volume of a developing amphibian limb-bud and still obtain a limb of normal form. There is, however, a threshold size below which an abnormal limb is formed. It is true that the abnormality takes the form of missing skeletal elements or whole digits, and thus has a quantitative aspect; but other qualitative differences between organisms do likewise. The point has been equally well made by increasing the volume of material in a developing fowl's leg. Here the result is a larger leg whose disproportionately larger fibula assumes the importance, and furthermore the anatomical relations, of this bone in *Archaeopteryx* (Wolff, 1958).

The Biological Significance of Embryonic Regulation

For embryology regulative behaviour has the important consequence of ruling out of court any general theory of development which demands that a precise, detailed, and irrevocable set of orders is given to various parts of the egg at or before fertilization, and that subsequent development is *only* the execution of these orders. But for those interested in homeostatic control systems in biology it poses a series of problems about regulating mechanisms that are not normally considered.

First we must notice that the effect of embryonic regulation is not in the ordinary sense to restore the *status quo ante*. Since the system which is regulating is simultaneously and independently changing its properties with time, any such restoration would be retrogressive. What regulation in fact accomplishes is a normal end result, despite a distortion of the process by which it is achieved. Yet there is one sense in which we can say, if we wish, that regulation here is restoring or tending to restore an earlier state. This is if we take the whole reproductive cycle into consideration and regard the developmental process as leading to the formation of a replica of the adult of the last generation. In this sense only regulation serves to restore an earlier prevailing state.

Such a view does not, however, rob the processes of regulation of their problems. How is it that an embryo, half of whose substance has been removed, can reorganize the other half to give a complete individual? Is the analogy perhaps not with a simple regulator but rather with the kind of control system which has as its model the 'lock-on' radar system described in Chapter 2. Here there are similarities but also an important difference. If we watch a lock-on radar performing on a series of aircraft which cross from one horizon to the other, the fact that it is functioning by first searching, then finding, and then holding its target aircraft, is effectively established because the different aircraft will take different paths across the sky, and however they move the radar follows them. It is thus a natural assumption that the radar really is following the aircraft and is not accidentally tracking the same path.

But now let us suppose that one watches such a radar set while a series of aircraft fly over, each following very nearly the same course. The outside observer is then unable to say with any confidence whether the radar set is guided only by the particular target or is also using the knowledge that all targets follow such a course to simplify its problem. Similarly when we watch successfully accomplished embryonic regulation we cannot be sure which of two possible controlling mechanisms has been at work. It could be that the developing system is continuously monitoring its progress by comparing it with a timetable-like programme laid down for it. In this case the departure of performance from the programme could generate an error signal and evoke a restoring response precisely tailored to the situation. On the other hand, successful regulation may only mimic such a control system; there may be no error signal of this sort, but only a general set of instructions on cell behaviour which happen to be able to take certain kinds of developmental accident in their stride. If we examine closely some of the limitations or failures of the regulative ability of those embryos which regulate most easily, we get some clue to which model is appropriate.

For example, it is not always true that a frog or newt egg divided into two along the plane of the first cleavage division will respond by the formation of identical twins. In a certain, not very large, proportion of cases, the plane separating the products of the first cleavage division passes through the egg in such a way that two non-equivalent cells are formed. If these cells are now separated and allowed to develop independently, only one of them provides a normal embryo, or regulates: the other fails to do so. In other words it is not quite correct to see in the amphibian egg a wholly uncommitted developmental system capable of regulating any sort of damage whatsoever. On the contrary, it is like a match which, if split along its length, will provide in principle at least two functional matches each capable of being struck, but if divided transversely can provide only one match capable of igniting, the other being a useless piece of wood. But if this is so then one is entitled to question the view that the regulative behaviour of these embryos is, in fact, a homeostatic device provided to help the egg cope with situations similar to those which the experimentalist provides.

Furthermore, the existence of mosaic eggs forbids us to regard the ability to regulate at any particular stage as an essential characteristic of animal development. How then did these striking regulative abilities evolve? Clearly not in anticipation of the assaults of experimental embryologists, but may it not safely be supposed that there

11*

is selective advantage in being able to make good less drastic accidental damage? If this were true, we should have to assume that for mosaic eggs there were compensating advantages of another sort, and this is quite reasonable. The difficulty that this view of regulation has to face is that the capacity to regulate in the laboratory so often far exceeds any demand that is ever made upon it in nature. In order to demonstrate that newt eggs can twin, it is necessary to perform technical tricks that it is safe to assume are mimicked in nature with a quite negligible frequency.

But though embryonic regulation may not often be called upon to reveal its full powers, natural selection has nevertheless offered it the recognition of using it to accomplish asexual reproduction in some species. The human identical twin is rare enough to be accounted an accident, but polyembryony is normal in some invertebrates and in the armadillos.

Most adult homeostatic devices must expect to be stretched to the full at some time or other and death is, in the last resort, the failure of one or more of them to cope with the demands made upon it. It is therefore surprising that embryonic regulation should be so excessively over-assured. Most embryologists would, in fact, be prepared to regard the external mechanical hazards to eggs and embryos as small indeed, and turn instead to endogenous sources of developmental abnormality as the important ones. Here we may, for convenience, consider two major categories. One consists of the contingent accidents that must intervene in the working of any very complicated system: this we can call noise in the meaning of modern information theory. The other category that must be considered is genetic abnormality.

Noise-type incidents in development may be of many kinds. For example, we must expect that, in addition to the morphogenetically-significant localized cell death to which Glückmann (1951) has drawn attention, there will inevitably be a cell mortality distributed throughout the whole embryo and occurring at all stages of its development, though not necessarily with an incidence that is uniform in space and time. Since neither we nor the embryo are in a position to predict which cell will die from, say, aneuploidy or a dominant lethal mutation, a general ability to regard any cell as expendable is certainly a device that might have selective value. It would be interesting to compare the radiosensitivities of embryos of highly regulative and of mosaic species, or of embryos of the same species at regulative and mosaic phases. One might predict that the mosaic form would show visible defects after a wide range of doses of ionizing radiation well below its

LD_{50},* but that regulative ones would show little effect until doses near the LD_{50}.

For all these reasons it seems wiser to regard the capacity to regulate as primarily an expression of genetic instructions. We might, of course, expect that it was a conservative force, since minor deviants of genetic origin from normal behaviour in embryonic cells would not be expressed because regulation would treat them as it would treat injuries of external origin. This prediction has interesting consequences for the spectrum of genetic abnormalities that might survive to scoring in mosaic and regulative eggs respectively. We should again expect many small phenotypic variants to show up in mosaic eggs, and the eggs themselves to have a lower threshold of genetic damage for lethality. Regulative eggs would show fewer trivial morphological departures from normal and a higher threshold for lethality. Such an argument, however, assumes quite arbitrarily that the norm to which regulation proceeds is not set by the genome of the fertilized egg, including such mutant alleles as it may have in expressible (dominant or homozygous recessive) form. This may well be true for any genes which have a short period of expression limited to embryonic life. It is not known to be true of any, and it remains one of the outstanding gaps in our understanding of morphogenesis that we do not know whether the processes of embryonic regulation and of post-embryonic morphological homeostasis—repair and regeneration—always tend to the morphological expression of the genotype of the cells involved, or if they sometimes tend to the creation of a more general species-specific form. This problem is, in practice, more easily posed for post-embryonic processes, notably regeneration.

Regeneration and Control of Adult Morphogenesis

Regeneration, or the replacement of lost parts in post-embryonic life, immediately invites comparison with embryonic regulation but, in fact, usually differs from it in significant ways. Biologically speaking there is more evidence for the selective advantage of regenerative capacity since animals do commonly lose extremities from accident or disease. Nor can one easily think that any capacity to regenerate, however extensive, is greater than its possessor will ever need. Yet in the somewhat haphazard distribution of regenerative ability in the animal kingdom, apparently largely unrelated either to likelihood of

---

* LD_{50} is a dose which kills 50% of the living organisms exposed to it within a specified time after exposure.

injury or to susceptibility to it once inflicted, we are faced again with the paradox that so desirable a characteristic is not more universally enjoyed. That selection does recognize the existence of regeneration is shown by those cases, amongst vertebrates and invertebrates, in which anatomical provision in the form of pre-existing fracture planes is made for the self-inflicted loss of parts which can be replaced (autotomy).

But while embryonic regulation is typically morphallactic, that is, it occurs at the expense of a 'remodelling' of the whole regulating system, practically all of whose parts suffer a change in fate, regeneration is commonly not. Morphallaxis does, it is true, enter into the regeneration of many invertebrates, but in other cases the new tissues are formed almost entirely from local cells at or near the site of loss or damage, a form of restitution called epimorphosis.

The control of these cells can be considered under three heads, of which the first covers the initiation of regenerative activity, or to put it another way, the repression of such activity until loss makes it desirable. In certain cases where organs are controlling the chemical composition of the blood (e.g. endocrine organs, liver, kidney) loss of part of the organ may lead at once to functional overloading of what remains. Here the failure of normal adult homeostasis provides a possible error signal for regeneration. And yet just because the normal organ *can* operate over a wide range of loads, the failure of the truncated one does not provide an adequate model for regeneration control. The error signal here must arise either from something directly monitoring the mass of tissue in the body or from something monitoring the abnormal functioning of the remaining tissue. Otherwise the removal of small parts, with only slight increases in functional load, would not result in replacement. Abercrombie (1957, 1959), Weiss (1955), and Goss (1964) have discussed organ growth control using models applicable to regeneration of this sort.

Epimorphic regeneration, as of limbs or tails, does not depend upon loss as such. This can be shown in several ways: the removal of organ rudiments in a mosaic stage of development need not lead to their replacement at a later stage when regenerative capacity is at its height. Neither need removal in later life, if the wounds caused by amputation are dressed in certain ways. In these cases it appears that the trauma of amputation involving, as it does, the section of nerve fibres, the death or wounding of cells, and the loss of body fluids is the initiator of regenerative activity; that is, it leads to an accumulation of undifferentiated cells at the wound site. On the other hand, 'regeneration' can sometimes be provoked without prior loss, as when newt limbs are formed in response to one of a number of stimuli applied to the flank (e.g., the implantation of nasal placodes or the diversion of the distal end of severed nerves).

These blastema cells proceed to undergo a morphogenesis strongly reminiscent of the primary development of the limb, and one is tempted to imagine that they do so under the control of the same genetic information as the embryonic cells possessed. Mitotic nuclear division, by which their nuclei are descended from that of the fertilized egg, might be held to have as its sole concern the handing on of unaltered genetic information. And yet before assuming that blastema cells work solely to the commands of their own genes, we must admit that other possibilities exist.

During primary embryogenesis an essentially one-dimensionally coded message is decoded, and the information it contains is used in the construction of a three-dimensional organism. In principle the two can be interchanged as sources of information. For example, when we know the base-sequences coding for blood group antigens we shall, in principle, be able to type individuals either from the antigens at their cell surfaces or from the DNA within their nuclei. Thus in replacing a lost part the blastema cells could call upon the morphogenetic information they require not only from their nuclei but also from the rest of the organism to which they belong.

In fact there is some evidence that blastema cells do use some information from outside themselves. In this they differ from most eggs which are normally, after fertilization, not dependent upon the environment for clues to development. We can ignore, of course, the supply of nutrients, or the satisfaction of other general metabolic dependencies which blastema cells share with other cells. These, though providing real prerequisites for regeneration, are not known to control its course. It is more difficult to be sure of the role of the nervous system which has often been shown to play an important, if non-specific, role in blastema formation. Because it is non-specific, i.e. independent of the central connections of the nerve fibres, we may assume that at most very little morphogenetic information is conveyed in this way.

On the other hand, there are cases where the contacts of the blastema cells with the healthy tissue of the stump left after amputation are known to determine the kind of tissue into which they will grow. Thus the presence of bone in the stump of a limb at the wound surface may cause skeletal tissues to appear opposite it in the regenerate, or the presence of fin tissues at the level of amputation of a newt tail may be a precondition of the formation of the same tissue in the regenerate (see Newth, 1958).

What we need, in order to determine the relative importance of nuclear information and extra-nuclear information in the determination of a cell's behaviour in development, is large numbers of observations in which regulation or regeneration take place in cell populations that have abnormal genetic (i.e. mutant) or abnormal extra-nuclear information at their disposal. We do possess evidence that the capacity of regulation or regeneration is not always adequate to cope with severe damage to extra-nuclear information sources. This is true of the newt egg divided so that its halves are not symmetrical, and of some of the regeneration situations discussed by Newth. In few cases only, have we well-established genes in animals suitable for experimental work of this kind. Where we have, the genetic constitution of the cell has been found to be all-important.

Summary

Animal cell populations destined to perform developmental tasks by giving rise to differentiated aggregates normally pass through phases during which dislocation of their spatial relations has little or no effect upon their later organization. The obvious explanation for this immunity to disruption is that the significant morphogenetic information upon which they will rely is still present only in either a protected or coded form. The capacity of the genetic code to be replicated at cell division and to remain intact during development exacts a rigid restriction of the repertoire of proteins that a cell can synthesize, but also makes possible a useful delay in determination during development. Cells can thus reserve their decision about their ultimate roles in the economy of a multicellular system until after some of the difficulties of early morphogenesis have been surmounted.

It is therefore possible to regard regulation in primary morphogenesis as a by-product of delay in restricting the potency of the cells involved, rather than a homeostatic process selectively favoured as an answer to environmental insults. Adult morphogenesis and regeneration on the other hand, presents a less clear-cut pattern of control in which the genetically determined properties of the participating cells are not always alone sufficient to guide morphogenesis to completion. Systemic and local somatic influences, derived at one remove from the genetic information in the egg, are also important in the control of regeneration.

References

Abercrombie, M. (1957). *Symp. Soc. Exp. Biol.*, **11**, 235.
Abercrombie, M. (1959). *Lect. Sci. Basic Med.*, **8**, 19.
Chun, C. (1880). *Die Ctenophoren des Golfes von Neapel*, Engelmann, Leipzig.
Curtis, A. S. G. (1957). *Proc. Roy. Phys. Soc. (Edin.)*, **26**, 25.
Driesch, H. (1892). *Zeit. Wiss. Zool.*, **53**, 160.
Drillen, C. M. (1964). *The Growth and Development of the Prematurely Born Infant*, Livingstone, Edinburgh.
Glückmann, A. (1951). *Biol. Rev.*, **26**, 59.
Goss, R. J. (1964). *Adaptive Growth*, Logos Press, London.
Heider, K. (1900). *Verhandl. Deutsch. Zool. Ges.*, **10**.
Mangold, O. and Seidel, F. (1927). *Roux. Arch. Entw. Mech. Organ.*, **111**, 593.
Maynard Smith, J. (1960). *Proc. Roy. Soc. (London)*, Ser. B, **152**, 397.
Newth, D. R. (1958). *New Biology*, **26**, 47.
Seidel, F. (1952). *Naturwissenschaften*, **39**, 355.
Tarkowski, A. (1964). *J. Embryol. Exp. Morphol.*, **12**, 575.
Twitty, V. (1955). In Willier, B. H., Weiss, P. and Hamburger, V. (Eds.), *Analysis of Development*, Saunders, U.S.A.
Weiss, P. (1955). In Butler, E. G. (Ed.), *Biological Specificity and Growth*, Princeton University Press, U.S.A.
Wolff, E. C. (1958). *Bull. Soc. Zool. Fr.*, **83**, 13.

12

The Preservation of Genetic Variability

SHEILA MAYNARD SMITH
and
J. MAYNARD SMITH
School of Biological Sciences, University of Sussex

It is a truism to say that the adaptations of animals and plants to their environment arose and are maintained by natural selection. This can be expressed by saying that the genotype, and the phenotype which it controls, is regulated by selection so as to correspond to requirements set by the environment. But it is not this aspect of the regulation of genetic systems which we intend to discuss here. Instead we shall discuss the processes which maintain the genetic variability of populations, and which cause a resistance to change in their genetic structure. The reason for this choice is that these processes have been referred to as examples of 'genetic homeostasis', and so an analogy has been drawn between them and the regulative processes which occur in the individual organism and which are the subject matter of physiology.

Homeostasis was first used as a general term for those processes which ensure the constancy of the internal environment. The term has since been extended to include processes which regulate development (Lerner, 1954), the genetic structure of populations (Lerner, 1954) and the numbers of animals (Wynne-Edwards, 1962). In this chapter we shall criticize the use of the term in population genetics. Since the term has a long and respectable history in physiology, its use in genetics would be justified only if there is a reasonably close analogy between the physiological and genetical uses, and this we do not believe to exist.

The types of observation which have led to the coining of the term 'genetic homeostasis' are described in the first section; the description is brief, since the correctness of these observations is not at issue. They show that there are features of the genetic structure of a population, measured as the frequency of particular alleles or combinations of alleles, which are resistant to change, and which, if they are changed by some distorting influence, tend to be restored as soon as that influence is removed. In the second section of this chapter, we shall review the mechanisms which can maintain stable intermediate gene frequencies. Our treatment of these mechanisms is more detailed than is necessary for our argument, because they are of interest whether or not we decide to describe them as homeostatic. Our reasons for wishing to abandon the term 'genetic homeostasis', or at least to restrict its use to certain particular cases, are given in the third section.

The Observational Evidence

Two main classes of observation have led to the concept of genetic homeostasis; the response of laboratory populations to selection, and in particular to the relaxation of such selection, and the maintenance of genetic heterozygosity in random-mating populations in the wild or in the laboratory. One example of each class will be described.

Artificial selection of varying degrees of severity has been practised on metrical characters such as abdominal chaeta number in *Drosophila melanogaster*, egg production in poultry or protein content in corn. Selection away from the average in either direction is often accompanied by a severe and embarrassing loss of fertility, with the result that many lines die out unless the intensity of selection is relaxed. Following such relaxation, the previously selected line frequently regresses very considerably towards the initial population mean. Results like this were obtained by Mather and Harrison (1949) when selecting for both high and low numbers of abdominal chaetae in *D. melanogaster*.

In the low selection line slow and irregular progress was made during the first 15 generations of selection; at this time the mean chaeta number had decreased by a little over five. After 34 generations of selection the mean chaeta number had decreased by 11 but sterility had set in and the population could only be kept in being by mass culture. Thereafter, all further 'low' lines selected from this population died out in a few generations.

In the high selection line, much greater progress was made in the first 20 generations, the mean chaeta number rising from 35 to 56, but at the same time fertility decreased so much that the workers had to resort to mass culture with relaxed selection. Over the next five generations the mean chaeta number fell back 80% of the way to the initial population mean. On resuming selection, the high level was reached once more in four generations but this time without extreme loss of fertility. Thus there was a tendency for the population to resist genetic change and to return to its initial state when selection was relaxed, but this tendency was finally overcome with a breakdown of the correlation between high bristle number and infertility.

A second type of experiment leads also to the conclusion that populations have, in a given environment, a preferred or optimal genetic structure. This is exemplified by Dobzhansky's work on *Drosophila pseudoobscura*. It was found that amongst wild flies caught in the Sierra Nevada there were three common types of structural arrangement of the third chromosome, which were named Standard (*ST*), Arrowhead (*AR*) and Chiricahua (*CH*). Other arrangements also existed but further reference will not here be made to them. A great deal of work was carried out by Dobzhansky and his coworkers using population cage techniques, in which flies are left to breed without artificial selection, samples being taken from the cage for examination at intervals. From these samples the frequency of the third chromosome arrangements can be estimated. These chromosome inversions segregate as units, and from the point of view of frequency and equilibrium determination they can be treated as if they were allelomorphs at an autosomal locus.

Among other results, it was found that in cages kept at 25°c there were considerable changes of frequency of chromosome arrangement. For example, if just Standard and Chiricahua arrangements were present, the frequencies approached the values of 70% *ST* and 30% *CH*, regardless of the initial frequencies, suggesting that a stable equilibrium existed at that point (Wright and Dobzhansky, 1946).

At 16·5°c there was no significant change in the frequencies of *ST* and *CH* in cages containing just these two arrangements; in one experiment the frequencies were 46% *ST*, 54% *CH* both initially and after six months, though there were fluctuations in the intervening period; in another experiment, the frequency of *ST* changed only from 83% to 86% in a period of five months. Thus the stability of the equilibrium depended on the environment.

Processes Maintaining Intermediate Gene Frequencies

In 1908 Hardy and Weinberg independently demonstrated that a large random mating, or panmictic population will, within one generation, come to an equilibrium with respect to one pair of autosomal alleles. In general, if p, q ($p + q = 1$) are the average frequencies of the autosomal alleles A, a respectively in a parental population breeding at random with no selection, mutation or migration, the frequencies of genotypes AA, Aa and aa among their progeny are p^2, $2pq$ and q^2 respectively.

The term random mating means that any given male in the group will not be influenced to mate with a female of a particular genotype except in so far as some genotypes will be more frequent than others, in other words the frequency of a particular type of mating depends only on the frequencies of the genotypes in the population. The effect is the same as if ova and spermatazoa from the population formed common gametic pools from which one of each is drawn at random, to unite and form each zygote. Although in practice an animal or plant is probably fertilized by another in its near vicinity, with regard to a large number of genetically determined characters the concept of panmixia is essentially valid. If human blood-group gene frequencies are ascertained from a parental population, genotype frequencies among the progeny are found to correspond very closely with Hardy–Weinberg expectations. Dobzhansky (1947) withdrew eggs laid in population cages and raised the larvae in optimal conditions in bottles; he then estimated the frequencies of the third chromosome arrangements among the gametic populations from which the larval zygotes were formed. He found that the observed larval frequencies of homozygotes and heterozygotes were in fair agreement with panmictic expectations though some selection against homozygotes was probable. To give a third example, Searle (1949) found that the frequencies with which ginger, tortoise-shell and non-yellow cats occur in London agree with the Hardy–Weinberg ratio.

The Hardy–Weinberg equilibrium is a neutral equilibrium; that is to say that if the gene frequencies are changed by migration or a differential death rate in one generation, the frequencies neither return to their original values, nor diverge further from them, but a new equilibrium is immediately established at the new gene frequency. If this were all, populations would be as passive as plasticine, and no one would be tempted to regard them as homeostatic; but in fact stable equilibria can be maintained by various mechanisms some of which will now be considered. Much of what follows is based on the work of Fisher and Haldane. *The Genetical Theory of Natural Selection* (Fisher, 1930) and *The Causes of Evolution* (Haldane, 1932) summarize this work and give other references.

Mutation

Rare diseases or conditions in which the disadvantageous effects of a gene are shown in the heterozygous individual can be kept in stable equilibrium by mutation. If an autosomal gene, A, has gametic frequency p (small), then under random mating the heterozygotes and unaffected homozygotes have frequencies $2p$ and $1 - 2p$ respectively at fertilization. If the relative fitnesses of genotypes Aa to aa are $1 - k{:}1$, then in each generation the loss of A genes is pk; if this is to be balanced by a mutation rate from a to A of μ/gene/generation, then

$$pk = \mu(1 - p) = \mu \text{ approximately}$$

Estimates of naturally occurring mutation rates of the order of 10^{-6} make it improbable that anything but rare conditions are maintained by mutation alone.

Heterosis

Selective advantages are much more powerful and more variable in their effects. Stable gene-frequency equilibria can be maintained by several means which include gametic selection, heterosis, cyclical selection, disruptive selection, advantages that differ between the sexes or are dependent on gene frequency, or by combinations of these with and without mutation. It will be assumed in what follows that the population is panmictic, has non-overlapping generations, and that the selective advantage or disadvantage acts between fertilization and reproduction. The advantage may be due to differential mortality or fertility or both but the following arguments hold if frequencies are always counted at the same stage of the life cycle.

If A and a are a pair of autosomal alleles with frequencies p_n and q_n respectively ($p_n + q_n = 1$) among the gametes forming the nth generation, and if the genotypes AA, Aa and aa have relative fitnesses $1 - K{:}1{:}1 - k$ respectively, then we have

Genotype	Zygotic frequency	Relative fitness	Frequency among breeding population
AA	$p_n{}^2$	$1 - K$	$p_n{}^2(1 - K) / \bar{W}$
Aa	$2p_nq_n$	1	$2p_nq_n \quad / \bar{W}$
aa	$q_n{}^2$	$1 - k$	$q_n{}^2(1 - k) / \bar{W}$

where $\bar{W} = 1 - Kp_n^2 - kq_n^2$ and is the mean fitness of members of the population at birth.

The gametic frequency of the gene A in the next generation is then given by

$$p_{n+1} = \frac{p_n^2(1 - K) + p_n q_n}{\bar{W}}$$

and the change in gene frequency is given by

$$\Delta p_n = p_{n+1} - p_n = \frac{p_n q_n}{\bar{W}} [kq_n - Kp_n]$$

$$= \frac{p_n q_n}{\bar{W}} [k - p_n(K + k)]$$

$$= \frac{p_n q_n}{2\bar{W}} \frac{\partial \bar{W}}{\partial p_n}$$

An equilibrium exists when $\Delta p = 0$; that is when

$$\left. \begin{array}{c} p = 0 \\ q = 1 \end{array} \right\}, \quad \left. \begin{array}{c} p = 1 \\ q = 0 \end{array} \right\}$$

or when

$$\left. \begin{array}{c} p = k/(K + k) \\ q = K/(K + k) \end{array} \right\}$$

The stability or instability of an equilibrium can be determined by considering the rate of change of Δp, or by investigating the sign of Δp at points close to the equilibrium value $\hat{p}$. If $p_1 = \hat{p} - \delta(\delta > 0)$ and $(\Delta p)_{p=p_1} > 0$, p will increase and tend towards the equilibrium value $\hat{p}$. The equilibrium is therefore stable against small displacements; if this is the only possible stable equilibrium, the frequency of A will tend to move towards $\hat{p}$ from any other initial value.

If K and k are both positive, the heterozygote is at an advantage relative to both homozygotes—a situation known as heterosis. In this case the equilibrium at $\hat{p} = k/(K + k)$ is stable and the equilibria at $\hat{p} = 0$ and $\hat{p} = 1$ are unstable; so if the gene a appears by mutation in a population consisting entirely of AA individuals and if it avoids chance elimination when it is very rare, it will spread and increase in frequency until the given stable equilibrium value is reached. A situation like this is probably the main cause for the maintenance in high frequencies among some African peoples of the gene controlling the production of sickle-cell haemoglobin. The sickle-cell homozygote is at a disadvantage owing to death from haemolytic crises: the normal homozygote is also at a disadvantage relative to the heterozygous individual as the former is more likely to die in early childhood from subtertian malaria than is the latter. The auto-somal gene controlling thalassemia and the sex-linked gene control-ling glucose-6-phosphate dehydrogenase (G6PD) deficiency are also both thought to give some protection against malaria when in hetero-zygous form. These three are all found in fairly high frequencies in parts of the world where malaria was, until recently, endemic, and they all affect the erythrocytes. The system is probably a complicated interlocking one with the essence of heterozygous advantage relative to the homozygotes applying throughout, though selection acting in opposite directions on the two sexes may also contribute to stability.

The stability of inversion polymorphism in *Drosophila pseudo-obscura*, mentioned above, has been shown to be due to the superior fitness of inversion heterozygotes (Wright and Dobzhansky, 1946). It is frequently difficult to show the existence of a heterotic effect, because a very small advantage of the heterozygote over the un-affected homozygote will maintain in appreciable frequency a gene with deleterious effects in the affected homozygote. For example, a gene, a, can be kept in stable equilibrium at a frequency of 5% if the relative fitnesses of genotypes $AA:Aa:aa$ are $1:1\cdot02:0\cdot64$.

If $k = 1$, $K > 0$, the gene a has lethal effects when homozygous and the stable equilibrium is then established with frequencies $\dfrac{1 - K}{1 + K} AA$, $\dfrac{2K}{1 + K} Aa$ in the breeding population.

When $K = 0$, $k > 0$, there is selection only against the homo-zygous genotype aa. The only stable equilibrium is when a is elimin-ated from the population or is maintained at a low frequency by mutation.

If $K < 0$, $k > 0$, the heterozygote has an intermediate relative fitness and again, omitting mutation, no equilibrium is possible except the elimination of the gene with the deleterious effects.

If K and k are negative, there is a position of equilibrium other than the fixation of one allele, but it is unstable; the gene with a frequency lower than the unstable equilibrium value will be eliminated.

If there are more than two alleles at a given locus, conditions for stable equilibrium have been given by Penrose, Maynard Smith and Sprott (1956). These conditions have been applied to the alleles for haemoglobins A, S and C in various small populations—in most cases the estimation of relative fitness values is open to great error as only very few SS or CC homozygotes are observed. In such circumstances it is not really very meaningful to say that these populations are or are not in equilibrium.

Different fitnesses in the two sexes

A single pair of alleles may also be kept in stable equilibrium without heterosis if the selective advantages are in favour of different homozygotes in the two sexes, or if the heterozygote is more fit than both homozygotes in one sex and is less fit than both homozygotes in the other sex (Owen, 1952, 1953; Haldane, 1962). If the genotypes AA, Aa and aa have relative fitnesses $1 - K:1:1 - k$ and $1 - L:1:1 - l$ in males and females respectively and if K, k are of opposite signs to L, l respectively though $K + k$, $L + l$ may be of the same or different signs, there may be three equilibria in addition to the trivial values $p = 0$ and $p = 1$. If these are denoted by $0 < \hat{p}_1 < \hat{p}_2 < \hat{p}_3 < 1$, either $p = \hat{p}_1$, $\hat{p}_3$ represent points of stability while $p = 0$, $\hat{p}_2$, 1 are points of instability, or vice versa. The former case can be achieved if the heterozygote is advantageous relative to the homozygotes in one sex, and is disadvantageous in the other sex. If the frequency of A is less than $\hat{p}_2$, equilibrium will be established at $p = \hat{p}_1$; if it is greater than $\hat{p}_2$, A will increase in frequency until the value $\hat{p}_3$ is reached. It would then be possible to shift the stable equilibrium frequencies, without change of selection coefficients, by, for example, a large migration or a severe epidemic.

Gametic selection

It has been assumed throughout the foregoing that each gamete was equally likely to be successful at fertilization—this is implicit in the idea of panmixia. However, there may be considerable selection at the gametic level; or other mechanisms, such as meiotic drive (Gershenson, 1928; Sandler, Hiraizumi and Sandler, 1959), as a result of which a heterozygote does not form gametes in a one-to-one ratio, may have a similar effect.

Intensive studies have been carried out by Dunn and his co-workers on a series of chromosomal abnormalities known as the t-alleles in house mice (Chesley and Dunn, 1936; Dunn and Suckling, 1956; Dunn, 1957).

The work started with the study of a laboratory stock of tailless mice which bred true. The line was shown to be heterozygous T/t', both the genes T and t' having lethal effects when homozygous. The T wild-type heterozygote, $T/+$, has a short-tailed phenotype while $t'/+$ produces a normal phenotype. Among the progeny of T/t' matings were a few normal-tailed mice which, on investigation, were shown to be heterozygous for one of several other alleles of T, i.e. they were t'/t^x.

It was found that in a number of lines the gametic segregation ratio from heterozygous males, either T/t^x or $+/t^x$, was markedly different from the expected 1:1 ratio. In some there was a great excess of t-bearing gametes and in others a deficiency. It was also found that t-alleles were widespread in wild populations; these are designated t^w to distinguish them from those which arose in laboratory stocks. Of 16 t^w-alleles, obtained from different wild populations, 13 have lethal prenatal effects when homozygous while the remaining three result in sterile, though viable, homozygous males. The proportions of t^w-bearing gametes from heterozygous males varied from 0·876 to 0·998, segregation being normal in heterozygous females.

Bruck (1957) outlined a deterministic model in which a stable equilibrium can be achieved in a large population by a balance of segregation advantage against selective disadvantage.

If m = the effective proportion of t-bearing gametes from heterozygous males, $\hat{q}$ = the equilibrium frequency of t-alleles amongst adults and the relative fitnesses of genotypes $++$, $+t$, tt are 1:1:0, then a stable equilibrium exists at

$$\hat{q} = \tfrac{1}{2}\left[1 - \sqrt{\frac{1-m}{m}}\right] \quad \text{provided that} \quad 2m > 1.$$

Lewontin and Dunn (1960) argue that mouse-breeding populations are small and consequently the effects of inbreeding and random gene fixation should be taken into account. They applied a stochastic model to a number of very small populations and showed that the t-allele tends to be eliminated. The speed with which this happens is very sensitive to the 'segregation ratio', m, and to the effective population size

$$N = \frac{4N_\male N_\female}{N_\male + N_\female}.$$

They suggest that the ubiquity of the t^w-alleles is due to continual 'infection' of normal populations by heterozygous males carrying a t^w-allele with a very high segregation ratio. They argue that alleles with low segregation ratios would quickly be eliminated; this could account for the differences in segregation ratio found between t^w-alleles and those t^x-alleles newly arisen in laboratory stocks.

If high values of m are used in Bruck's model, the expectation of adult heterozygotes, given by $\hat{h} = 2\hat{q}$, is much greater than the frequencies actually found. The model suggested by Lewontin and Dunn, allowing for inbreeding and random fixation in small populations, accounts for this discrepancy, and there is recent evidence which supports their hypothesis of small mating isolates (Anderson, 1964).

Cyclical selection

In all the foregoing cases a genotype has been considered as having a relative fitness value which remains constant throughout the time and environment considered. This is an over-simplification which may approach the truth if these fitness values are taken as averages over small fluctuations. However, equilibria can also be established when genotypes have relative fitness values that vary appreciably in time or space or both. Fitness values may also be dependent on genotype or gene frequency. We will consider first variations in fitness with time.

Consider an autosomal locus in a diploid panmictic population which is large enough to avoid the chance extinction of one allele. If u_n is the ratio of A to a in the gametes forming the nth generation and if the relative fitnesses of the genotypes AA, Aa and aa are $F_n:1:f_n$, then

$$u_{n+1} = \frac{u_n(1 + F_n u_n)}{u_n + f_n} \quad \text{(Haldane, 1924)}$$

Haldane and Jayaker (1963) have shown that both alleles will remain in the population provided that the geometric means of the F_r and the f_r are less than unity even though the arithmetic means may be greater than unity. This means that a gene with deleterious effects in the homozygote and, to a lesser degree, in the heterozygote could persist if an epidemic, famine or some other cause periodically violently reduced the fitness of the normal homozygote relative to the heterozygote. If there is a cyclical variation in relative fitness, then a cyclical increase and decrease of gene frequency would follow. For example, if AA, Aa and aa have steady relative fitnesses of $1\cdot2:1:0\cdot7$ for nine generations, but in the tenth generation, on the average, they become $0\cdot15:1:1\cdot1$, then an 'equilibrium cycle' of ten generations will become established with $u_0 = 5\cdot62$ (frequency of $a = 0\cdot151$), $u_9 = 32\cdot09$ (frequency of $a = 0\cdot030$) and $u_{10} = 5\cdot62$ once more. Such a population starting at some other frequency will approach this 'equilibrium' from either direction. The geometric and arithmetic means of the F_r (AA) are $0\cdot975$ and $1\cdot095$ respectively, while the geometric and arithmetic means of the f_r (aa) are $0\cdot732$ and $0\cdot74$ respectively.

Haldane and Jayaker also give the conditions under which a bivoltine population (i.e. a population which regularly has two generations a year) will be in equilibrium. Let the relative fitnesses of AA, Aa and aa be $1:1:f_1$ and $1:1:f_2$ in the spring and autumn generations respectively and let the gametic gene frequencies from which the spring and autumn generations are formed be q_n and q'_n respectively $[q = (u + 1)^{-1}]$. Then, if $f_1 > f_2$, stable equilibrium is established at

$$\hat{q} = \frac{1 - f_2 - (1 - f_1 f_2)^{1/2}}{f_2(f_1 - 1)}$$

$$\hat{q}' = \frac{f_1 - 1 - (1 - f_1 f_2)^{1/2}}{f_1(1 - f_2)}$$

provided that $f_1 f_2 < 1$ and $f_1 + f_2 > 2$.

These requirements are very stringent when the selection coefficients are small and it may be that this type of equilibrium only operates where selection is fairly severe; for example, if $f_1 = 1\cdot1$, $0\cdot9 < f_2 < 0\cdot909$, but if $f_1 = 2\cdot0$, $0 < f_2 < 0\cdot5$.

If, say, f_2 is zero, i.e. *aa* homozygotes do not survive the winter, they (*aa*) will have to be more than twice as fit as the other genotypes in the spring generation in order to maintain the equilibrium which is then given by

$$\hat{q} = \frac{f_1 - 2}{2(f_1 - 1)}$$

$$\hat{q}' = 1 - \frac{2}{f_1}$$

Different fitnesses in different places

Suppose that the geographical range of a species is divided up into a number of regions, in which different genotypes are at a selective advantage. If the distances moved by individuals between conception and mating are large enough to carry alleles from one region to another but not so large as to lead to effective panmixis, then the allele favoured by selection at any one point will be diluted by an inflow of alleles from other regions. This can lead to a stable distribution of gene frequencies. Particular cases have been considered by Haldane (1930, 1948).

Alternatively, different alleles may be established in two subpopulations of a species although they cannot both survive in a single population. The simplest explanation of the Rhesus blood-group D-d polymorphism is that it is a transient one, consequent on the mixing of two populations homozygous for different alleles. Maternal–foetal incompatibility of the Rhesus type results in the elimination of heterozygotes; the more frequent allele is thus established at the expense of the rarer one.

Fitness a function of gene frequency

A stable intermediate gene frequency can be maintained if the fitness of genotypes depends on their frequency in the population. Thus in the case of two alleles A and a, if AA is fitter than aa when A is rare, and aa fitter than AA when a is rare, then whichever is the rarer allele will increase in frequency to some stable intermediate equilibrium. This will be true if the heterozygote Aa is intermediate in fitness at all frequencies, or equal to one homozygote at all frequencies, or even equal to the less fit homozygote at all frequencies.

The classical example of such a situation is Batesian mimicry (Ford, 1953). If a species is polymorphic for forms which mimic a distasteful species and forms which do not, then when the mimic form is rare it will be the fitter of the two, but if it is too common relative to the model, it will cease to be protected by its mimicry, and it may be the less fit of the two forms. Haldane (1949) has pointed out that a somewhat similar situation may arise in the case of disease resistance.

Disruptive selection may also lead to a stable equilibrium by increasing the fitness of genotypes as their frequency decreases. This possibility has been demonstrated experimentally by Thoday and Boam (1959). Maynard Smith (1962) has pointed out that the possibility arises only if there are two or more 'ecological niches' open to a species, and if the density of the population is separately regulated in each niche. He considered the case of an environment divided into two ecological niches, for example different food plants. It is supposed that the factors responsible for maintaining the population density act independently in the two niches, and ensure that the total density N is made up of n_1 individuals in niche 1 and n_2 individuals in niche 2. In the simple case of a single pair of alleles, with additive inheritance of fitness in both niches, the relative fitnesses will be:

	AA	Aa	aa
Niche 1	$1 + K$	$1 + \frac{1}{2}K$	1
Niche 2	1	$1 + \frac{1}{2}k$	$1 + k$

If mating is at random, and if there is no habitat selection (that is, the probability that an individual will live in a particular niche is independent both of the genotype of the individual and of the niche in which it was raised) then a stable equilibrium exists when the frequency of allele A is given by

$$p = \frac{n_1}{n_1 + n_2} + \frac{n_1 K - n_2 k}{(n_1 + n_2)Kk}$$

If an equilibrium is to be possible, p must lie between 0 and 1. This will clearly be the case if $n_1 K = n_2 k$, so an equilibrium is possible for any values of n_1 and n_2 provided that the selective advantages are nicely adjusted to the sizes of the niches. But if K and k are small an equilibrium is in practice unlikely. For example, if $n_1 = n_2$ the requirement for an equilibrium to exist becomes

$$\frac{K - k}{Kk} < 1$$

Thus if $K = 0{\cdot}1$ (10% selective advantage), k must lie between $0{\cdot}09$ and $0{\cdot}11$; but if $K = 1$, there will be a stable equilibrium provided k is greater than $0{\cdot}5$.

Stable equilibria maintained by disruptive selection require independent regulation of the population density in the different niches, and rather intense selection pressures. The stability of the equilibria arises because when a genotype is rare the competition in the niche to which it is adapted is less, and consequently its mean fitness is higher.

Group selection

In all the cases so far considered, fitness has been ascribed to individuals, not to groups of individuals. We come now to a process which appears to require that selection should act not only at the individual but also at the group level.

We are here considering characters, such as size or bristle number, with continuous variation within a population. The observed distributions and kin correlations for such characters are best explained by the hypothesis that they are largely controlled by several gene pairs at different loci, the small individual gene effects being additive between loci. As has been mentioned earlier, in these cases selection is often in favour of the intermediate. Phenotypic variation of this kind could be maintained by processes already discussed in previous sections such as heterosis at individual loci or selection for different genotypes in different environments together with migration. However, Mather (1943) has suggested that these characters are controlled by blocks of linked genes and it is this mechanism which appears to require the action of group selection for its establishment.

Suppose a metrical character, say size, is influenced by a number of allele pairs, A, a; B, b; C, c; D, d. Capitals are used to denote large size, not dominance. Heterozygotes are assumed to be intermediate in size and the gene effects are additive between loci. Suppose also that fitness depends on size, optimal fitness being associated with intermediate size, i.e. with genotypes with equal numbers of 'large' and 'small' genes, while both extremes are relatively unfit.

Then genotypes such as

$$\frac{A}{A} \frac{B}{B} \frac{c}{c} \frac{d}{d}, \quad \frac{a}{a} \frac{B}{B} \frac{c}{c} \frac{D}{D}, \quad \frac{A}{a} \frac{B}{b} \frac{C}{c} \frac{D}{d},$$

would all be the same size and would be optimally fit. A population consisting entirely of either of the first two genotypes would breed true and all individuals would have optimal fitness. But should the environment change so that a different size was advantageous, such a population would be unable to evolve until one or more new mutations had occurred. In contrast, a population starting from individuals of the third genotype would in later generations give rise to a wide range of segregant types, many of low fitness, yet would be able to evolve in response to changing environments.

Mather points out that the twin desiderata of stability of phenotype and potential variability can be achieved by close repulsion linkage between genes for large and small size. If the two common chromosomes in the population are $AbCd$ and $aBcD$, both homozygotes and heterozygotes are of maximal fitness but recombinant gametes such as $ABcD$ would give rise to individuals of lower fitness. The linkage groups would therefore be maintained by selection against genotypes carrying recombinant chromosomes, but the population could evolve through recombination without waiting for new mutations. Much of this would also be true of a population in which the two common chromosomes were $ABcd$ and $abCD$, but in the latter case the release of variability through recombination would be greater than in the former. Single recombination between the repulsion linkage groups can do no more than put three genes for large size in one chromosome and three genes for small size into the other, whereas single recombination between $ABcd$ and $abCD$ can produce chromosomes carrying four genes for large size or four for small size. The greater the number of coupling linkages, the greater the rate of release of variability and consequently the less stable the phenotype of the population. From this point of view the repulsion linkage groups would clearly be a better proposition than the coupling groups.

The equilibrium between chromosomes $AbCd$ and $aBcD$ would be a neutral one. But it would be stable if there were some degree of fitness heterosis associated with one of the pairs of alleles or with a closely linked locus. What is more difficult to understand is how a situation of this kind would arise in the first place. In Mather's model, and in more general discussions of the evolution of genetic systems (e.g. Darlington, 1939), it is supposed that groups with the ability rapidly to produce new genetic variants survive because they can evolve adaptions to new circumstances. The implication is that polygenic blocks exist because populations or species with such blocks have survived when others have become extinct, and not because some individuals have survived when others have died.

But there is another reason why groups with high genetic variance might be favoured by selection: in social animals a phenotypically-variable group may be better adapted than a phenotypically-uniform one. This is most obviously true in social insects, with their worker, soldier and reproductive castes, but it is probably also true of human communities and perhaps of other animal societies. It has for example been suggested (Kalmus, 1964a) that in a food-gathering society, although colour-blind individuals would be worse at finding most things, they might be better at finding some types of food, and hence groups with a few colour-blind individuals would be at an advantage.

The selective mechanisms responsible for the maintenance of phenotypic variance in groups are not easy to see. In social insects the phenotypic differences depend on differences in nutrition in larval life. Thus selection has favoured a genotype giving a sharply differentiated response to varying nutritions, because family groups with such genotypes have survived better than groups with a more uniform phenotype. (For discussion of the evolution of social insects see Sturtevant, 1938; Williams and Williams, 1957; Hamilton, 1964; Kalmus, 1964b.)

It is not clear whether selection favouring polymorphic groups could maintain a genetic as opposed to an environmentally-induced polymorphism. It is useful to distinguish two processes, kin selection and group selection (Maynard Smith, 1964), both of which can cause the evolution of 'altruistic' behaviour, whereby individuals evolve characters which increase the chances of survival of other members of the species; to avoid the value judgement implied by the word altruistic, Williams and Williams (1957) have suggested that such behaviour be called 'donor' behaviour. Kin selection occurs because individuals have genes in common with their relatives; for example, an individual which sacrifices its life to save the lives of more than two siblings will be acting so as to perpetuate its own genotype. Kin selection requires that individuals should live near their relatives, but not that the population be divided into small reproductively-isolated groups. If kin selection is to maintain a genetic polymorphism, then the advantage to the group relative to the disadvantage to the individual of a particular character must be greatest when the character is rare. It is not easy to think of examples in which one would expect this to be the case, although the example quoted above of colour-blindness is a possible case. A recent analysis of alarm notes in birds (Maynard Smith, 1965) led to the conclusion that either the 'donor' habit of giving an alarm call would spread to all members of a population, or it would disappear; there is no intermediate stable equilibrium.

Group selection differs from kin selection in that it requires that the population be divided into small partially isolated groups, so that an individually disadvantageous but socially desirable gene can be established by genetic drift in all members of a group. Such a process obviously cannot give rise to genetically variable groups.

There is one case in which the fitness of family or larger groups may be increased by genetically caused variation, and that is in the division of labour between the sexes. Despite the difficulties of group selection, it seems certain that the sexual processes of meiosis and fertilization, and other methods of the exchange of genetic material, owe their existence to the long-term evolutionary advantages they confer. Given that there are two sexes, or initially two mating types, selection can produce a division of labour between them, at first in the nature of the gametes, but often also in the form and function of the adults. In those cases in which males have an essential function to perform other than fertilization, this acts as a protection against losing the evolutionary advantages of sex through the acquiring of parthenogenesis.

It seems, therefore, that selection can establish genotypes which respond differentially to environmental conditions, as in the social insects; that it can cause the evolution of altruistic traits even in the absence of isolated groups; that it may occasionally establish a stable polymorphism for an altruistic trait. It is unlikely, however, often to maintain genetic variance by favouring the immediate survival of genetically variable groups at the expense of genetically uniform ones. It is of course quite possible that groups which are genetically variable because of selection acting at the individual level (e.g. superior fitness of heterozygotes, cyclical selection, increased fitness of rare phenotypes) will also be fitter as groups.

The Concept of Genetic Homeostasis

It is, or ought to be, assumed that the mechanisms of physiological and developmental homeostasis have evolved because individuals which possessed them survived when others did not, and further, that the increased chances of survival arose because of the regulative effects of the mechanisms in question. For example, it is assumed that mammals shiver because those which shivered in the past survived, and that they survived because shivering helps to regulate temperature. There is an obvious analogy between the adaptations of animals and of machines (Kalmus, 1963; Kapp, 1954). There is likely to be a resemblance between a structure which has evolved by natural selection because it performs a particular function effectively, and one which has been designed to perform the same function. The evolution of clocks presents no difficulty to a selectionist; a modern Paley would be forced to seek for evidences from complex organic structures for which no function could be conceived.

It is our thesis that the word 'homeostasis' should be used only for mechanisms which have evolved *because of their regulative function*. Only if this is the case is the resemblance between different mechanisms likely to be close enough for the use of a common term to be illuminating. For example, an introductory chapter on self-regulating machines has been included in this book, in the hope that concepts such as error regulation which are helpful in the design of such machines may be helpful in understanding natural regulating mechanisms. This is a reasonable hope only if the function (i.e. the property which has been favoured by natural selection) of the natural mechanism being studied is in fact to regulate something.

In the case of population genetics, it follows that the word homeostasis should be used only for genetic mechanisms which have evolved by inter-group selection. The frequency of a particular allele, or block of alleles, is a property of a population; an individual does not have a gene frequency. The entity whose future survival is affected by a gene frequency is again the population. Consequently a particular genetic system can be regarded as an example of genetic homeostasis only if it has evolved because in the past groups with that system have survived when others have not. In such cases it may be justifiable to say that the 'function' of a particular genetic mechanism is to ensure the constancy of the genetic make-up of a population. Thus Mather, if he accepts our account of his views on polygenic blocks, could properly refer to them as an example of genetic homeostasis. But it is doubtful whether Lerner (1954) ought to have used the term as the title of a book, since in most cases the processes he describes are the consequence of selection at an individual level.

It is not easy to decide what features of the genetic system of a species are due to group selection. Sexual processes, and mechanisms for the exchange of genetic material generally, are the most obvious and important examples. In unicellular organisms, in which sexual differentiation arose, reproduction by cell division is directly contrary to the sexual (or cannibalistic) fusion of two cells. It is only at a later evolutionary stage that sex and reproduction come to be intimately associated. The advantages of sex appear to be at the population level, although one suspects it may also confer advantages at the individual level through hybrid vigour. Sexual processes are regulative in the sense that they make more efficient the evolutionary adjustment of populations to their environments; but it is a little odd to regard them as homeostatic if their function is to facilitate evolutionary change. The evolution of plants suggests that constancy and immediate fitness are often best achieved by the abandonment of sex. Certainly most of the processes described above which maintain the variability of populations are the consequences of individual selection. It is therefore no accident that categories such as feedback or error regulation seem singularly inappropriate to their analysis.

To conclude, we would argue that consistency requires that the word homeostasis be used only for mechanisms which have the effect of ensuring the constancy of some property of organisms or populations, and which have evolved because of this regulative effect. Waddington (1957) has criticized the use of the term 'developmental homeostasis' because the target of regulation is not a steady state but a state changing in time in a particular way (a 'creode'); he has suggested 'homeorhesis' as an alternative. But there is a real analogy between homeorhesis and physiological homeostasis, in that both are the consequences of selection for regulative ability.

But mere constancy, or a tendency to return to an initial state, is not enough; a deformed balloon resumes its original shape.

Similar objections can be raised to the use of the word homeostasis in ecology. There are two somewhat different views which can be taken concerning the processes which limit the numbers of animals. One view, put forward in this book by Wynne-Edwards, is that the properties of animals, particularly of their behaviour, which limit population size, have evolved by a process of group selection; that is to say, groups which limit their density behaviourally are fitter than those which outrun their food supply and starve. If this view is true, the concept of homeostasis is applicable. The alternative view is that, in so far as behaviour patterns which limit population density exist, they have evolved because they increase the fitness of the individual and not of the group to which it belongs. If this view is true, then the processes limiting animal numbers are similar to most of those which maintain the genetic variability of populations, and the concept of homeostasis is not readily applicable.

References

Anderson, P. K. (1964). 'Lethal alleles in *Mus musculus*: local distribution and evidence for isolation of demes.' *Science*, **145**, 177.

Bruck, D. (1957). 'Male segregation ratio advantage as a factor in maintaining lethal alleles in wild populations of house mice.' *Proc. Nat. Acad. Sci. U.S.*, **43**, 152.

Chesley, P. and Dunn, L. C. (1936). 'The inheritance of taillessness (anury) in the house mouse.' *Genetics*, **21**, 525.

Darlington, C. D. (1939). *The Evolution of Genetic Systems*, Cambridge University Press, Cambridge; 2nd ed. 1958, Oliver and Boyd, Edinburgh.

Dobzhansky, T. (1947). 'Genetics of natural populations. XIV. A response of certain gene arrangements in the third chromosome of *Drosophila pseudoobscura* to natural selection.' *Genetics*, **32**, 142.

Dunn, L. C. and Suckling, J. (1956). 'Studies of genetic variability in populations of wild house mice. I. Analysis of seven alleles at locus T.' *Genetics*, **41**, 344.

Dunn, L. C. (1957). 'Studies of genetic variability in populations of wild house mice. II. Analysis of eight additional alleles at locus T.' *Genetics*, **42**, 299.

Fisher, R. A. (1930). *The Genetical Theory of Natural Selection*, Clarendon Press, Oxford; republished 1958, Dover Publications Inc., New York.

Ford, E. B. (1953). 'The genetics of polymorphism in the lepidoptera.' *Advan. Genet.*, **5**, 43.

Gershenson, S. (1928). 'A new sex ratio abnormality in *Drosophila obscura.*' *Genetics*, **13**, 488.

Haldane, J. B. S. (1924). 'A mathematical theory of natural and artificial selection.' Part 1, *Trans. Camb. Phil. Soc.*, **23**, 19.

Haldane, J. B. S. (1930). 'A mathematical theory of natural and artificial selection.' Part 6, *Proc. Camb. Phil. Soc.*, **26**, 220.

Haldane, J. B. S. (1932). *Causes of Evolution*, Longmans, Green and Co., London.

Haldane, J. B. S. (1948). 'The theory of a cline.' *J. Genet.*, **48**, 277.

Haldane, J. B. S. (1949). 'Disease and evolution.' *Ric. Sci.*, **19**, 68.

Haldane, J. B. S. (1962). 'Conditions for stable polymorphism at an autosomal locus.' *Nature*, **193**, 1108.

Haldane, J. B. S. and Jayakar, S. D. (1963). 'Polymorphism due to selection of varying direction.' *J. Genet.*, **58**, 237.

Hamilton, W. D. (1964). 'Genetical evolution of social behaviour.' *J. Theoret. Biol.*, **7**, 1.

Hardy, G. H. (1908). 'Mendelian proportion in a mixed population.' *Science*, **28**, 49.

Kalmus, H. (1963). 'Axioms and theorems in biology.' *Nature*, **198**, 240.

Kalmus, H. (1964a). *Diagnosis and Genetics of Defective Colour Vision*, Pergamon Press, Oxford.

Kalmus, H. (1964b). 'Symposium on social organisation of animal communities. Origins and Genetical features.' *Symp. Zool. Soc. London*, **14**, 1.

Kapp, R. O. (1954). 'Living and lifeless machines.' *Brit. J. His. Philos. Sci.*, **5**, 18.

Lerner, I. M. (1954). *Genetic Homeostasis*, Oliver and Boyd, London.

Lewontin, R. C. and Dunn, L. C. (1960). 'The evolutionary dynamics of a polymorphism in the house mouse.' *Genetics*, **45**, 705.

Mather, K. (1943). 'Polygenic inheritance and natural selection.' *Biol. Rev. Camb. Phil. Soc.*, **18**, 32.

Mather, K. and Harrison, B. J. (1949). 'The manifold effects of selection.' *Heredity*, **3**, 1–52, 131–162.

Maynard Smith, J. (1962). 'Disruptive selection, polymorphism and sympatric speciation.' *Nature*, **195**, 60.

Maynard Smith, J. (1964). 'Group selection and kin selection.' *Nature*, **201**, 1145.

Maynard Smith, J. (1965). 'The evolution of alarm calls.' *Am. Naturalist*, **99**, 59.

Owen, A. R. G. (1952). 'A genetical system admitting of two stable equilibria.' *Nature*, **170**, 1127.

Owen, A. R. G. (1953). 'A genetical system admitting of two distinct stable equilibria under natural selection.' *Heredity*, **7**, 97.

Penrose, L. S., Maynard Smith, Sheila and Sprott, D. A. (1956). 'On the stability of allelic systems, with special reference to Haemoglobins A, S and C.' *Ann. Hum. Genet.*, **21**, 90.

Sandler, L., Hiraizumi, Y. and Sandler, Iris. (1959). 'Meiotic drive in natural populations of *Drosophila melanogaster*. I. The cytogenetic basis of segregation-distortion.' *Genetics*, **44**, 233.

Searle, A. G. (1949). 'Gene frequencies in London's cats.' *J. Genet.*, **49**, 214.

Sturtevant, A. H. (1938). 'Essays on evolution. II. On the effects of selection on social insects.' *Quart. Rev. Biol.*, **13**, 74.

Thoday, J. M. and Boam, T. B. (1959). 'Effects of disruptive selection. II. Polymorphism and divergence without isolation.' *Heredity*, **13**, 205.

Waddington, C. H. (1957). *The Strategy of the Genes*, George Allen and Unwin Ltd., London.

Weinberg, W. (1908). 'Über den Nachweis der Vererbung beim Menchen.' *Jahrb. Ver. Vaterl. Naturk. Württemb.*, **64**, 368.

Williams, G. C. and Williams, D. C. (1957). 'Natural selection of individually harmful social adaptations among sibs with special reference to social insects.' *Evolution*, **11**, 32.

Wright, S. and Dobzhansky, T. (1946). 'Genetics of natural populations. XII. Experimental reproduction of some of the changes caused by natural selection in certain populations of *Drosophila pseudo-obscura.*' *Genetics*, **31**, 125.

Wynne-Edwards, V. C. (1962). *Animal Dispersion*, Oliver and Boyd, Edinburgh.

IV

GROUPS and POPULATIONS

13

Seasonal Regulations in the Honeybee Colony

J. B. FREE

Rothamsted Experimental Station, Harpenden, Herts

The success of the social organization of honeybee (*Apis mellifera* L.) colonies over the past several million years (Kalmus, 1966) and their survival in different environments is reflected in their distribution in the subarctic and tropical regions as well as throughout the temperate zones. The honeybee colony owes much of this success to the fact that, during the winter or other unfavourable seasons when it cannot collect forage, it can survive on food stored in its combs. Little or no brood is reared then and the colony consists of a cluster of several thousand morphologically identical workers (sexually undeveloped females) and a single sexually developed female, the queen. Unlike bumblebees and wasps whose queens alone survive the winter, the honeybee colony has a large worker force immediately able to collect nectar and pollen when conditions become favourable, as when plants flower in spring. As the season progresses, the number of worker honeybees increases, drone rearing begins and later new queens may be reared and sometimes colonies reproduce.

This Chapter is concerned with the amount and type of forage collected, the number and type of brood produced, colony reproduction, and with the regulation of the environment within the hive, particularly in unfavourable circumstances. All these factors are undoubtedly regulated and interrelated but, because of the difficulty of controlling one independently of the others, most of our information is from observations only, and it is only in the last decade or so that much experimental work has been done.

351

Spring Growth

It is convenient to consider first an average colony in a temperate climate in early spring. It will be relatively small and will contain a small amount of brood being reared on the pollen and honey collected and stored during the previous year; the amount reared depends on the colony's size and the quantity of pollen stored (Allan and Jeffree,

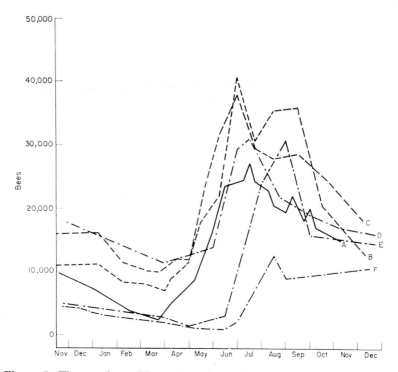

Figure 1. The number of bees throughout the year in one colony (A) in 1945 (——), two colonies (B, C) in 1946 (— — — —) and three colonies (D, E, F) in 1947 (·—·—·). (After Jeffree, 1955.)

1956). The brood is on the central combs and on either side, and sometimes also above and below, there are combs containing honey and pollen. In early spring only a small proportion of the population is occupied with brood rearing and most workers are relatively idle. However, with the increase in the number of flowers in the spring, the number of foragers collecting food also increases, until a limit is reached at which no more bees can be spared from rearing the brood. If the supply of forage were limitless, a colony's income would depend entirely on the number of its foragers. The rate at which a colony's foraging force increases depends on the amount of brood reared and the longevity of the workers. Therefore, to increase rapidly the amount of food collected in the spring, a colony should ideally have plenty of brood and workers with a long life expectancy. However, colonies in the spring have relatively little brood and the average life expectancy of the workers is at its shortest. Relatively few of the workers present have been reared on the colony's reserves of stored food during the winter, and most are old, overwintered bees that will probably survive only a few days of continuous foraging. In fact, for a time in spring, more individuals die than emerge, so the colony size decreases to its minimum for the year. The time of this population minimum depends both on the colony and the environment. Thus Jeffree (1955) found that the minimum occurred later in 1947, when spring was also later, than in 1945 or 1946 (Figure 1). In 1947 it was also later in small than in large colonies. However, the spring decrease in population is only temporary and, as the previous year's bees die, there is a greater proportion of young bees which have a greater life expectancy so that the total population, the number of foragers and the food collected all increase.

Interactions between Foraging and Brood Rearing

The amount of food a colony collects affects the activities of its bees. In a normal colony the duties undertaken by a worker depend largely on its physiological condition, which in turn tends to depend on its age (Rösch, 1925; Lindauer, 1952; Sakagami, 1953). Thus newly emerged workers clean cells, making them ready to receive eggs and food; within two or three days their brood-food glands develop and produce a secretion with which the larvae are fed; later their wax glands become active and they can secrete wax and build combs; they become foragers when two to three weeks old and may collect nectar and pollen or, less often, water and propolis (resinous cement); before beginning to forage, they may pack pollen loads into cells and relieve foragers of nectar at the hive entrance. Each receiving bee, in turn, gives food to one or more bees and so food passes from the vicinity of the hive entrance to the storage cells, or to the main places of consumption, which are the brood nest and the regions where wax is being produced and new comb is being built. Transfer of food within the hive is both extensive and rapid in summer; Nixon and Ribbands (1952) fed some foragers of a colony with 20 ml of sugar syrup containing radioactive phosphorus and a day later found that

12*

most bees of the colony were radioactive. Thus the workers of a colony soon become aware of changes in nectar supplies. It is less obvious how changes in incoming pollen supplies are appreciated, as foragers deposit their loads directly in cells, although household bees pack them down. The pollen storage cells tend to be on the edge of the brood area of a comb so perhaps the nurse bees become aware of changes in the amount of pollen present by personal inspection.

Pollen is the sole source of protein for bees and the amount collected can influence brood rearing, particularly when supplies are scarce (Nolan, 1925; Todd and Bishop, 1941). When natural pollen is lacking, beekeepers can encourage brood rearing by feeding bees with 'substitute' protein food (Haydak, 1945; Spencer-Booth, 1960). Feeding colonies with sugar syrup also increases brood rearing (Free and Spencer-Booth, 1961), but, as such feeding also increases pollen collection, it is difficult to distinguish between the effect of the syrup and the increased pollen collection. There is little evidence that increased nectar in the absence of increased pollen influences brood rearing.

When food supplies influence brood rearing, the workers, once they are aware of the changed situation, must on their part somehow influence the behaviour of the queen who is the sole useful egg laying individual in the colony. Queens do not feed themselves but are fed by workers of similar ages to those feeding brood (Allen, 1955), and because their guts do not contain pollen the protein they require for egg production is assumed to come from the workers' hypopharyngeal glands. Workers probably control the number of eggs the queen lays by the amount of food they give her; they certainly feed her less shortly before the colony swarms (Allen, 1960). Workers probably also regulate the amount of brood produced by eating some, or on occasions all, of the eggs laid, as there is a considerable discrepancy between the number of eggs laid and the number of larvae produced, particularly during unfavourable weather (Merrill, 1924; Myser, 1952); Villumstad (1962) found eggs were produced several weeks before larvae appeared after the winter break in brood rearing so perhaps different factors govern egg production and egg retention in the early part of the year. An increase in food supply causing increased egg laying may be only temporary and, by eating or neglecting some of the eggs and young larvae, the workers can adjust the amount of brood to suit current circumstances. Similarly, when a colony has more older larvae than it can feed, those in a small area are fed but the rest are neglected and die (Gontarski, 1953).

The rate at which a queen lays eggs may also be influenced by the

number of cells the workers have cleaned and prepared for them. If
there is a shortage of cells, an increase in nectar will provide for
increased wax and cell production. Ribbands (1953) suggested that
when storage cells are too few, the bees retain newly gathered nectar
in their honeystomachs and that its assimilation increases wax production and comb building, so helping to relieve the cell shortage.
However, as the comb is not built outside the area the cluster can

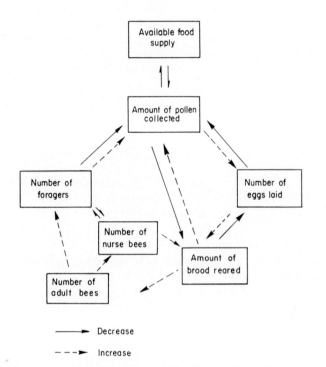

Figure 2. Relationship between foraging and brood rearing.

cover this will apply only when the cluster is large enough to enclose
extra cells; when there is no space left in the cluster to build new cells
those vacated by newly emerged bees may be used for storage, so
diminishing brood area and egg laying.

The increased brood rearing that usually follows increased forage
in the spring will increase the population, and so again the number
of potential foragers is increased. In turn, the amount and type of
forage collected is considerably influenced by the amount of brood
present. Although it is possible to construct a flow diagram (Figure 2)
illustrating such effects, it must be emphasized that these are the

net results of the overall activities of some thousands of workers, many of which may be acting against the general trend or undoing the work of others. Filmer (1932) found that in spring the proportion of pollen to nectar gatherers in colonies increased with the amount of brood present. Free (1966a) found the amount of pollen collected by colonies could be increased or decreased by adding or removing brood, and that these changes reflected changes in the foraging behaviour of the individual bees. In experiments in which foragers were secluded by wire screens from the brood and were only able to smell it, they collected less pollen than when they had access to it but more than when neither brood nor its odour was present; when foragers were allowed to smell brood and to touch bees confined with brood, they collected more pollen than those only able to smell brood but less than foragers with direct access to brood. Thus foragers seem to be stimulated to collect pollen by contact with brood, by the smell of brood, and indirectly through other bees. Eggs, larvae and pupae stimulated pollen collection but larvae were the most effective.

Possibly increasing the amount of brood increases the demand for nectar in the brood area and encourages the passage of nectar from the foragers at the hive entrance to the nurse bees, so the foragers give their loads to the household bees more quickly and spend less time in the hive and more in the field.

Successful colony growth enables the queen to reach her maximum rate of egg laying, which may be about 1,500 per day, but depends on her age and differs with individual queens (Nolan, 1925; Moeller, 1958). The number of bees flying from growing colonies increases approximately proportionally with colony size (Free, 1960). When egg laying is at its maximum, the larva:worker ratio decreases and the proportion of bees available for foraging increases. The reports that large colonies store proportionally more honey in relation to their size than small ones (Sharma and Sharma, 1950) probably mean that full size colonies were being compared with growing ones. Moeller (1958) found a correlation between the rate of egg laying in colonies and the amount of honey they produced. Presumably the amounts of honey stored by mature colonies of different sizes are directly proportional to their populations.

So far in this Chapter it has been assumed that during spring and summer the supply of forage is unlimited. This is obviously not so as the supply fluctuates with the flowering of various crops. The object of many beekeeping practices is to obtain full size colonies in time for the flowering of the major nectar crops (see Nolan, 1925).

The accessibility of a food supply varies with such environmental factors as temperature, weather and day length; the ability of foragers to inform other members of their colony of the source of a good supply of food (Frisch, 1923, 1946) enables them to exploit it rapidly. When a crop is not particularly attractive foragers do not recruit other workers. The attractiveness of a crop depends on many factors,

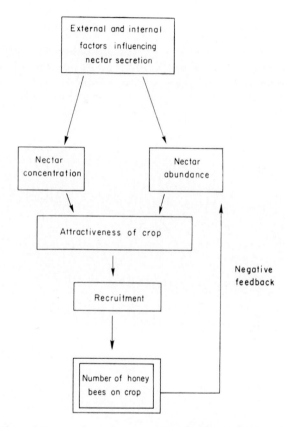

Figure 3. Regulation of the number of nectar-gathering honeybees on a crop.

such as the quantity of nectar and pollen, the sugar concentration of nectar, the quality of pollen, the density of the flowers, the number of competing insects (including honeybees) and the distance from the hive (Figure 3). The threshold level of attractiveness of a crop depends on the other competing crops present and when there are few competing crops, foragers recruit other bees to crops whose nectar sugar concentration is far below what would be attractive at other times. Despite these regulatory mechanisms very different proportions of the foraging populations of neighbouring colonies may visit the same crops (e.g. Synge, 1947), presumably reflecting chance differences in the number of bees from each colony that first found them. Consequently changes in the amount of a particular food does not affect all the colonies in an area equally.

Longevity of Bees

Day-to-day fluctuations in forage probably have little influence on brood rearing (Nolan, 1925) although larger fluctuations may do so. A decrease in brood rearing and subsequently in the number of adults

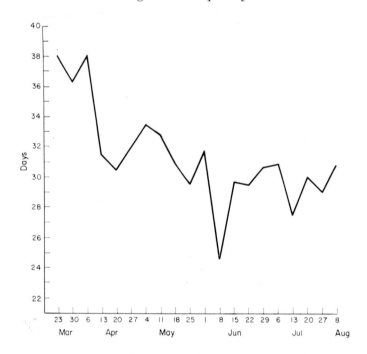

Figure 4. Mean longevity of bees emerging at various times during the spring and summer. (After Free and Spencer-Booth, 1961.)

produced is compensated to some extent by an increased longevity. Normally, the longevity of bees decreases in accordance with their date of emergence from early spring to mid-summer and then increases again (Figure 4). Free and Spencer-Booth (1959) found that bees that emerged in March lived 37 days, whereas those that emerged in June lived only 28 days; many bees that emerged in August, September and October survived the winter. A colony reaches its maximum size in June or July (Jeffree, 1955) and consequently its larva:worker ratio will have decreased, so the short lives of bees emerging in mid-summer are probably associated with less demand for brood rearing which encourages them to begin to forage younger. Ribbands (1952) found that the older the bees were when they began foraging the longer they lived, and Mauermayer (1954) observed that the longer bees could forage each day the shorter their lives. Therefore, foraging is probably more arduous and more hazardous than hive duty, and affects longevity more. Probably both an increase in longevity and a decrease in brood rearing are caused directly by a decrease in available forage, although the decreased brood rearing will also favour increased longevity.

The increased longevity does not, of course, completely compensate for decreased brood rearing, so that food supply will ultimately influence the number of foragers available to collect it, and after prolonged unfavourable circumstances a colony's population may become too small to store sufficient honey and pollen for the winter.

The greatly increased longevity of bees that overwinter arises from a difference in their physiology compared with that of 'summer' bees and contrasts with the relatively slight prolongation of life associated with lack of work alone. Thus most bees in the winter cluster have well-developed 'fat bodies', containing protein as well as fat, and their hypopharyngeal glands remain fully developed (Lotmar, 1939; Maurizio, 1946; Haydak, 1957). Bees with the life expectancy of 'winter' bees can be produced by feeding abundant pollen to young bees in summer, but the treatment has little effect on older bees (see Maurizio, 1961).

Swarming

Even when the food supply and other external conditions remain constant, the numbers of a colony's brood can fluctuate and affect foraging. Thus brood rearing often decreases before swarming thus anticipating (see pages 27 and 155ff) the social requirements. Swarming is likely to occur when a colony becomes overcrowded (Simpson, 1957; Simpson and Riedel, 1963), i.e. when it has grown too big for its nest or hive space. In such circumstances any cells vacated by newly emerged bees may quickly become filled with food and consequently foragers may have difficulty in passing on their loads and so remain longer in the hive and add to the congestion. The queen lays many

fewer eggs (Nolan, 1925), probably partly because of less space for
eggs and partly because the workers feed her less (Huber, 1792;
Taranov-Ivanova, 1946; Allen, 1960). Presumably with less brood
the stimulus to forage also decreases and the workers live longer,
temporarily increasing the adult population and causing more congestion. If the situation were not relieved by swarming the congestion
would eventually be relieved by more bees dying than being reared.
Also food stores would be used and there would be more cells to receive eggs. Brood rearing would increase until there was again
congestion.

When a colony of social insects swarms (i.e. reproduces itself) one
or more queens accompanied by workers leave the colony and eventually settle in a new home. Swarming by honeybees differs from that
of other social insects in that it is the old queen, often thin because
she is no longer laying well, who accompanies the swarm, and the
new queen stays behind and heads the old colony. The workers that
leave with the swarm from a crowded honeybee colony are not predetermined long beforehand (Martin, 1963; Simpson and Riedel,
1963), although those that leave the hive when an uncrowded colony
swarms possibly are; the factors determining which bees go with the
swarm are unknown. Because of decreased brood rearing in the parent
colony many of the swarming bees have well developed fat bodies
and hypopharyngeal glands that serve as protein stores for the
swarm, and many have full honey stomachs and so can secrete wax
and build comb as soon as the swarm settles in a new home.

Undoubtedly the initial small amount of comb in a new colony
limits brood rearing but, as more cells are built, more eggs can be laid.
Assuming that the daily rate of cell construction and the number
used for food storage and egg laying remain constant, the first new
workers will emerge three weeks after the first eggs were laid and an
approximately equal number will emerge daily for the next three
weeks. However, provided the comb building potential of the swarm
has not been reached, as soon as cells are being vacated by newly
emerged workers the queen can, theoretically, double her egg laying
rate because there will be twice the number of cells. However, such
an increase is unlikely because nurse bees will be few; the original
workers of the swarm will be dying rapidly, and the death rate will
exceed the birth rate for a while, as happens to colonies in the
spring. Similar dwindling may occur in the swarm's mother colony;
it takes the bees a maximum of eleven days to rear a new queen and
it is another one to two weeks before she has mated and is ready to
lay eggs. Therefore, egg laying will be resumed later in the parent

colony than in the swarm, but in the parent colony egg laying will not be restricted by lack of comb space.

Queen Replacement

When the queen of a colony dies suddenly the situation is similar to that in the parent colony from which a swarm has departed, except that there will have been no decrease in brood so the associated dwindling of the population will be less. The absence of the queen alone, without any subsequent decrease in brood results in less pollen being collected (Free, 1966a). However, a queen usually keeps near the centre of her colony and is well protected from enemies, so that under natural conditions it is unlikely that she will die suddenly, although she may be killed accidentally by a beekeeper. When a queen becomes old or otherwise begins to fail, her colony usually ensures its continuity by rearing another queen. The onset of queen rearing is generally followed by less brood rearing (Simpson, 1959) perhaps because queen larvae are fed at the expense of the worker larvae. When queen rearing occurs during summer, it may be accompanied by swarming (Simpson, 1959) but otherwise the queen is replaced by a new one without colony division. Sometimes the old queen continues to lay eggs until her successor begins to do so and may even continue for some time afterwards. Thus supercedure of the old queen by the new has relatively little effect on brood rearing and the population only dwindles temporarily.

The mechanisms by which bees know their queen is defective or absent have recently been studied intensively. The queen's mandibular glands produce volatile substances (pheromones, Karlson and Butenant, 1959) that inhibit workers from rearing queens (Butler, 1954; Callow, Chapman and Paton, 1964). Queens heading colonies that are rearing new queens produce less of these substances than usual (Butler, 1960) and presumably too little to inhibit the workers. Distribution of the pheromones may also be hindered in large or overcrowded colonies and so this effect is further diminished. The new queens that the workers rear produce enough of these substances after they are mated and so correct the deficit. When queen rearing is accompanied by swarming, presumably the old queen, who heads the swarm, can inhibit queen rearing again in her smaller colony (Figure 5).

The same materials that inhibit queen rearing also inhibit the workers' ovaries from developing (Butler and Fairey, 1963). In colonies preparing to swarm some workers have ovaries more developed than normal (Perepelova, 1926) but usually they develop sufficiently to lay eggs only in those colonies that have become hopelessly queenless, such as those in which a queen has become lost on her mating flight and there is no brood young enough from which to rear a replacement queen (see Sakagami, 1959). Possibly the immature queens being reared produce something that compensates in part for the lack of the queen's inhibitory material. Undoubtedly many more pheromones will be discovered during the next few years but already the resemblances of the chemical interactions between

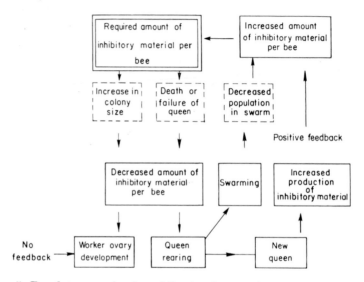

Figure 5. Regulatory mechanisms following decrease in amount of inhibitory material available per bee.

the individuals of a honeybee colony to the hormonal interactions between the tissues of an organ (see Chapter 8) are quite striking. A more general comparison between the regulatory functions of hormones and pheromones has been made by Kalmus (1965).

In queenless colonies only a relatively small proportion of the workers lay eggs (Perepelova, 1928a and Sakagami, 1959) and some laying workers are licked and examined by others as if they were queens (Perepelova, 1928b; Park, 1949; Sakagami, 1954, 1959); perhaps their presence in some way limits the ovary development and egg laying of other workers. However, bees with developed ovaries may also be mauled by others (Sakagami, 1954; Hoffman, 1961). These observations are difficult to reconcile, but it seems that laying workers help maintain the cohesion of a queenless colony, if only because the drone brood they produce stimulates foraging and nursing.

Drone rearing will shorten the workers' lives and hasten the colony's end, but any males produced may help other colonies in the vicinity to survive. Whether in fact such altruistic potentialities are ever selected is controversial (see Kalmus, 1964). The question of what units are selected in general is one of the perplexing problems in evolution (see pages 330 and 340ff). If a virgin queen cannot mate she eventually lays unfertilized eggs which, like workers' eggs, produce drone brood. The males produced in such circumstances may be particularly beneficial by helping to redress the drone shortage in the locality and enable other virgins to mate.

Drone Production

Nearly all colonies headed by mated queens produce some drones during the year, but the numbers range from a few hundred to several thousand (Rosser, 1934; Allen, 1958). In Britain drone rearing starts in April or May and continues until September; the peak period is usually in June and early July. The amount of drone brood fluctuates more than the amount of worker brood (Nolan, 1925; Allen, 1958), so drone production is probably more sensitive to nectar and pollen increase. Drone brood may be discarded in unfavourable circumstances or only a very small proportion of drone brood may be reared to maturity (Weiss, 1962).

Drones are reared in larger cells than workers, and whereas the eggs a queen lays in worker cells are fertilized those she lays in drone cells are not. Drone production may be limited by a shortage of drone cells (Allen, 1963) and beekeepers tend to discourage the building of drone comb, and thus the rearing of drones, as being wasteful. In most hives drone cells are to be found only at the margins of the combs or used to repair parts of damaged worker combs. However, when colonies are allowed to build comb naturally about 13% of the cells produced are drone (Weiss, 1962), and drone-cell production reaches its annual peak in April and May (Free, 1966b). The size of a colony seems to influence its tendency to build drone comb, as in colonies under natural conditions the drone cells tend to be along the bases of the central combs and in the outside combs (H. J. Wadey, personal communication). The proportion of drone to worker cells built also depends on the number of drone cells already present (Free, 1966b).

It has been suggested that drone cell production is associated with the vigour of the queen and that drone production may be associated with swarming, but there is little evidence for this. At times of the year when drones are normally not reared, the production of drone cells can be initiated in a colony that is queenless and rearing queens, but not in one that is only queenless (Free, 1966b). It is not known whether queen rearing, when a queen is present, is effective in similar circumstances.

A colony that has reared drones seems to tolerate them provided sufficient food is being collected. In autumn when food becomes scarce the workers drive the older drones away from the combs and may even drag them out of the hive. Meanwhile, younger drones are tolerated on the brood combs and also fed by the nurse bees (Free, 1957). It is obviously beneficial to a colony to stop rearing drones and to discard mature ones when colony reproduction is over for the year, but the factors resulting in drone expulsion are not understood, although it is common experience that queenless colonies will retain their drones when others are discarding them. Even if their retention in queenless colonies is associated with a lack of inhibitory substances produced by the queen, their expulsion from normal colonies cannot be explained by the opposite assumption, because when forage is scarce more workers remain at home and the distribution of inhibitory substances is, if anything, hindered. Perhaps the destruction of worker brood and drones in unfavourable weather share a common cause.

When nectar suddenly becomes scarce the out-of-work foragers often attempt to rob each other's colonies and guards are alerted to challenge and repel intruders, which are recognized by their alien odours (Butler and Free, 1952; Kalmus and Ribbands, 1952; Ribbands, 1954) which arise through difference in food supply. In these circumstances drones are quickly expelled whereas their expulsion tends to be prolonged when the nectar supply gradually becomes less (Bro. Adam, personal communication) and robbing is less likely. Kalmus (1941) found that workers from different colonies, feeding at the same dish of sugar syrup, tolerate each other as long as food is plentiful but when it becomes scarce they fight. Possibly drones have an alien odour tolerated by workers only as long as there is sufficient forage and their antagonism to strange odours has not been aroused by the odours of robber bees from other colonies. If so, drone retention in a queenless colony may be associated with the colony's diminished tendency to defend itself against robbers.

Temperature Regulation in Summer

It is not known whether a decrease in environmental temperature has any direct effect on drone expulsion but it seems unlikely because much of the success of the honeybee colony depends on its ability to regulate its temperature and humidity independently of its external environment, and so to survive conditions in which individual bees die. The brood nest in summer is usually kept at 34–35°C and any deviations from this are soon corrected (see for example, Gates, 1914; Himmer, 1932) even by unprotected clusters in the open (Büdel, 1958). Hess (1926) and Wohlgemuth (1957) found that at outside temperatures of 51 and 50°C the temperature in the brood nest was only 36 and 38°C.

Bees and brood inevitably produce metabolic heat and the additional workers are forced to remain in the colony during cold, unfavourable weather and at night undoubtedly provide additional help in maintaining the colony's temperature. However, when the outside temperature approaches the optimum brood nest temperature, the bees need to take one or more compensatory actions (Figure 6). First,

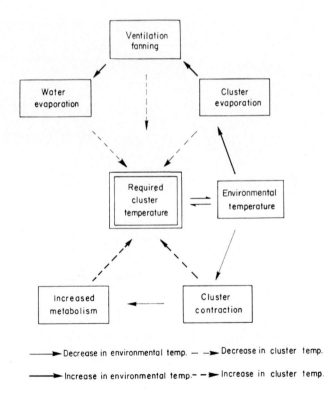

Figure 6. Compensatory mechanism for change in environmental temperatures.

as the temperature rises the bees move further apart on the combs and some leave the hive and cluster outside. As it gets still hotter groups of bees on the alighting-board of the hive face its entrance and by fanning their wings draw a current of air out of the hive. Similar groups just inside the entrance and even on the brood combs help to maintain the flow (Himmer, 1932; Hazelhoff, 1941). When there is an opening at the top of the hive, fresh air is sucked downwards

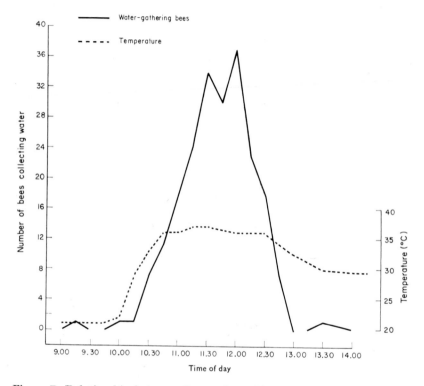

Figure 7. Relationship between the number of bees collecting from a water site and the temperature around the hive. (After Lindauer, 1955.)

through it; when there is no top ventilator the air is drawn in through one part of the entrance and forced out through another. Provided the outside air is cooler than that in the hive the colony is effectively cooled. Fanning bees are mostly of preforaging age (Sakagami, 1953); the reason why some bees, but not others, fan is unknown. Because the metabolism of the bees increases with rise in temperature, the hive contains more carbon dioxide and less oxygen as the colony becomes hotter; fanning is induced more readily by introducing carbon dioxide into the hive (Bigot, 1953) than by introducing pure nitrogen, so probably an increase in carbon dioxide is more relevant than oxygen lack (Hazelhoff, 1941). However, carbon dioxide content is not the only factor causing ventilation fanning, for when hives are hot it continues after the excess carbon dioxide is removed.

In association with ventilation, fanning bees evaporate water to cool the colony. They not only spread minute drops of water in the cells but by folding and unfolding their tongues, cause small drops of regurgitated water to be drawn out into thin films between their tongues and other mouthparts and so to be evaporated (Chadwick, 1922, 1931; Parks, 1929; Lindauer, 1955). Water is not stored in the hive but is collected when needed (Figure 7). When there is no water, bees sometimes use dilute nectar for the purpose. Water collectors inform other bees of the source of their loads by dances (Park, 1923). When there is a danger of a colony overheating, those foragers that return home with water or dilute nectar have their loads eagerly received by other bees whereas foragers with concentrated nectar have difficulty in finding bees that will accept it (Lindauer, 1955), (Figure 8). Consequently the collection of water or dilute nectar is encouraged and collection of concentrated nectar discouraged.

Kiechle (1961) found that the honey stomach contents of bees in a small part of a hive he had artificially heated were dilute, whereas in the rest of the hive they were concentrated. Therefore, water or dilute nectar had obviously not been passed in stages from the bees that received it at the hive entrance to bees that evaporated it, for if this were so the honey stomach contents would have become more concentrated approaching the area where cooling was necessary. Bees that receive water from foragers must carry it directly to where it is needed; perhaps they also evaporate it there.

Colonies with brood need water or dilute nectar to dilute stored honey and prepare brood food. Water carriers are particularly numerous after cool or rainy weather during which bees have been unable to collect nectar and have had to use stored honey (Kiechle, 1961), and there are probably always some workers collecting water. Kiechle (1961) found that at those times when there was small demand for water only a few foragers collected it between trips for nectar and pollen; when water was wanted urgently these bees knew where it was and could inform others. Perhaps when the nectar circulating among the bees of the colony is concentrated some foragers are induced to collect water instead of nectar or pollen. When water is no longer needed, the foragers returning with it find difficulty in getting their loads accepted by others and so are discouraged from collecting more, but loads of concentrated nectar are readily received again (Lindauer, 1955).

Despite the numerous sources of water vapour in a hive the brood nest is usually kept at a relative humidity of about 40% (see Büdel, 1960). Bees concentrate nectar by regurgitating a drop and drawing it out as a thin film beneath the tongue in the same way as they expose water or dilute nectar when cooling the hive; when much nectar

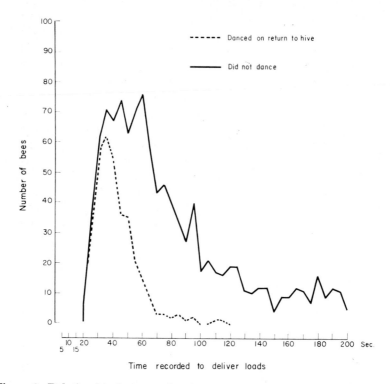

Figure 8. Relationship between the time water carriers take to deliver their loads and their tendency to dance. (After Lindauer, 1955.)

is being concentrated in a colony the relative humidity increases and the bees rectify this by ventilation fanning. Even when the outside temperature is higher than the optimum for brood rearing, the warm outside air is drawn in by ventilation fanning to rid the hive of water produced by cooling. Presumably the bees fanning just outside the hive entrance continue to do so only when the air they are drawing from inside the hive is either warmer or has a higher relative humidity than the outside air. At high temperatures isolated individual bees live longer in humid than in dry air, because desiccation is the limiting factor; but desiccation is unlikely in a colony where bees survive brief periods of heat better at low relative humidities because they can cool themselves more readily by evaporation, see Figure 9 (Free and Spencer-Booth, 1962).

With inadequate ventilation fanning, the air inside a hive would soon become saturated and the bees would be unable to cool the atmosphere by evaporating water; alternatively, with inadequate

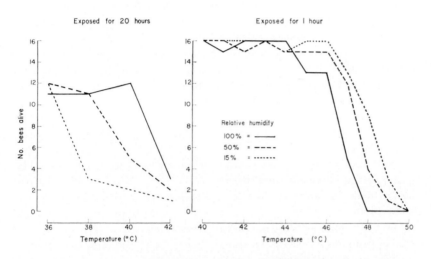

Figure 9. Effect of exposing bees to various temperatures and humidities. (After Free and Spencer-Booth, 1962.)

evaporation, and the outside temperature higher than 35°C, ventilation fanning could only aggravate the situation. In either circumstances the colony would soon die.

In continued hot weather, many foragers would be continuously collecting water and many housebees carrying and evaporating it; less nectar and pollen would be collected and less brood reared, so colonies would become smaller and less able to regulate their temperature. Thus continued heat is detrimental to the economy of a colony; perhaps the fact that nectar and pollen are often abundant in such circumstances compensate for this.

Temperature Regulation in Winter

In temperate climates brood rearing ceases in autumn and the colony lives on its accumulated food reserves until it can forage again in the spring. When bees are kept in small groups of up to 200 at 20°C or below, an increased proportion clusters in a compact group as the temperature falls, or as the number of bees is increased, and nearly all are clustered at 10°C (Free and Spencer-Booth, 1958). The temperature in the centre of a broodless cluster in winter usually fluctuates between 20 and 30°C (Gates, 1914; Wilson and Milum, 1927; Corkins, 1930); such a cluster consists of a closely packed mass of bees with active ones inside and more sluggish ones outside. As the temperature falls the cluster becomes more compact thus decreasing its surface and its heat loss (Corkins, 1930), (Figure 6). A compact cluster contains about 21 bees per cubic inch (Jeffree, 1959). However, the heat loss is not directly proportional to the cluster size as it is complicated by the inclusion of parts of combs within the cluster and these act as cooling fins whose thermal conductivity depends on their contents. The contents of the combs surrounding the cluster, the hive structure and the amount of ventilation also determine the rate at which heat is lost from the cluster (see Büdel, 1960, and Simpson, 1961).

Individual bees become immobilized by cold at 9 or 10°C. The temperature at which they die depends on the duration of exposure; individuals can survive two or three days at 0–10°C (Kalabuchov, 1933), but only three hours at −3°C, or one hour or less at −4°C (Free and Spencer-Booth, 1960). Temperatures in the cluster decrease from the centre to the edge. The edge of the cluster is kept at 9–10°C or more (Himmer, 1926); obviously it must not get much colder or the bees would become immobile, fall from the cluster and soon die when exposed to the lower temperatures on the hive floor. Presumably, the proportion of bees producing additional heat and the density and thickness of bees in the centre and outside shell, all affect the temperature gradient between the inside and outside of the cluster and help to prevent the centre from becoming too hot or too cold while maintaining the periphery above the minimum temperature for survival.

In addition to checking heat loss, a colony can react to cold weather by generating more heat. The temperatures of resting bees are similar to those of their surroundings at 30–40°C, but at lower air temperatures an active bee can be as much as 20°C above air temperature (Pirsch, 1923; Himmer, 1925). The total heat generated by a cluster enables it to survive cold that would kill individuals and so there is some resemblance to the temperature control of homoiothermic animals (see p. 87ff). When even small groups of 25–200 bees are exposed to cold they keep their temperatures above that of air temperature and so increase their chance of survival (Figure 10); both the temperature of the group and its longevity increase with its size. When the air temperature falls, the amount of food consumed increases, showing that bees can respond to cold by increasing their metabolism (Free and Spencer-Booth, 1958). Although it has long

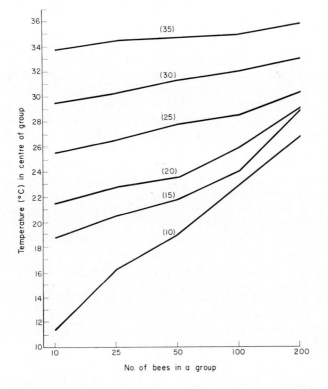

Figure. 10. The ability of small numbers of bees to maintain their temperatures above the environment. Temperatures of groups at various environmental temperatures (given in brackets) three hours after the start of the experiments. (After Free and Spencer-Booth, 1958.)

been assumed that colonies generate more heat in the cold (e.g. Phillips and Demuth, 1914; Himmer, 1926), this has been demonstrated only recently (Free and Simpson, 1963), when colonies were kept in a room where the temperature could be regulated between +20 and −40°c. The carbon dioxide production per bee was least at +10°c and increased both above and below this temperature (Figure 11). The increase above 10°c is obviously not part of the temperature-regulating mechanism and presumably reflects the greater activity of colonies at summer rather than at winter temperatures, but the increase below 10°c must indicate greater metabolism and heat production and is best understood as regulatory.

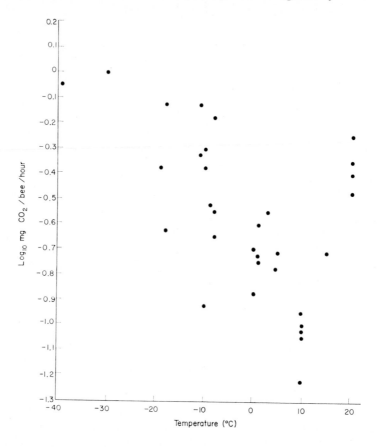

Figure 11. The carbon dioxide output of three colonies kept at various environmental temperatures. (After Free and Simpson, 1963.)

Cluster concentration probably reaches its limit at about +5°c, see Figure 12 (Wedmore, 1953), and compensation at colder temperatures must be by the cluster generating more heat. The temperature in the centre of a broodless cluster is often inversely related to the outside atmosphere (e.g. Gates, 1914; Hess, 1926; Himmer, 1926), but this relationship tends to be brief and presumably results from bees producing more heat than is needed. Over longer periods the temperatures tend to be more closely related to outside temperatures (Corkins, 1930; Lavie, 1954). Moreover, because of the insulation provided by the hive and combs, there is inevitably a time lag between a change in the outside temperature and the response of the colony.

It is difficult to understand exactly how heat production in a cluster is regulated. The bees forming the periphery of a cluster would be the first to perceive a temperature change but bees exposed to temperatures such as those that occur at the periphery produce little

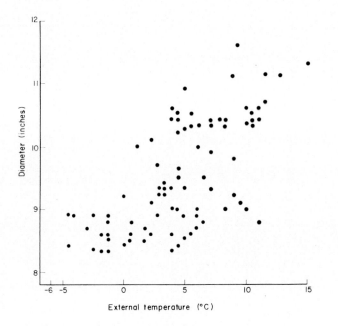

Figure 12. The cluster diameter at different external temperatures. (After Wedmore, 1953.)

heat (Allen, 1959). They must either communicate the temperature change in some way to the bees in the centre who then metabolize faster, or themselves move into the centre and generate it. Bees in the centre and periphery certainly change places but it is not known whether there is enough exchange to account for the heat compensation that follows a sudden temperature change. Winter bees have a thermal preference of 28–33°C (Heran, 1952) and so should prefer the cluster centre; acclimatization to the colder temperatures at the outside of the cluster would hardly affect their thermal preference, although it might help the bees there to survive the cold (Free and

Spencer-Booth, 1950). Bees at the outside of a cluster face inwards; perhaps the cluster contraction observed when the air temperature falls reflects the efforts made by the outside bees to reach the centre.

Because of their smaller ratio of surface area to volume, larger colonies conserve heat more efficiently than smaller ones, so probably

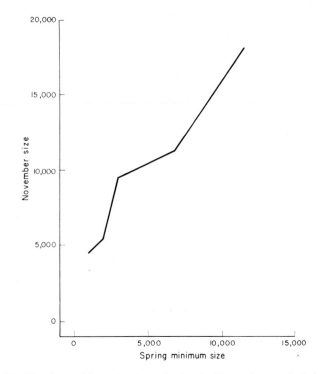

Figure 13. Number of bees in six colonies in November and the following spring. (After Jeffree, 1955.)

bees of large colonies need to produce less metabolic heat and so tend to survive longer (Figure 13), see Jeffree (1955); also large colonies are presumably more efficient in maintaining their brood nest temperatures when brood rearing begins early in the year. The amount of this early brood and the time when it is produced varies greatly (Jeffree, 1956; Villumstad, 1962) and can substantially affect colony growth for the rest of the year. However, little is known of the factors determining the initiation of brood rearing for the year, or whether it is affected by the temperatures in the cluster centre.

References

Allen, M. D. (1955). 'Observations on honeybees attending their queen.' *Brit. J. Animal Behav.*, **3**, 66–69.

Allen, M. D. (1958). 'Drone brood in honeybee colonies.' *J. Econ. Entomol.*, **51**, 46–48.

Allen, M. D. (1959). 'Respiration rates of worker honeybees of different ages and at different temperatures.' *J. Exp. Biol.*, **36**, 92–101.

Allen, M. D. (1960). 'The honeybee queen and her attendants.' *Animal Behav.*, **8**, 201–208.

Allen, M. D. (1963). 'Drone production in honeybee colonies (*Apis mellifera* L.).' *Nature*, **199**, 789–790.

Allen, M. D. and Jeffree, E. P. (1956). 'The influence of stored pollen and of colony size on the brood rearing of honeybees.' *Ann. Appl. Biol.*, **44**, 649–656.

Bigot, L. (1953). 'Les particularités de l'atmosphère interne des ruches.' *Apiculteur*, **97**, 39–46.

Büdel, A. (1958). 'Ein Beispiel der Temperaturverteilung in der Schwarmtraube.' *Z. Bienenforsch.*, **4**, 63–66.

Büdel, A. (1960). 'Bienenphysik.' In Büdel, A. and Herold, E. (Eds.), *Biene und Bienenzucht*, Ehrenwirth, Munich, pp. 115–180.

Butler, C. G. (1954). 'The method and importance of the recognition by a colony of honeybees (*A. mellifera*) of the presence of its queen.' *Trans. Roy. Entomol. Soc. London*, **105**, 11–29.

Butler, C. G. (1960). 'The significance of queen substance in swarming and supersedure in honeybee (*Apis mellifera* L.) colonies.' *Proc. Roy. Entomol. Soc. London, Ser. A*, **35**, 129–132.

Butler, C. G. and Fairey, E. M. (1963). 'The role of the queen in preventing oogenesis in worker honeybees.' *J. Apic. Res.*, **2**, 14–18.

Butler, C. G. and Free, J. B. (1952). 'The behaviour of worker honeybees at the hive entrance.' *Behaviour*, **4**, 262–292.

Callow, R. K., Chapman, J. R. and Paton, P. N. (1964). 'Pheromones of the honeybee; chemical studies of the mandibular gland secretion of the queen.' *J. Apic. Res.*, **3**, 77–89.

Chadwick, P. C. (1922). 'Ventilation.' *Am. Bee J.*, **62**, 158–159.

Chadwick, P. C. (1931). 'Ventilation of the hive.' *Gleanings Bee Culture*, **59**, 356–358.

Corkins, C. L. (1930). 'The metabolism of the honeybee colony during winter.' *Bull. Wyo. Agric. Exp. Sta.*, No. 175.

Filmer, R. S. (1932). 'Brood area and colony size as factors in activity of pollination units.' *J. Econ. Entomol.*, **25**, 336–343.

Free, J. B. (1957). 'The food of adult drone honeybees (*Apis mellifera*).' *Brit. J. Animal Behav.*, **5**, 7–11.

Free, J. B. (1960). 'The pollination of fruit trees.' *Bee World*, **41**, 141–151, 169–186.

Free, J. B. (1966a). 'Factors determining the collection of pollen by honeybee colonies.' *Animal Behav.* (in press).

Free, J. B. (1966b). 'The production of drone comb by honeybee colonies.' *J. Apic. Res.* (in press).

Free, J. B. and Simpson, J. (1963). 'The respiratory metabolism of honeybee colonies at low temperatures.' *Entomol. Exp. Appl.*, **6**, 234–238.

Free, J. B. and Spencer-Booth, Yvette (1958). 'Observations on the temperature regulation and food consumption of honeybees (*Apis mellifera*).' *J. Exp. Biol.*, **35**, 930–937.

Free, J. B. and Spencer-Booth, Yvette (1959). 'The longevity of worker honeybees (*Apis mellifera*).' *Proc. Roy. Entomol. Soc. London, Ser. A*, **34**, 141–150.

Free, J. B. and Spencer-Booth, Yvette. (1960). 'Chill coma and cold death temperatures of *Apis mellifera*.' *Entomol. Exp. Appl.*, **3**, 222–230.

Free, J. B. and Spencer-Booth, Yvette. (1961). 'The effect of feeding sugar syrup to honeybee colonies.' *J. Agri. Sci.*, **57**, 147–151.

Free, J. B. and Spencer-Booth, Yvette. (1962). 'The upper lethal temperatures of honeybees.' *Entomol. Exp. Appl.*, **5**, 249–254.

Frisch, K. von (1923). 'Über die Sprache der Bienen.' *Zool. Jber.* (3), **40**, 1–186.

Frisch, K. von (1946). 'Die Tänze der Bienen.' *Öst. Zool. Z.*, **1**, 1–48.

Gates, B. N. (1914). 'The temperature of the bee colony.' *Bull. U.S. Dep. Agr.*, No. 96, 1–29.

Gontarski, H. (1953). 'Zur Brutbiologie der Honigbiene.' *Z. Bienenforsch.*, **2** (1), 7–10.

Haydak, M. H. (1945). 'Value of pollen substitutes for brood rearing of honeybees.' *J. Econ. Entomol.*, **38**, 484–487.

Haydak, M. H. (1957). 'Changes with age of the appearance of some internal organs of the honeybee.' *Bee World*, **38**, 197–207.

Hazelhoff, E. H. (1941). 'De luchtverversching van een bijenkast gedurende den zomer.' *Maandschr. Bijent.*, **44**, 16.

Heran, H. (1952). 'Untersuchungen über den Temperatursinn der Honigbiene (*Apis mellifica*) unter besonderer Berücksichtigung der Wahrnehmung strahlender Wärme.' *Z. Vergleich. Physiol.*, **34**, 179–206.

Hess, W. R. (1926). 'Die Temperaturregulierung im Bienenvolk.' *Z. Vergleich. Physiol.*, **4** (4), 465–487.

Himmer, A. (1925). 'Körpertemperaturmessungen an Bienen und anderen Insekten.' *Erlanger Jber. Bienenk.*, **3**, 44–115.

Himmer, A. (1926). 'Der soziale Wärmehausalt der Honigbiene I. Die Wärme im nicht brütenden Wintervolk.' *Erlanger Jber. Bienenk.*, **4**, 1–50.

Himmer, A. (1932). 'Die Temperaturverhältnisse bei den sozialen Hymenopteren.' *Biol. Rev. Cambridge Phil. Soc.*, **7**, 224–253.

Hoffmann, I. (1961). 'Über die Arbeitsteilung im weiselrichtigen und weisellosen Kleinvölkern der Honigbiene.' *Z. Bienenforsch.*, **5** (8), 267–278.

Huber, F. (1792). *Nouvelles Observations sur les Abeilles I*, Paschaud, Geneva.

Jeffree, E. P. (1955). 'Observations on the decline and growth of honeybee colonies.' *J. Econ. Entomol.*, **48**, 723–726.

Jeffree, E. P. (1956). 'Winter brood and pollen in honeybee colonies.' *Insectes Sociaux*, **3**, 417–422.

Jeffree, E. P. (1959). 'The size of honeybee colonies throughout the year and the best size to winter.' *Rept. Cent. Assoc. Beekeepers*, April, 1959.

Kalabuchov, N. I. (1933). 'Contribution to the study of dormancy and "anabiosis" in the honeybee, *Apis mellifera*.' *Zool. Zh.*, **12** (4), 121–153.

Kalmus, H. (1941). 'The defence of a source of food by honeybees.' *Nature*, **148**, 228.

Kalmus, H. (1964). 'The evolution of albuism.' *New Scientist*, **20**, 550–551.

Kalmus, H. (1965). 'Possibilities and constraints of chemical tele-communication.' *Proceedings of 2nd International Congress of Endocrinology*, pp. 188–192.

Kalmus, H. (1966). *Rates of Evolution in the Voices of Time*, Brazillier, New York.

Kalmus, H. and Ribbands, C. R. (1952). 'The origin of the odours by which honeybees distinguish their companions.' *Proc. Roy. Soc. (London), Ser. B*, **140**, 50–59.

Karlson, P. and Butenant, A. (1959). 'Pheromones (Ectohormones) in insects.' *Ann. Rev. Entomol.*, **4**, 39–58.

Kiechle, H. (1961). 'Die soziale Regulation der Wassersammeltatigkeit im Bienenstaat und deren physiologische Grundlage.' *Z. Vergleich. Physiol.*, **45**, 154–192.

Lavie, P. (1954). 'L'enregistrement thermique continu dans les populations *d'Apis mellifera*, au cours de l'hivernage.' *Insectes Sociaux*, **1**, 39–48.

Lindauer, M. (1952). 'Ein Beitrag zur Frage der Arbeitsteilung im Bienenstaat.' *Z. Vergleich. Physiol.*, **34**, 299–345.

Lindauer, M. (1955). 'The water economy and temperature regulation of the honeybee colony.' *Bee World*, **36**, 62–72, 81–92, 105–111.

Lotmar, R. (1939). 'Der Eiweiss-Stoffwechsel im Bienenvolk während der Ueberwinterung.' *Landw. Jahrb. Schweiz.*, **53**, 34–70.

Martin, P. (1963). 'Die Steuerung der Volksteilung beim Schwärmen der Bienen. Zugleich ein Beitrag zum Problem der Wanderschwärme.' *Insect Soc.*, **10**, 13–42.

Mauermayer, G. (1954). 'Untersuchungen über die Beziehungen zwischen Arbeitsleistung und Lebensdauer bei Arbeiterinnen der Honigbiene.' *Arch. Bienenk.*, **31**, 31–41.

Maurizio, A. (1946). 'Beobachtungen über die Lebensdauer und den Futterverbrauch gefangen gehaltener Bienen.' *Beih. Schweiz. Bienenztg.*, **2**, 1–48.

Maurizio, A. (1961). 'Lebensdauer und Altern bei der Honigbiene (*Apis mellifica* L.).' *Gerontologia*, **5**, 110–128.

Merrill, J. H. (1924). 'Sealed and unsealed brood.' *Am. Bee J.*, **64**, 424–5.

Moeller, F. E. (1958). 'Relation between egg-laying capacity of queen bee and populations and honey production of their colonies.' *Am. Bee J.*, **98**, 401–402.

Myser, W. C. (1952). 'Ingestion of eggs by honeybee workers.' *Am. Bee J.*, **92**, 67.

Nixon, H. L. and Ribbands, C. R. (1952). 'Food transmission within the honeybee community.' *Proc. Roy. Soc. (London), Ser. B*, **140**, 43–50.

Nolan, W. J. (1925). 'The brood-rearing cycle of the honeybee.' *Bull U.S. Dept. Agric.*, No. 1349, 1–56.

Park, O. W. (1923). 'Behaviour of water-carriers.' *Am. Bee J.*, **63**, 553.

Park, O. W. (1949). 'The honeybee colony—life history.' In Groot, R. A. (Ed.), *The Hive and the Honeybee*, Dadant, Hamilton, Illinois.

Parks, H. B. (1929). 'Water storage by bees.' *Iowa State Apiarist Rept.*, 1928, pp. 53–56.

Perepelova, L. I. (1926). 'Biology of laying workers.' *Opuit. Pas.*, (**12**), 8–10.

Perepelova, L. I. (1928a). 'Biology of laying workers. I. Relationship of the age of bees to the development of workers.' *Opuit. Pas.*, (**1**), 6–10.

Perepelova, L. I. (1928b). 'Biology of laying workers. VI. The time taken by a laying worker to lay an egg.' *Opuit. Pas.*, (**2**), 59–61.

Phillips, E. F. and Demuth, G. S. (1914). 'The temperature of the honeybee cluster in winter.' *Bull. U.S. Dep. Agric.* No. 93, 1–16.

Pirsch, G. B. (1923). 'Studies on the temperature of individual insects, with special reference to the honeybee.' *J. Agric. Res.*, **24**, 275–287.

Ribbands, C. R. (1952). 'Division of labour in the honeybee community.' *Proc. Roy. Soc. (London), Ser. B*, **140**, 32–43.

Ribbands, C. R. (1953). *The Behaviour and Social Life of Honeybees*, Bee Research Association, London.

Ribbands, C. R. (1954). 'The defence of the honeybee community.' *Proc. Roy. Soc. (London), Ser. B*, **142**, 514–524.

Rösch, G. A. (1925). 'Untersuchungen über die Arbeitsteilung im Bienenstaat. I. Die Tätigkeiten im normalen Bienenstaate und ihre Beziehungen zum Alter der Arbeitsbienen.' *Z. Vergleich. Physiol.*, **2**, 571–631.

Rosser, J. H. (1934). 'On drone rearing.' *Bee World*, **15**, 58.

Sakagami, S. F. (1953). 'Untersuchungen über die Arbeitsteilung in einem Zwergvolk der Honigbiene. Beiträge zur Biologie des Bienenvolkes, *Apis mellifera*, L.' *I. Jap. J. Zool.*, **11**, 117–185.

Sakagami, S. F. (1954). 'Occurrence of an aggressive behaviour in queenless hives, with considerations on the social organisation of honeybees.' *Insect Soc.*, **1**, 331–343.

Sakagami, S. F. (1959). 'Arbeitsteilung in einem weisellosen Beinenvölkchen.' *Z. Bienenforsch.*, **4**, 186–193.

Sharma, P. L. and Sharma, A. C. (1950). 'Influence of numbers in a colony on the honey-gathering capacity of bees.' *Indian Bee J.*, **12**, 106–107.

The impetus for their reinvestigation came from the new disciplines of ecology and ethology, and the underlying motives consisted in the main of a desire to show that adaptations, deemed protective on anecdotal or intuitive grounds, do indeed protect. An essential part of these studies has been the elucidation of the more subtle properties of the adaptations. Two examples will suffice in this context.

1. Procryptic insects closely mimic objects in their environment, such as twigs, dead leaves, or aerial rootlets, and the resemblance may be very exact. Similarly, in Batesian and Müllerian mimicry, the convergence between the superficial facies of the convergent animals may be extraordinarily detailed. In both cases, it was assumed that predators are confused by these similarities. One of the arguments used against the proponents of mimicry theory was that the predators of these insects are unlikely to have the powers of sensory discrimination needed to impose selection pressures capable of causing the evolution of such precise imitations (McAtee, 1932a, b). Two sorts of solution to this difficulty seemed possible. Carter (1946, 1948) suggested that the responses of birds to food-objects are triggered by inborn releasing mechanisms; generalizing from Lorenz's work on the social behaviour of jackdaws, he supposed that a given species of predator would be responsive to no more than a very few of the configurational cues given by its food, or by inanimate and mimicked objects in the environment. Carter then showed that this apparent support for the critics of mimicry theory can be negated if it is supposed that different predator species in the same stimulus situation have inborn responses to different configurational aspects of the stimulus. In principle, a population of many species of predator could create a selection pressure which would result in resemblances more complex than could be imposed by any single species of predator alone.

It proves that this attractive hypothesis is probably not necessary to explain the facts (Blest, 1956). The inborn responses of birds are not so configurationally limited as was at first supposed; their visual equipment is excellent; and they can learn and retain remarkably fine visual discriminations in experimental situations. It may also be added that many of the more spectacular cases of mimicry are found in the tropics, where the role of the visually sophisticated small primates as insect predators has probably been underestimated.

2. The chemical defences of the arthropods—defensive sprays, and various kinds of noxious secretion—are well known. Until recently, there has been no reason to suppose that the detailed constitution of

these secretions has any special significance; their virtue might be supposed to reside solely in their general unpleasantness. Recently, however, their efficacy and chemical constitution have been studied in several laboratories. Two particular cases are worth quoting. Eisner and his colleagues (Eisner, Meinwald, Monro and Ghent, 1961) have shown that the defensive spray of the whip-scorpion *Mastigoproctus giganteus* consists of 84% acetic acid, 11% water and 5% caprylic acid. The most noxious material is clearly the acetic acid but the small percentage of caprylic acid seems not to be present merely as a biochemical accident. It is lipophilic, and promotes both the spreading of the spray-droplets over the cuticles of arthropod predators, and the penetration of the acetic acid through them. Eisner and his coworkers have suggested that a consideration of the minority constituents of other insect secretions might reveal similar patterns of chemical adaptation.

Recently, Remold (1963) has studied the defensive secretions of pentatomid bugs, which are composed of saturated and unsaturated aldehydes of chain length C_6–C_{10}, and n-tridecane. Of these compounds, only the aldehydes are toxic when applied to the cuticles of test insects. Tridecane is non-toxic, but it accelerates the penetration of the cuticle by the liquid, probably by altering the chemical bonding of the cuticular lipids and increasing the cuticular permeability.

These investigations, however, are concerned with fairly straightforward questions: how do presumptive adaptations work? and can they be shown to be explicable in terms of the fine action of natural selection? Essentially, the problems do not differ from those presented by mimicry.

In the case of Batesian mimicry, a palatable insect achieves a close resemblance, albeit superficial, to a distasteful one; in the case of Müllerian mimicry, two or more unpalatable species converge to achieve a common appearance. Both types of mimicry require either that predators learn by experience which of their potential prey species are edible and which are distasteful, or they demand that predators have built-in patterns of responsiveness which are specifically adjusted to normal features of their prey objects. The former assumption appears to be largely true for avian predators, while the latter may prove to be true for the responses of insectivorous lizards (Muhlmann, 1933–34; Mostler, 1934–35; de Ruiter, 1952, 1955; Sexton, 1964). All these problems can be studied with reference to hypotheses about natural selection of very simple kinds. The selection pressures which are postulated are *direct*, and the relationships which are assumed to pertain between individual predators and prey, and between the prey and its natural environment, are supposed to be relatively invariant in time. Although the lives of insects are broken up into successive stages which are often morphologically and biologically very distinct, it is not unnatural to assume that the selective forces acting on an adult butterfly, say, at the beginning of its adult life will not be greatly different from those imposed upon it shortly before its death a few days or weeks later. There is a tendency, too, to see mimicry in terms of the dramatic examples described, for instance, by Eltringham (1916), in the large neotropical butterfly genus *Heliconius*, or of the close resemblances between certain neotropical danaine butterflies and the pericopid moths which fly with them in the same habitats.

Recent work has tended to reveal more complex possibilities. Those which will be mentioned in this chapter can be summarized as follows:

Complex adaptations, tailored very specifically to the ethological weaknesses of particular predators.

Adaptations whose incidence and efficiency depend upon interactions between the prey and its environment.

Adaptations which are limited or directed by the sexual or social behaviour of the prey itself, or by similar interactions between several species of prey in the same habitat.

Adaptations which are generated by the changes of structure or behaviour which both prey and predator undergo with time.

These categories are no more than convenient, and they disguise what is, perhaps, a more fundamental distinction: that between conventional *direct* selection, which can account for all of the effects to be discussed under the first three headings and *group*, or, more properly, *kin* selection, which must be invoked to explain some of the phenomena under the fourth heading. The circumstances under which kin selection can reasonably be used as an explanatory concept have recently been discussed by Hamilton (1964) and Maynard-Smith (1964).*

---

* *Kin* selection requires only that a character present in certain individuals of a community should favour the survival of siblings which do not possess it. *Group* selection implies that groups of individuals which are reproductively in partial isolation one from another differ in the presence or absence of a character which affects their fitness. Selection, in this case, acts upon the groups as units, and for it to do so, the population must be divided into partially isolated breeding units. The conditions which permit kin selection to occur are the less rigorous.

Tailored Adaptations

One very striking example of a series of physiological and behavioural adaptations which have been 'tailored' to meet a particular situation has been studied in some detail by Roeder and Treat; this is the acoustic and behavioural responses of moths to the echolocation pulses of flying bats. The echolocation systems of bats have been reviewed by Griffin (1958), and the behaviour of flying moths in relation to them by Roeder (1963); an extensive series of publications describes the structure and physiology of the moth tympanic organ which is responsible for receiving acoustic stimuli and transmitting information about them to the central nervous system (Treat, 1955; Roeder and Treat, 1957, 1961; Roeder, 1962a, b, 1963). The features of this system of interaction between prey and predator which are of particular interest lie in the evolution by the moths of a receptor system of a very simple kind, which is responsive to the range of ultrasonic frequencies used by bats in echolocation, and a repertoire of behavioural responses which is adapted to the bats acoustic relationships with their potential prey.

The tympanic organ is sensitive to frequencies as high as 150 kc/s, and its property of rapid adaptation to a continuous pure tone means that it functions most efficiently as a pulse detector. The pulses of free-flying bats can be detected at distances of 100 ft or more. There is little frequency discrimination, but considerable sensitivity to intensity differences, which, at least in principle, could provide the flying moths with information about the distance and flight-path of an oncoming bat. Bats do not react to flying moths at distances greater than some 12 ft and at this short distance it has been determined that the response of the tympanic organ would in any case be saturated, so that it could no longer indicate the direction of the predator. Between 100 ft and 15 ft, however, the movements of an approaching bat could probably be assessed by its victim.

In response to the pulses of an approaching bat, a flying moth can do one of several things, which appear to depend partly on the intensity of the ultrasonic pulses which it receives, and partly on vagaries of its own physiological state.

a. *Low sound intensities* (i.e. the bat is at a large distance from the moth). Moths so stimulated change their course, and start to fly in a relatively straight line away from the source of sound, upwards, laterally, or downwards.

b. *High sound intensities* (i.e. the bat is at a short distance from the moth). Moths may respond in several ways:

(i) *Complex, non-directional manoeuvres.* The moths may go into a power-dive, spiral, or zig-zag, or perform a series of sharp turns or loops. These manoeuvres may be alternated with periods of passive fall, during which the wings remain folded.

(ii) *Passive fall.* The moths may respond simply by folding their wings and falling passively.

(iii) *'Answering back'.* Some arctiid and ctenuchid moths are equipped with metathoracic tymbal organs, whose mechanics and acoustics have been described by Blest, Collett and Pye (1963). Although many species have been shown to use their tymbals to generate ultrasonics when they are handled (Blest, 1964) others have been shown in preliminary experiments to answer pulses from a bat simulator, or from live bats, with short bursts of ultrasound from the tymbals. The response is only given to acoustic stimuli by moths in free flight, and the tymbal response does not persist beyond the cessation of the stimulus (Roeder, in preparation; Pye, Darwin and Blest, in preparation). Dunning and Roeder (1965) have shown that the catching performance of bats presented with mealworms thrown from a 'gun' was adversely affected by the simultaneous presentation of moth sounds, but not by competing bat echolocation signals of similar intensity. The effect was statistically striking, but the mechanism is, as yet, unknown. It is unlikely that simple jamming of echolocation is involved, for the system is highly resistant to acoustic interference (Griffin, 1958).

The adaptive significance of the response to low sound intensities is immediately obvious. It seems likely that the main feature of the non-directional manoeuvres, their variability and lack of stereotypy, makes them as unpredictable to an attacking predator as they are to a human observer. In addition, there is a special advantage attached to folding the wings and falling passively. Measurements of the proportion of sound reflected from the wings of a flying moth show that at ultrasonic frequencies the greater part of the echo is reflected from the spread wings (Roeder, 1963). By folding its wings, therefore, a moth can become acoustically invisible to a bat which has just begun to detect it at the periphery of its field of acoustic vision.

This is a rather complex example of a tailored system, although its individual features are simple enough. A less elaborate example is provided by the evolution of eyespot patterns on the wings of many insects (see Figure 1); these patterns are usually concealed when the insects are at rest, and displayed suddenly to attacking predators. Experimental evidence suggests that these imitations of the appearance of the vertebrate eye are 'exploiting' the inborn responses

13*

of passerine birds to their own predators. The eyes of predators (e.g. owls, cats, etc.) are among the most important of the stimuli releasing mobbing and evasive responses from small passerines (Hinde, 1954).

Figure 1. The eyespot display of *Automeris hamata* (Saturniidae, hemileucinae).

Figure 2. Anterior view of the pupa of an unidentified noctuid moth from Panama, in its natural site. The 'head' somewhat resembles that of common species of small *Anolis* found in the same habitat.

The whole situation reminds one forcibly of a statement made many years ago by Sumner that 'the cause must have had eyes'. The existence of these inborn releasing mechanisms in a major class of predator has undoubtedly provided the selection pressures for the widespread evolution of eyespot patterns by the insects (Blest, 1957a, b); they are in fact 'tailored' to the inborn releasing mechanisms of small passerines. They are certainly ineffectual against White-faced Monkeys and Geoffroy's Pinché, both of which are potential predators of the ocellated species of *Automeris* in neotropical forest (unpublished observations). A number of caterpillars and pupae also bear eyespots, and are perhaps, mimicking snakes and lizards (cf. Figure 2), and so do some fish, in which they probably serve to deflect the attacks of predators to less vulnerable parts of the body.

Adaptations whose Incidence and Efficiency Depend on Interactions between the Prey and its Environment

In the broad sense, it is obvious that no protective adaptation functions in complete isolation from the sum of relationships which its owner has with its environment. There are, nevertheless, instances in which an apparently isolated adaptation can only be seen to make sense in terms of the overall biology of a species.

At the most naïve level, generalized cryptic coloration (that is, a general resemblance to some feature of the natural environment, such as dead leaves or bark, which does not involve a precise procryptic imitation) must necessarily bear some relationship to the gross features of the natural environment. In neotropical forests, casual observation suggests that cryptic populations of noctuids and other families of small moths of similar size and appearance tend to contain more greenish species in forests with a high density of epiphytes, than in forests where the density of epiphytes, and particularly of mosses, is low. A precise study of this type of phenomenon would be of some interest (Blest, 1963a).

But cryptic coloration in an adult insect has widespread consequences for its biology. Whereas aposematic forms, that is, those which advertise their distastefulness by bright colours, are often gregarious and afford the inexperienced predator who must learn that they are unpalatable the optimum chance to do so, cryptic species must usually be dispersed in their habitat. Experiments by de Ruiter (1952, 1955) have emphasized the adaptive reasons for dispersal. Captive jays were allowed to search for stick caterpillars scattered at random among a greater number of the twigs which they resembled. The first capture of a stick caterpillar usually followed its accidental discovery. The jays would then start picking up twigs at random. If the density of caterpillars to twigs was high, the jays were rapidly reinforced by a high number of successes, and could, eventually, learn to distinguish caterpillars from twigs. Faced with a low density of caterpillars, however, an accidental capture was followed by a few random pecks at twigs, after which the birds, unless they found another insect, lost interest.

These experiments show that there must be a considerable selection pressure to ensure the dispersal of cryptic species; not only must it affect the social behaviour, or lack of it, of the adult forms, but it is likely to affect the reproductive potential, and egg-laying behaviour, and also the whole course of early development. In general, we may expect that a cryptic species will lay its eggs singly and scatter them over a wide area, so that its larvae, particularly if they themselves are cryptic, will have a low population density. Newly emerged adults are vulnerable, and it is usual for larvae which will give rise to cryptic adults to disperse before pupation. This is particularly striking in the case of such moths as *Pseudosphinx tetrio*, whose aposematic larvae feed conspicuously and semi-gregariously on the poisonous frangipani, but disperse widely before pupating. The adults are cryptic and palatable.

Adaptations Limited or Directed by the Sexual or Social Behaviour of the Prey Itself, or by Interactions between Several Prey Species in the Same Habitat

The evolution of a visual protective adaptation may lead to a conflict between selection imposed by predators, and selection arising from other features of the biology of the species. In diurnal Lepidoptera with visually-dominated courtship behaviour, it is possible to imagine a selective conflict between the evolution of cryptic coloration on the one hand, and of adequate epigamic signals on the other. Similar difficulties attend the initial stages of the evolution of a Batesian mimic: an inefficient and incomplete resemblance to the model carries the risks of enhanced conspicuousness. Schmidt (1958) has discussed this problem, and points out that the initial populations of an evolving Batesian mimic must be small in relation to those of their models if the incipient resemblance is to carry an advantage, and that evolution could be expected to proceed rapidly. He remarks on the absence in the literature of descriptions of incomplete Batesian mimics, and concludes that they must be rare. There are, however, such obvious examples as the temperate species of ant-mimicking spiders in the genus *Myrmarachne*, whose mimetic efficiency seems feeble compared to that of their more specialized tropical congeners.

However, it seems to be true that incipient or incomplete resemblances are common in Müllerian complexes, and may, perhaps, be more frequent than the complete and spectacular resemblances which dominate the literature (Blest, 1963b, 1964).

A complex case of selective competition has been described by Stride (1956). *Danaus chrysippus* L., a danaid butterfly, is mimicked by the females only of *Hypolimnas misippus* L., a nymphalid. There are, in Africa, three principal colour forms of *D. chrysippus*, to which correspond three 'matching' variants of *H. misippus*, as indicated in the following Table.

	D. chrysippus	*H. misippus*
1	Typical	Typical
2	f. *dorippus* Klug	f. *inaria* Cramer
3	f. *alcippus* Cramer	(f. *alcippoides* Butler)

While the first and second forms of the two species probably represent a true mimetic association, the form *alcippoides* of *H. misippus*, while it resembles *D. chrysippus* form *alcippus* in possessing a predominantly white hindwing, is too scarce to be regarded as a true mimic. Furthermore, in regions such as Achimota, on the Gold Coast, where all the individuals of *D. chrysippus* are of the form *alcippus*, the populations of *H. misippus* achieve no more than a slight suffusion of their hindwings with white scales.

It cannot be supposed that *H. misippus* lacks the genetic potentiality to achieve full mimicry of *D. chrysippus* form *alcippus*: the mere existence of form *alcippoides* shows that a white hindwing can, in fact, be achieved. A study of the male courtship behaviour of *H. misippus* suggests a reason for this mimetic failure. The early stages of the male courtship behaviour are under the control of visual stimuli from the female, of which the hindwing attributes exert the stronger effect. While altering the colour of the female hindwing from its natural orange–brown diminished its effectiveness as a releasing stimulus, the substitution of white had a more dramatic effect; it actively inhibited the male courtship response. Thus there is a selective block preventing the evolution of a true mimic of *D. chrysippus* form *alcippus*. It is not proven why, in adaptive terms, white should have become an inhibitory stimulus, but the preponderance of white pierine butterflies in the same habitats occupied by *Hypolimnas* may have caused selection to operate against 'wasted' courtship flights directed to inappropriate objects. If this is the case, the situation can be interpreted as an example of selective competition between the mimetic demands of the environment, and the sexual demands of the species.

Adaptations Generated by the Changes in Structure or Behaviour which Prey and Predator Undergo with Time

Discussions of the relations between insects and their predators have tended to treat the various properties of the prey as if they were constant in time. It does not take any very deep consideration of the problem to reveal that this is not, in fact, the case. The contribution that a given individual makes to the reproductive potential of its population changes with age; a fertilized female moth, for example, progressively depletes the store of new individuals contained within her as she lays her eggs; a virgin moth probably does not maintain her power to disperse sex attractant and becomes progressively less likely to attract a mate as she grows older. The protective coloration of adult moths is achieved by a covering of scales and hairs which are peculiarly vulnerable to mechanical damage. The mere activity of flight itself is sufficient to reduce the efficiency both of protective resemblances and of aposematic patterns. Indeed, the damaging consequences of flight may have contributed to the selection pressures which have caused individual species of tropical moths to restrict their flight times often to very brief periods of each night.

The effects on a population of the removal of young or of old individuals, therefore, are not the same; the sacrifice of a series of individuals of increasing age will have progressively less significance for the population to which they belong. The sacrifice of post-reproductive individuals of a non-social species, will, within these terms of reference, be selectively without significance (Medawar, 1955; Williams and Williams, 1957; Williams, 1957). That selective importance can, in fact, be attributed to the capture of post-reproductive prey follows from the inconstancy in time of the behaviour of individual predators. The argument may best be illustrated with the following example.

Consider two species of non-feeding adult insect, whose life-spans are limited by their metabolic reserves, one of which is procryptic, and the other aposematic. Both are subject to predation by a species of predator whose choice of prey objects is determined by learning processes. The interaction between prey and predator will result in the following two distinct selective processes.

a. The longer that a cryptic insect survives after reproducing, the greater the chance that it will be found by a predator. It is known that birds, at least, must learn to find cryptic prey, and the lesson must be periodically reinforced by success. Survival, therefore, prejudices the safety of other individuals, which may still be reproductive. Kin selection should shorten post-reproductive survival, and direct selection should shorten the reproductive life-span, should the overall ecology of the species permit it to do so.

b. The converse of this argument applies to the aposematic insect. Birds also learn to avoid conspicuous but unpalatable prey, and they may even do so by single-trial learning (Mostler, 1934–35). The post-reproductive survival of an aposematic insect, then, should favour the survival of its siblings, for it is available to train a predator to avoid them without prejudicing the reproductive potential of the population. Kin selection should lengthen the post-reproductive survival of the aposematic species; direct selection may be expected to lengthen the overall life-span.

While selective actions of these kinds may be postulated on purely logical grounds, they are peculiarly difficult to test, for the ecology of most species introduces competing and interacting selective effects which might be expected to mask the results of a critical comparison.

A differential effect of kin-selection on life-spans could only be demonstrated where the following conditions exist.

(i) The adult forms to be compared naturally coexist in a fairly uniform habitat.

(ii) They do not feed as adults.

(iii) There are no gross differences between the species in the availability of their food-plants, and in the length of time which must be available to them for mating and egg-laying.

(iv) The species are of comparable size.

These conditions are met for a number of neotropical saturniid moths of the subfamily hemileucinae, and an attempt has been made to test the hypothesis by comparing the behaviour and life-spans of five species in seasonal tropical evergreen forest on Barro Colorado Island in the Panama Canal Zone (Blest, 1963b). The moths and their behaviour patterns are listed in Table 1.

One of the species, *Lonomia*, is a procryptic leaf mimic; two *Automeris* combine leaf mimicry with hindwing eyespot patterns, *Dirphia (Dirphiopsis)* and *Dirphia (Periphoba)* are unpalatable to White-faced Monkeys and to Geoffroy's Pinché and possess aposematic displays.

The comparisons which were practicable were limited by the

Table 1

Species	Coloration	Protective Devices
1 *Lonomia cynira* Cramer	Procryptic leaf mimic	Immobility to disturbance
2 *Automeris junonia* Walker	Forewings: leaf mimicry Hindwings: eyespots	Eyespot display
3 *Automeris boucardi* Druce	Hindwings: eyespots	Eyespot display
4 *Dirphia* (*Dirphiopsis*) *eumedide* Stoll	Aposematic with banded abdomen	Display of abdomen, with discharge of foul-smelling meconium
5 *D.* (*Periphoba*) *hircia* Cramer	Drab with hairy and banded abdomen	Display of abdomen with oriented discharge of foul meconium

circumstances under which the wild moths were captured at ultra-violet light-traps. Males which arrived at the lights immaculate, and with their fringe-scales undamaged were presumed to have hatched on the night of capture, but it cannot be assumed that they were virgin. The adult male life-tables shown in Figure 3, therefore, may

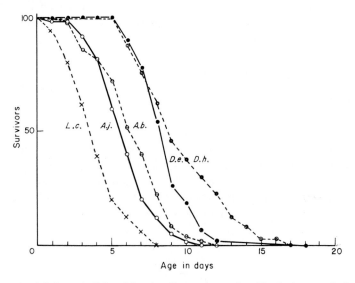

Figure 3. Adult male life-tables for *Lonomia cynira* (*L.c.*) *Automeris junonia* (*A.j.*) *A. boucardi* (*A.b.*) *Dirphia* (*Dirphiopsis*) *eumedide* (*D.e.*) *D.* (*Periphoba*) *hircia* (*D.h.*). Each curve is derived from a sample of 50 moths. (From Blest, 1963b.)

comprise both virgin and mated moths. The moths from which they were obtained were stored in plastic boxes under constant illumination from capture until death.

The hypothesis predicted that the life-spans should be ranked in the order: *Lonomia* < *Automeris* < *Dirphia*, and this expectation is fulfilled. This result, in itself, does not require the action of kin selection. For example, there is, presumably, some advantage to be gained from reducing the larval life to as short a period as possible; a restricted larval life-span might limit the metabolic reserves of the adult moths, which in turn might be supposed to limit their life-spans. The mere possibility that mating and oviposition could be completed shortly after hatching would be sufficient, in principle, to allow a reduction of the adult life-span by the conventional processes of direct selection acting on the larvae.

That this alternative and simpler explanation is inadequate is suggested by the following two lines of evidence.

1. The behavioural characteristics of the adult moths are also adjusted to their modes of protective coloration. The stability of the adult behaviour in time can be inferred from that of the rhythmic response ('rocking') which is performed when the moths settle (Blest, 1960), and which can be precisely measured. The behaviour of the cryptic species is markedly unstable in time, and the individual moths become more excitable and prone to fly as they age. These changes are absent, or at least less dramatic, in the aposematic species, which undergo relatively little change in their flight excitability with age. Thus, even if the aposematic and cryptic types were to start with equal metabolic reserves, the cryptic forms would use them up faster. It seems that kin selection may have acted upon the adult behaviour patterns to extend the life-spans of the aposematic moths, and to restrict those of the cryptic insects.

2. Most relevant of all, if a comparison is made between the post-reproductive survival of mated females, it is found that the majority of cryptic females die within 24 hours of completing oviposition; aposematic females tend to survive for several days afterwards (Blest, 1963b).

The special argument outlined above may be extended to a consideration of the well-known dichotomy between the fragility of cryptic lepidoptera, and the 'toughness' of aposematic forms. It is usually argued that the toughness of aposematic moths and butterflies is an adaptation which allows them to survive 'sampling' by inexperienced predators. But it is, surely, an advantage to be tough, even if cryptic, for predators differ widely in their capture-efficiency, and may be very inefficient when taking prey from the extremes of their own habitats.

Cryptic fragility may, in fact, be a device which hinders an attacking predator from learning the visual characteristics of its prey. Some insectivorous birds take time to reduce their prey to a state in which it can be swallowed. During this period of preparation, fragile lepidoptera rapidly lose their scales, while aposematic lepidoptera do not. Given these assumptions, it follows that kin selection could generate cryptic fragility, even if it is supposed that evolution is taking place in a population whose predators are 100% successful in capturing their prey after they have recognized it.

In summary then, evolutionary interactions between prey and predator can operate at several different levels of selective complexity, ranging from simple relationships, in which it is assumed that selection is constant throughout the lives of both individual prey and individual predator, to complex relationships, in which it can be shown that the changes which prey and predator undergo in time are selectively important in their own right, and tend to generate indirect effects. These more complicated evolutionary situations have been scarcely investigated, and are perhaps the most fruitful field for future study.

References

Blest, A. D. (1956). 'Protective coloration and animal behaviour.' *Nature*, **178**, 1190–1191.

Blest, A. D. (1957a). 'The function of eyespot patterns in the Lepidoptera.' *Behaviour*, **11**, 209–256.

Blest, A. D. (1957b). 'The evolution of protective displays in the Saturnioidea and Sphingidae.' *Behaviour*, **11**, 257–309.

Blest, A. D. (1960). 'The evolution, ontogeny and quantitative control of the settling movements of some New World saturniid moths, with some comments on distance communication by honey-bees.' *Behaviour*, **16**, 188–253.

Blest, A. D. (1963a). 'Relations between moths and predators.' *Nature*, **197**, 1046–1047.

Blest, A. D. (1963b). 'Longevity, palatability and natural selection in five species of New World saturniid moth.' *Nature*, **197**, 1183–1186.

Blest, A. D. (1964). 'Protective display and sound production in some New World arctiid moths.' *Zoologica*, **49**, 161–181.

Blest, A. D., Collett, T. S. and Pye, J. D. (1963). 'The generation of ultrasonic signals by a New World arctiid moth.' *Proc. Roy. Soc. London, Ser. B*, **158**, 196–207.

Carter, G. S. (1946). 'Mimicry, protective resemblance, and animal behaviour.' *Sci. Progr. (London)*, **135**, 547–551.

Carter, G. S. (1948). 'Colour and colour vision in animals.' *Nature*, **162**, 600–601.

Cott, H. B. (1940). *Adaptive Coloration in Animals*, Methuen, London.

Dunning, D. C. and Roeder, K. D. (1965). 'Moth sounds and the insect-catching behaviour of bats.' *Science*, **147**, 173–174.

Eisner, T., Meinwald, J., Monro, A. and Ghent, R. (1961). 'Defence mechanisms of arthropods. I. The composition and function of the spray of the whip-scorpion *Mastigoproctus giganteus* (Lucas) (Arachnida, Pedipalpida).' *J. Insect Physiol.*, **6**, 272–298.

Eltringham, H. (1916). 'On specific and mimetic relationships in the genus *Heliconius* L.' *Trans. Entomol. Soc. London*, 101–148.

Griffin, D. R. (1958). *Listening in the Dark*, Yale University Press, New Haven.

Hamilton, W. D. (1964). 'The genetical evolution of social behaviour.' *J. Theoret. Biol.*, **7**, 1–52.

Hinde, R. A. (1954). 'Factors governing the changes in strength of a partially inborn response, as shown by the mobbing behaviour of the Chaffinch.' *Proc. Roy. Soc. London, Ser. B*, **142**, Parts 1 and 2, 306–331, 332–358.

Maynard-Smith, J. (1964). 'Group selection and kin selection.' *Nature*, **201**, 1145–1147.

McAtee, W. L. (1932a). 'Effectiveness in nature of the so-called protective adaptations in the animal kingdom, chiefly as illustrated by the food habits of nearctic birds.' *Smith. Misc. Coll. Washington*, **85**, 7, 1–201.

McAtee, W. L. (1932b). 'Protective adaptations of animals.' *Nature*, **130**, 961–962.

Medawar, P. B. (1955). 'The definition and measurement of senescence.' *Ciba Found. Colloq. Aging*, **1**, 4–15.

Mostler, G. (1934–35). 'Beobachtung zur Frage der Wespen-Mimikry.' *Z. Morphol. Oekol. Tiere*, **29**, 381–454.

Muhlmann, H. (1933–34). 'Im Modellversuch kunstlich erzeute Mimikry und ihre Bedeutung fur den "Nachahmer".' *Z. Morphol. Oekol. Tiere*, **28**, 259–296.

Poulton, E. B. (1890). 'The colours of animals.' *Intern. Sci. Ser. London*, **68**, xiii, 360.

Remold, H. (1963). 'Scent-glands of land-bugs, their physiology and biological function.' *Nature*, **198**, 764–768.

Roeder, K. D. (1962a). 'The behaviour of free-flying moths in the presence of artificial ultrasonic pulses.' *Anim. Behav.*, **10**, 300–304.

Roeder, K. D. (1962b). 'Ultrasonic interactions of bats and moths.' In Bernard, E. E. and Kave, M. R. (Eds.), *Biological Prototypes and Synthetic Systems*, Vol. 1, Plenum Press, New York, pp. 54–57.

Roeder, K. D. (1963). 'Echoes of ultrasonic pulses from flying moths.' *Biol. Bull.*, **124**, 200–210.

Roeder, K. D. and Treat, A. E. (1957). 'Ultrasonic reception by the tympanic organ of noctuid moths.' *J. Exp. Zool.*, **134**, 127–158.

Roeder, K. D. and Treat, A. E. (1961). 'The detection of bat cries by moths.' In Rosenblith, W. (Ed.), *Sensory Communication*, M.I.T. Technology Press, Cambridge, Massachusetts.

Ruiter, L. de (1952). 'Some experiments on the camouflage of stick caterpillars.' *Behaviour*, **4**, 222–232.

Ruiter, L. de (1955). 'Countershading in caterpillars.' *Arch. Neerl. Zool.*, **11**, 285–342.

Schmidt, R. S. (1958). 'Behavioural evidence of the evolution of Batesian mimicry.' *Anim. Behav.*, **6**, 127–138.

Sexton, O. J. (1960). 'Experimental studies of artificial Batesian mimics.' *Behaviour*, **3–4**, 244–252.

Sexton, O. J. (1964). 'Differential predation of the lizard *Anolis carolinensis*, upon unicolored and polycolored insects after an interval of no contact.' *Anim. Behav.*, **12**, 101–110.

Stride, G. O. (1956). 'On the courtship behaviour of *Hypolimnas misippus* L. (Lepidoptera, Nymphalidae), with notes on the mimetic association with *Danaus chrysippus* L. (Lepidoptera, Danaidae).' *Brit. J. Anim. Behav.*, **4**, 52–68.

Treat, A. E. (1955). 'The response to sound in certain Lepidoptera.' *Ann. Entomol. Soc. Am.*, **48**, 272–284.

Williams, G. C. (1957). 'Pleiotropy, natural selection, and the evolution of senescence.' *Evolution*, **11**, 398–411.

Williams, G. C. and Williams, D. C. (1957). 'Natural selection of individual harmful social adaptations among sibs, with special reference to social insects.' *Evolution*, **11**, 32–39.

15

Regulation in Animal Societies and Populations

V. C. WYNNE-EDWARDS

Natural History Department, Marischal College,
Aberdeen

Population Homeostasis

In *The Origin of Species* Darwin (1859) pointed out that animals and plants in general have a capacity to increase their numbers by geometrical progression, although in practice they do not continuously do so. The stock of individuals occupying a given region tends instead to fluctuate irregularly in size as the years pass, often with less than a tenfold extreme departure from its basic level; and in stable habitats it is the constancy of numbers that tends most often to emerge as the striking characteristic of long-period observations.

Darwin named four 'checks to increase', which appeared to him to be responsible for preventing populations from expanding indefinitely. These were (i) the amount of food available which 'gives the extreme limit to which each can increase', (ii) the effects of predation by other species, (iii) adverse climatic effects, and (iv) the toll of epidemics. His views afterwards became widely accepted and deeply entrenched, and until very recent years the populations of living organisms have been quite generally pictured by biologists as 'striving to the utmost to increase in numbers' in the face of these four recurrent and normally insuperable restraints. As recently as 1954 Lack (1954, p. 4) reaffirmed that 'the best introduction to the subject [of the natural regulation of animal numbers] is that given in Chapter 3 of *The Origin of Species*'.

Against this, evidence has been independently accumulating over the last 40 years and more, through experiment and field observation, that supports the view that animals in natural environments are by no means necessarily or ceaselessly striving to expand in numbers. Numerous representative species belonging to most of the great phyla have been shown to possess self-regulatory adaptations, which give them the initiative in restricting their own population densities without any need for the intervention of Darwin's four checks on increase. Darwin's intuitive deductions on this subject are not in other words being borne out by the test of experiment and there can no longer be any justification for uncritically accepting them.

The new evidence indicates that means of regulating their own numbers have evolved in animals as a typical homeostatic process with negative feedback, all controlled within the populations themselves. As might be expected, homeostasis at the population level, as at the physiological one, has reached its best and most conspicuous development in the most advanced types of animals, such as the arthropods and vertebrates.

Darwin was correct in pointing to the food resource as the most common 'ultimate' factor in determining the possible upper limit of population density. As the normal result of this, each habitat in which animals live has a certain carrying capacity for the species concerned, a capacity which often varies with the seasons and from one year to the next. The homeostatic machinery appears to operate so as to keep the number of mouths to be fed as nearly as possible in harmony with the existing availability of food, adjusting the population density to match significant changes in resource levels. The adjustments can be brought about in a variety of ways. For instance, physiological stress can exert a negative or inverse effect on the reproductive input to the stock (i.e. on recruitment); or it may effect emigration, or mortality. Whatever the particular mechanism being used, such homeostasis generally depends on regulatory processes whose intensity of action varies on a sliding scale and is described as *density dependent*.

Food resource levels, and even population levels, are sometimes rapidly reduced by uncontrollable outside events, most often climatic in character (e.g. drought, cold). When this happens on a big enough scale a large imbalance can be created between existing numbers of animals and the local carrying capacity, which it may take some time to correct. In highly unstable habitats like those of the arctic and desert margins, and equally in cultivated land where there is characteristically a deliberately disruptive regime of ploughing, harvesting and crop rotation, wild animal populations may exist for much of the time very much as Darwin imagined them, in an erratic pioneering state of building up their strength in the wake of repeated disasters; and it may be relatively seldom that they reach a level at which any homeostatic population-limiting device need be called into play.

In the less extreme natural environments of the temperate zones, still large but far more regular seasonal changes commonly occur in food availability. These not infrequently culminate in periods of superabundance, especially in the spring and summer, which animals generally exploit for breeding. Breeding usually produces a large, well-timed and often quite short-lived upsurge in population density; but the temporary increase in demand that results from it may still be far from sufficient to fully exploit the summer food peak, and part of it may consequently go to waste. Only in the most constant and serene environments, in marine habitats and the rain forests of the tropics for instance, and among species that breed all the year round, is anything like a perfect static balance likely to be indefinitely maintained.

Seasonal variations in food availability, as pointed out, generally result in corresponding changes in carrying potentials. The fact that animals are not necessarily exploiting their food resources to the full all the time does not affect the general premise that food is the ultimate factor deciding the ceiling of numbers that can be supported in each succeeding generation. Its determining influence will be felt again as soon as the superabundance ends, and demand for food in the habitat begins to catch up with the supply. At that point the homeostatic machinery might be expected to come into play to ensure that demand is prevented from outrunning the resources available. The question arising here is what the factors are that determine the best exploitation rate. Under a system of natural selection, in other words, at what point on the sliding scale of supply and demand will the optimum population density be found?

Optimum Density of Population

Animal populations exploit their food resources under conditions that are closely paralleled in various human enterprises, such as the fisheries and primitive forms of agriculture. In these it is notoriously easy to demand too much of the natural resource and, by over-exploiting it, to bring about a lowering of productivity. In fishing, for any species there is a maximum harvest or yield that one can take and sustain, season after season, without making inroads into the stock which impair its recuperative powers. It is the same with natural pastures; the density of livestock has to be kept down to a level dictated by experience, otherwise the productivity of the pasture soon begins to fall away, the sward grows thin, and the stones begin to show through. So often nature's standing crops in their pristine state have given the appearance of being limitless resources; and time after time human experience has shown how tempting it is to get rich in the short term by over-exploiting them, and yet how essential to manage them with restraint if they are to yield the far greater long-term profits that accumulate through sustained production.

It is clear that animals whose populations are ultimately limited by the growth and productivity of living foods face situations of the same kind. The maximum tenable density for them will consequently be the one that prevents desperate shortages of food from arising, where these can result in damage being inflicted which lowers the capacity of the food to regenerate and produce future crops. Population growth has generally therefore to be checked, prematurely as it were, and a ceiling imposed on population density some time before the available food has been used up. How far, in any such situation, it is safe to go will depend very much on the type of resource being exploited. With perennial herbaceous plants including many palatable grasses almost the whole of the annual leaf, flower or seed production can sometimes be utilized for food without impairing vegetative regeneration. But where the food consists of evergreen foliage, or of potentially long-lived animals like fish, birds or mammals, the consumers have generally got to limit their demands to a small fraction of a relatively enormous standing stock. To give a quantitative illustration, experiments with red deer (*Cervus elaphus*) on the Hebridean island of Rhum conducted by the Nature Conservancy have shown that one-sixth of the population (taken from all age groups over one year old) can be killed each year without depleting the herd.

There are some types of food, many of them 'non-living', of which detritus, carrion, and the honey produced by flowers are examples, where the whole production can be consumed without detriment to the source. But, like any typical food resource, honey for instance is produced at a finite rate per unit area, and there is still an important advantage in the homeostatic control of the numbers of honey consumers, because this safeguards their chance of success in obtaining sufficient rations to complete their life-cycle and to reproduce. If overcrowding was not prevented, malnutrition would result. It is in fact widely accepted that the elaborate social systems of the Hymenoptera, the group that includes the honeybees themselves, have been evolved primarily to regulate the reproductive output of the population, in relation to currently available resources (cf. Richards, 1953, p. 74). This has in turn brought numerous other regulatory devices into existence (Free, p. 351).

The writer's theory of population regulation in animals (Wynne-Edwards, 1962) takes as its starting point this common basic requirement, to curb the growth of numbers and not exceed the maximum density consistent with sufficient and sustained yields of food; and it shows how this appears to be more or less successfully realized through an intricate web of adaptations, which together provide the working mechanism of population homeostasis.

The most easily detected adaptations result in some kind of spacing-out process which almost always depends on individual behaviour. A system of *exclusive territories*, for instance, each held by an individual, a pair, or a social group, leads to a pattern of land-tenure (which can, for example in fish, be under water), not basically different from those of human farming communities. Provided the behaviour of the territory owners automatically compels them to claim an area of sufficient size, and to resist all attempts by rival claimants to compress it below a determinate minimum, the population density can be prevented from rising above any arbitrary limit which natural selection has fixed as being favourable. If there are too many contestants to be accommodated with territories in the habitat on this basis, those that fail to establish themselves are identified *ipso facto* as being surplus to its capacity. This can lead to their immediate expulsion and off-loading from the habitat; in other cases they may be allowed to remain on sufferance but strictly as nonbreeders. Variants on this situation have been described in a number of birds and reptiles, for example in the red grouse (*Lagopus scoticus*) in north-east Scotland by Jenkins, Watson and Miller (1963); in the Australian magpie (*Gymnorhina tibicen*) by Carrick (1963); and in the rainbow lizard (*Agama agama*) in Nigeria by Harris (1964).

There are innumerable variant forms of property tenure, some of which are referred to later in this Chapter. Invariably they concern individual rights, and are more or less inseparably related to the rank or status of the individuals concerned. In the more gregarious kinds of higher animals a *personal hierarchy* tends to establish itself among the members of each group. The top-ranking, dominant members have the right to preempt every desirable situation at the expense of their subordinates; and when this is applied to obtaining preferred access to communal food resources, it can work, in a way that is

closely analogous to the property-holding systems already mentioned, as another device for cutting off from further use of the
resources an arbitrarily determined surplus element, in this case
forming the tail of the hierarchy.

Hierarchies of the peck-order type (Schjelderup-Ebbe, 1922) are
necessarily based on personal acquaintance between members of the
group concerned. There are other types of association, exemplified
for instance in a school of herrings, a flock of sea-gulls or starlings, or
a dancing swarm of gnats, in which all the members of the group
appear to be equal in status, and to be distinguished only collectively
from non-members, or from members of other groups. Usually,
though not necessarily, they involve such large numbers of individuals
that complete mutual acquaintance would be impossible. Here population stress may perhaps in some cases condition all members alike,
uniformly influencing their behaviour and even their reproductive
output; but in other cases inequalities of status are undoubtedly exposed by adverse conditions, and the machinery of the hierarchy invoked to cut the surplus off.

There are thus two types of 'internal' mechanism capable of acting
as *proximate* factors (see p. 398) in limiting the intensity of exploitation of food resources (and, in many species, of limiting also the quota
of the population attaining reproductive status at any given time).
These are, respectively, systems of property tenure, and systems
involving personal status. In many situations, as has already been
said, they are largely or completely combined. Each can offer desirable prizes or goals to be competed for, and won or lost, by the
aspiring occupants of any habitat. By their possession the individual
can become established as a member or resident; without them his
status is inferior and his future precarious. Each of them provides a
conventional kind of substitute, which effectively displaces all direct
competition for the actual food itself. Food still remains the *ultimate*
factor, behind the scenes, on which the real carrying capacity of the
habitat depends. But the evolution of these conventional substitutes
as immediate or proximate objectives of competition allows, as we
have seen, a density ceiling to be fixed for any population at the level
best favoured by natural selection.

The Origins of Social Behaviour

Consequences of the greatest theoretical and practical significance
flow from the fact that the proximate factors controlling population
have a conventional basis.

The prizes of competition derive their conventionality from the fact that, although they take the form of real possessions like territories, nest-sites, resting places, lairs, dens and burrows or of superior positions in an abstract hierarchy, they confer rights to something totally different—namely sharing the food resources, and often breeding as well. Not only are the prizes themselves conventional, they are competed for by equally conventional methods. Birds achieve much of their territorial defence by singing. Male dogs decide their mutual status by growling, raising their manes, and urinating. Although success or failure in interspecific competition can, and often does, become a matter of life and death, of leaving offspring or leaving none, the contestants do not fight to the death in order to establish their claims. Decision is ordinarily reached merely by threats of power, often without any drop of blood being shed. Moreover even the uncouth display of teeth and claws is readily superseded by less unseemly methods of domination, through the power of voice, scent, visual adornment and elegance.

The dependence of population homeostasis on these substituted conventional kinds of competition is the central feature of the writer's theory, because *it gives a key to the meaning of society itself*. Social organization is seen to have evolved to provide the indispensable stage and scenery, the artificial conditions and arbitrary rules, without which such improvised competition could neither operate nor be effective. On this hypothesis a completely new perspective of sociality opens before us, and for the first time we are able to give *society*, as a biological term, a single, comprehensive, functional definition. It appears simply as 'an organization of individuals capable of providing conventional competition among its members' (Wynne-Edwards, 1962, p. 132).

Seen for the first time in this light, society may appear barely recognizable. As a social animal himself, man has usually concentrated his attention on the agreeable elements of cohesion and fellowship, as the true hallmarks of society; he has turned his back on the discordant undertone of rivalry that a moment's reflexion will show is equally inherent in it, and has tended not to connect competition with sociality at all. In truth, the basic quality of human society is an adherence to common codes of convention. These set the group above the individual and characteristically call forth the dominance of leaders, the submission to their will of those that are led, and the desire for one's own side to win. The height of antisociality is self-seeking by criminal methods, that is, by methods that contravene the social code; but getting ahead in 'fair' competition with one's fellows, and helping one's own team to righteous victory are socially praiseworthy.

In its primitive essentials therefore man's social behaviour does not differ at all from that of other animals. Instead it provides a convincing example of the inseparability of brotherhood and rivalry in the social fabric, and the all-pervasiveness of conventional competition. But in at least one philosophically interesting respect man differs from most other species, for he has the choice of disobeying the social code. This is probably shared in a diminished degree by some at least of the other mammals which, like the familiar horse and dog, are capable of doing wrong and knowing it. For many animals compliance with the code seems automatic, even in the face of death.

Balancing Recruitment and Loss

In nearly constant environments, such as there are in parts of the tropics, especially in aquatic habitats and in rain forests, some of the animals (including many of the vertebrates) can as a rule be found breeding all the year round. Their population densities may as a result remain almost constant for long periods. Other species have synchronized annual periods for breeding (see Kalmus, p. 169). The latter is the normal situation in all regions where habitats are subject to predictable yearly alternations of summer and winter, or wet and dry seasons. Even there, taking one year with the next, populations may still be very stable, annually pursuing a regular rise and fall of reproduction and mortality. In more fluctuating climates greater differences tend to occur between one year and another in the population peaks and minima attained, although over a decade or longer period there is still no persistent trend up or down.

In every situation where constancy, integrated over suitably chosen time intervals, is maintained, the income to the population must obviously be equal to the loss. Recruitment comes primarily from reproduction; but there is generally no direct connection between this and mortality, which stems from completely separate causes and often falls largely at other seasons of the year. If in the long run both are equal, the most obvious explanation seems to be that one is somehow regulated to match the other: one would then be an independent and the other a dependent variable. On the question which is which, both alternatives have been postulated at various times by zoologists.

On the one hand there is the classical view that reproduction rates in animals have in general been adjusted to the hazards of their life-cycles. Parasites with highly specific requirements as to their hosts, and minimal powers of locomotion, have, as Leuckart (1863) emphasized a century ago, prodigious fecundities; whereas animals that live long and care for their young have correspondingly small fecundities. On this view, natality has been broadly matched to mortality, at least in the long term. The opposite proposition is that natural selection promotes 'fitness', and fitness in the individual must be measured by the relative share he succeeds in contributing to the next and subsequent generations; it can be claimed therefore that natural selection is bound to have elicited in every species the highest possible reproductive rate, measured by the numbers of offspring that in their turn survive to become adult. On this view it is of course reproduction that goes full speed ahead, and mortality that must vary dependently to equal it.

The first of these alternatives appears to be a well-founded inference based on the fecundities observable in numerous different types and groups throughout the animal kingdom; the second is a theoretical postulate of apparently incontrovertible logic. Clearly they cannot both be right (see p. 397).

If the hypothesis outlined in this chapter is correct, however, both can be dismissed as wrong, because they are too superficial and based consequently only on half-truths. The original question is too simply conceived. The equation needs to be further broken down into its component elements. Recruitment arises not only from reproduction but from immigration. Mortality comes not only from outside agencies such as predators, disease and accidents, but from the inside consequences of conventional competition, including expulsion from the habitat, deprivation, cannibalism and the like. On both sides of the equation there are clearly 'dependent' terms, capable of being varied and regulated through the appropriate homeostatic machinery.

The equation describing population size can be expanded to take account of this in the following way:

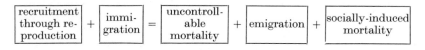

When set out thus, it becomes clear that the true independent variable is the first term on the right-hand side, the 'uncontrollable mortality' which arises from the external agencies. All the other four terms can be brought under intrinsic control through homeostatic processes.

Starvation, it needs to be said, is not always due to uncontrollable causes, but can be socially induced, in the sense that established individuals with sufficient social status are frequently able to maintain an adequate level of nutrition at times when their subordinates starve. It has recently been shown (Jenkins, Watson and Miller, 1963) that even predation can be promoted or diminished by the social system, and that it falls more heavily (in the species investigated—the Scottish red grouse) on outcasts than it does on established individuals. Under similar conditions of social deprivation, an otherwise tolerable burden of parasites may equally prove pathogenic.

To achieve a balance, the homeostatic machinery has thus to compensate for the losses falling outside its control, by adjusting one or more of the other four terms in the equation. Where these uncontrollable losses are perpetually vast, as tends to happen in parasitic animals, the scale of recruitment from reproduction (i.e. fecundity) must be correspondingly enormous. The other terms in the equation, while they may still be very important, for this reason tend to be eclipsed and escape notice. It is this that gives verisimilitude to the classical view outlined above. Yet even among parasites there still exists an optimum population density in terms of the carrying capacity of their habitat, in this case the host, and adaptations have generally evolved which prevent the latter becoming overburdened and thus killed too quickly. In the case of pathogenic organisms, loss of virulence and other similar devices may also help in keeping the sources of sustenance alive.

Intergroup Selection

The fallacy in the second postulate, that selection must automatically produce a maximum reproductive output, arises from the fact that whatever is of survival value to the stock or group as a whole must in the long run override, and in the end effectively forestall, selection for the immediate but antisocial advantage of the individual. Between different local stocks, intergroup selection can take place, favouring those groups that prove adaptable and make efficient and conservative use of their habitat; and this allows their social systems to be selected and evolve as units in their own right. Social systems can come to impose obligations of self-sacrifice on the individual in the interests of group survival. This has been a factor of great importance in evolution which has hitherto either been overlooked, or dismissed by modern Darwinists as untenable, although its genetic basis presents no great difficulty (cf. Wynne-Edwards, 1963).

The working of population homeostasis entirely depends on group adaptations such as the social hierarchy, which are not and cannot be the property of isolated individuals but only of a collective group. Their modification and perfection must necessarily come about through the kind of selection that favours the ones that work best and confer the longest survival on the stock that exhibits them. If there were no selection for the fitness of the group or social order as such, effectively repressing the selection at a lower level of antisocial advancement in the individual, then not only would fecundity always be maximized, but by the same logic so would the destruction of rivals. Uncurbed self-promotion among individuals is incompatible with adherence to a conventional code, and no social species can therefore survive it.

In practice the curbs are necessarily effective, and the fecundity of the individual in particular is consequently bound to be immune to quick-acting selection. This could be achieved simply through a polygenic inheritance of fecundity, which would render it stable and resistant in exactly the same way as characters like body-size and social dominance.

As a natural reality, intergroup selection receives impressive support from the remarkable variety and complexity of adaptations evolved to maintain the integrity of local breeding groups and prevent their adulteration. In primitive human communities we can recognize the typical hostility to strangers, the tendency to proliferate exclusive local languages and customs, and even to marry someone physically resembling one's relatives. But probably the most telling instances are those of the homing wanderers and migrants, be they insects, fish, amphibia, turtles, birds, seals or bats, which have evolved almost unbelievable powers of precise navigation (Kalmus, 1964) in order to permit them to return from wherever they find themselves, and retain their citizenship as breeders in a local ancestral clan.

Homeostatic Mechanisms

Social adaptations that facilitate the homeostatic control of population density are readily demonstrable in natural environments. To single out two of them, there are conventional systems of property ownership, which exist in many divisions of the animal kingdom; and can be shown to exclude surplus individuals; and second, established ownership or dominance as a qualification required for sexual maturation, together with the sex-inhibition of individuals that fail to become established, have also been demonstrated many times in species of fish, reptiles, birds and mammals.

Both these adaptations limit the population, or its reproductive segment, to a determinate quota and they suppress any remainder of leftover individuals. This alone, quite apart from all other homeostatic devices, can have a strong regulatory influence on population density. To give an illustration, the Royal Society of Edinburgh during the decade 1954–63 continued its established practice of recruiting 25 new members a year, and, notwithstanding the chance element in human mortality, the total membership each autumn varied by less than 1% on either side of the mean value of 752. A gannet colony, in which the number of nest-sites is limited by tradition—the breeding females each lay one egg and life-expectancy is long and actuarially predictable—would for the same reasons automatically achieve a stable population.

The quota system is but one among many adaptations for regulating the recruitment from reproduction. Fertility, meaning the number of eggs laid or young born, has been shown in an increasing number of cases to be influenced by social stress. Examples can be found in insects, including *Drosophila*, *Lucilia*, *Tribolium* and *Sitophilus*, in mammals such as mice, muskrats and deer, and in certain fish and birds (Wynne-Edwards, 1962, pp. 495–515). In the higher vertebrates it is rather common for reproduction to begin on a generous scale, and then later, as the season progresses and contemporary conditions for the rearing of young become known, brakes are applied and the output is progressively cut down to an appropriate final size. In mammals, pregnant females may resorb their foetuses; in birds, adults may desert their eggs. Juvenile mortality can be greatly affected in many of the higher animals by the solicitude of the parents. In a surprising diversity of species, almost all of them in 'advanced' groups such as arthropods and vertebrates, parents are known from time to time to destroy their own eggs or young. At a still later stage, the age at which maturity is first reached and breeding begins is variable and density-dependent; thus when survival of adults has been good or food is scarce, young recruits are held back or driven away, but when adult density is low in relation to the resources available, the normal period of adolescence is curtailed and parenthood accelerated.

In imposing these restraints, economic stress does not necessarily depress all individuals to a roughly uniform extent at any one time, as one would expect if there was no 'group effect'. Instead the same end-result can be achieved by restricting the percentage of adults that embark on reproduction at all, and later of those that carry it through to the finish, so that the effective line is drawn between individuals that achieve full or normal success and those that achieve nothing. This makes it perfectly clear that the tax is levied on the population or group as the unit, and not on each member separately, and it brings out once again the social character of the homeostatic mechanism which controls numbers. It provides a parallel to the adaptation in territorial species which insists that holdings must be of at least a certain size, and prevents overcrowding by driving supernumeraries away; or a parallel again to the adaptation of the hierarchy, which in a similar way can separate haves from have-nots, and distribute economic restrictions unevenly. In each case, the interests of those individuals constituting a surplus are sacrificed, with the result that the group can maintain its establishment in health and vigour.

Epideictic Phenomena

Like all homeostatic systems, population homeostasis requires a feedback of information, showing whether it is in equilibrium and if not how far it has strayed above or below the balance point. The kind of yardstick required is one that can weigh population density, ultimately against food supplies but proximately in terms of current demands for conventional property and status.

The need appears to be invariably met through some kind of social operation. In the simplest situations prizes such as territories and stances which are limited in number are competed for, and automatically the more contestants there are the fiercer the contest. Population pressure is gauged and relieved in one process, which results in the unsuccessful being disqualified. But there are other more numerous situations in which the pressure is sampled first in order to provide the feedback; and the latter then proceeds to react, for example through the individual's endocrine system (p. 176), and to condition his response in a separate, subsequent adjustment of the homeostatic balance. The preliminary stress phase may be brief, but more often it is protracted and builds up until the time becomes ripe for a density change.

The need for demonstrations of population strength and pressure has, I believe, given rise to a very large class of social phenomena, so conspicuous as to be obvious to naturalists and laymen alike, but hitherto incapable of valid explanation. As one would expect, they are manifest most of all at times when the sequence of the life-cycle is about to produce, as part of the normal course of events, a major alteration in homeostatic balance, and especially therefore just before breeding and seasonal migration. Before breeding begins it is important to secure an efficient deployment or *dispersion* of the population in occupiable habitats, and then to determine how large a reproductive input is justified under present conditions in each local area. Emigration produces an opposite change, relieving population pressure that has previously existed, typically as a climax to mounting social tension.

Population feedback phenomena are said to be *epideictic* (Wynne-Edwards, 1962, p. 16). This is a word that comes from the Greek and it suggests demonstrating something by means of an actual sample. In its present technical use it implies that a sample of the population is being displayed, in such a way that its density can be sensed by the participants through the social stress and stimulation that the display evokes. A familiar example of an epideictic display is provided by the assembly and flight demonstrations of swallows as the summer draws to a close. A little later, while migration is in progress, millions of travelling birds must find suitable places in which to break their long journeys for a week or two en route, until the fat reserves that serve them as fuel tanks are refilled. Then epideictic displays may be essential at all hours of the day at the stopping places to guard against congestion and food shortages, and to regulate the massive traffic flow.

More spectacular than these are the social assemblies of locusts gathering their strength for a migratory eruption. Migratory-phase locusts are born in special generations, often after many years of population quiescence. They are dressed in a different, much more striking livery than their solitary-phase parents, are intensely gregarious and from birth imbued with restless activity. As young hoppers they form into bands, then into armies. After the last moult they can fly and, in peaks of excitement, repeatedly take wing in brief massed manoeuvres. These increase in duration and size and continue day by day while the locusts are still voraciously provisioning themselves in preparation for their mission. At some stage a threshold is reached, 'D' day arrives, and the immense expedition, stretching for miles and darkening the sky, takes off never to return.

Nearer home, assemblies of starlings gather nightly at customary points to roost, and can often be seen in their thousands performing exhilarating manoeuvres against the evening sky. If the hypothesis of homeostasis is correct, these again are epideictic displays, providing an automatic census tally to the massed inhabitants of a

communal feeding territory, sometimes over a hundred square miles
in extent. Each bird is primed or conditioned by the stress of social
stimulation. Day by day the population is weighed against its economic situation in order that a balance can be preserved; and when
hard weather comes and food supplies are curtailed, 'weather movements' are triggered which discharge the population excess.

In the breeding season many species mate, and are dispersed in
their habitats as mated pairs. There is then a profound and important
division of labour between the sexes and the whole responsibility for
population homeostasis and dispersion ordinarily falls on the males.
The males exercise every epideictic function and can be described as
the epideictic sex. In birds it is the cocks that sing, have bright
plumage and hold territories in rivalry with other cocks; in mammals,
males have the tusks, horns and manes, the extra size and strength,
the vocal sacs, scent glands and other symbols of power and domination over rival males. Everywhere that sexual dimorphism occurs
in the animal kingdom it is the same, the males vie with males (unless
in rare cases the roles of the sexes have been reversed) in ritualized
competition. In a few species of frogs (e.g. *Rana catesbiana*, the
American bull-frog) the males not only exclusively possess a deep
loud voice, but they have ear-drums twice the area of those of the
females; in the cicadas it is essentially the same.

Some of the most splendid male adornments occur in bird species
in which there is a quite different pattern of breeding dispersion, and
where the seasonal epideictic display takes place on a traditional
arena. Here, especially in the early morning, all the males from the
surrounding communal territory gather day after day to spar and
joust or join in bizarre concerts and dances. Famous exponents are
the ruff (*Philomachus pugnax*), the blackcock (*Lyrurus tetrix*) and
the flame-coloured cock-of-the-rock (*Rupicola rupicola*). Females
appear at the arena, usually briefly, until they have been fertilized,
but otherwise the males are left to themselves. Our hypothesis suggests that the males, conditioned by the strength and intensity of
their contest, automatically react to fertilize an appropriate quota
of females and no more. This is incidentally a prediction it seems
possible to verify and an attempt is at present being made to pursue
it.

It is revealing to turn to yet a third dispersionary pattern, in the
colony-nesting birds. Here, no matter what order or family they
belong to, both sexes are almost invariably alike and in some cases
both are also specially adorned for the season. Their conventional
holding is a nest-site, which they are obliged to leave at least in order

to feed. One bird must always stay to guard the site, but either will suffice because they both exhibit the necessary status symbols. Both sexes participate wherever there are special epideictic displays.

One of the most important by-products of the social-homeostasis hypothesis is this indication that the classical views on dimorphism and sexual selection are in the main erroneous, and that the pheno-menon is primarily connected with the social machinery of population control. Even courtship itself lends support to this interpretation, for it consists essentially in overcoming an initial resistance to mating, and it is successful only when the ardour of the male (sometimes it has to be mutual) is sufficient. Why has such an initial barrier be-tween the sexes evolved at all? Our answer would be that it gives the male control over the fertilization of the female; and if, under the epideictic stress to which he is constantly exposed at that time, his ardour fails, the female will remain unfertilized and by that the out-put from reproduction will be diminished.

Interspecific Integration

The account so far given has been concerned exclusively with the regulation of numbers within a single species, in which homeostasis is normally linked to the food supply. But many important food organisms are exploited by a variety of consumers, and vice versa, so that in a typical ecological community the web of food-chains is consequently complex and interwoven. There would be little point in one species evolving a highly perfected homeostatic system for con-serving the food supply, if this merely gave an opportunity to prodigal and unsophisticated competitors to overtax the same resources, to the disadvantage of all.

The productivity of any food-resource is measurable as the yield per unit time and area; and where the product is shared in the same period and area between several species, their demands are comple-mentary. If the share of one is increased, those of the others must be diminished; and according to generally acceptable definitions the species are said to be in competition. Biologists have sometimes pictured interspecific competition as leading to a wholly uncom-promising and relentless campaign to gain ground at the expense of equally militant competitors. Natural selection it has been argued must necessarily favour the more efficient exploiter of the habitat; so that no two competitors with identical needs could ever coexist for more than a transitory period, because it is inconceivable that their efficiencies could be balanced on a knife-edge. One would swiftly gain the ascendancy and oust the other. This is the well-known principle of Gause (1934).

It is important to remember that, in the end, natural selection is concerned with only one issue—the survival of the stock. Chances of future survival are often increased by the occupation of more ground, by the fostering of diversity and adaptability, and of varied ecotypes. They are similarly prejudiced by the persistent attrition and encroachment of competitors. But there is nothing in principle opposed to establishing a *modus vivendi* which offers a chance of survival to two or more competing users.

In the plant world this is a familiar occurrence. In a Malayan rain-forest, one acre of uniform ground may occasionally contain over 200 species of trees, a number so large that it almost precludes the possibility that any two trees should be the same. In this community it would be absurd to claim that all the species are occupying different ecological niches. Many of them are manifestly interchangeable. It is characteristic of ancient forest floras in many parts of the world that they should harbour a diversity of trees, among which very surprisingly none has evolved as a dominant, after continuous coexistence perhaps from the mezozoic era, 50 million years ago. In these serene habitats, survival can no doubt depend almost indefinitely on relatively slender threads, and there has been no selection consequently for persistent aggression and take-overs. Instead the situation has developed into a harmony of equals, a majestic *entente cordiale*.

Interspecific harmony is not an exclusive attribute of plant communities, although among animals it appears to be less common to find such near-identity in the environmental demands of two species in the same community. But cases are not difficult to find, for instance among birds or ants, in which two species divide up the ground into a single mosaic of territories, so that where one is in occupation the other is excluded. A celebrated case in birds relates to the eastern and western meadowlarks (*Sturnella magna* and *neglecta*) in North America; they have a wide overlap in range where they show this 'interspecific territoriality' (Lanyon, 1956). There are many others (for a survey, see Wynne-Edwards, 1962, p. 391). The simultaneous presence of two or more sibling species in the same habitat is on the whole rather common in insects, and it would often be very difficult to find any ecological differences between them. Familiar examples of this can be found in many parts of the world in the common yellow wasps (*Vespula*), of which two or three species may even be attracted to the same pot of jam in rural parts of Britain.

There is at the same time a strong tendency for species with competing interests to diverge, and develop specializations that separate them ecologically into different habitats. As far as food is concerned the tendency of habitat specialization is to narrow down the range of diet; and if this were driven to the limit, each kind of animal would be left in the end depending on very few or perhaps only one staple food species, of which by luck it might be the sole consumer. This is indeed quite a common situation; but it has the disadvantage of linking the fortunes of the user inescapably with those of the species consumed. Group survival is no doubt better assured by having a diversity of food sources, so that there are alternatives to draw on if one of them temporarily fails. A compromise appears to be generally established by natural selection between dangerously narrow food-specialization on the one hand and a highly competitive overlap with other exploiters on the other.

The effect of this compromise is that the great majority of animals do not have the exclusive use of the particular produce on which they depend; to a greater or less extent they share their needs with competing species. Some interspecific system of mutual 'management' or conservation of food resources seems therefore indispensible; and the question is, can we see any detectable signs of compromise or co-operation between species that overlap in this way? Is there, in particular, any evidence of epideictic activity in which two or more competing species participate collectively?

These questions lead to the most interesting topic of multispecific social groups, very familiar and common in the higher vertebrates, for the existence of which no really convincing explanation has ever hitherto been suggested. Over much of the savanna country of Africa, for example, it is common to find more or less loose associations of grazing animals, comprising perhaps two or three species of antelopes, along with wildebeest, warthogs and zebras. All of them are dependent on the same grassland for food, and the dietary overlap between them is probably very extensive. A close parallel to this is found in the association of several species in bird flocks, normally comprising (as with the big game herbivores) the same groups in the same small areas each day. These occur for example among the forest insect-hunters, such as titmice (*Parus*) and kinglets (*Regulus*), seed and bud eaters such as siskins and redpolls (*Carduelis* spp.), and ground-feeding sparrows and buntings. In Africa there is an interesting situation among the numerous family of weavers (Ploceidae), in that two or more related species often associate in non-breeding flocks, and then resemble each other so closely as to be superficially indistinguishable; but when the breeding season approaches, the males at least moult into highly distinctive plumages. It certainly looks as though they merged their identities in a common social organization for the asexual part of the year. It is reminiscent of the sympatric *Vespula* wasps, referred to earlier, of which several species can scarcely be told apart with the naked eye, even in the hand; or again of the herring and sprat (*Clupea* spp.), which mix indistinguishably together in feeding nursery shoals, when they are all small and similar in size.

The feeding flocks of birds sometimes form up for only part of the day, often at a regular time. Just as common in birds is the association of mixed flocks at nightfall at a customary roosting place. Various members of the thrush family (Turdidae) associate with starlings in this way, and in North America most of the grackles and American blackbirds (Icteridae) do the same. Exactly analogous social roosting associations are found in bats, both insectivorous and fruit-eating, in various parts of the world; these are established of course by day. Even insects have in a few known cases evolved the same habit; among the Hymenoptera for example two or more cases of multispecific communal roosts have been reported (see Wynne-Edwards, 1962, p. 310).

Another well-known alliance that appears to be fundamentally of the same type occurs between foxes and badgers, which to a great extent share a common system of earths or underground dens and may inhabit them even at the same time. Like many other mammals that hunt alone, individual foxes or badgers do not have exclusive territories of their own. They belong instead to social organizations based on these local traditional dens, which only certain individuals at any given moment possess and share the right to use. Numerical equilibrium is almost certainly effected through the den system, which provides the basis for social competition and the essential power of exclusion if there is a population surplus. The mutual association in a single system of these two partially competing predatory mammals, which there is reason to think are about as companionable as dogs and cats, appears on any other basis a complete paradox.

Probably the most impressive evidence for multispecific homeostasis lies in the frequency of mixed breeding colonies, in which two or more species exploiting the same environment are more or less intimately associated. In Britain this is particularly notable in seabirds. On the east coast from Yorkshire northwards there are traditional bird cliffs, typically occupied by an alliance of herring gulls, kittiwakes, guillemots, razorbills, puffins and shags, all of which share as their staple summer foods the immensely prolific sand-eels (*Ammodytes*), herring and sprats. In spite of a tendency to specialize in the type of site they prefer for nesting, some in holes and crevices, others on ledges that are wide or narrow, high or low, they contrive to make do and stick together, and not to oust those species whose needs are not ideally met, nor overflow their customary places until one or other has usurped the whole cliff or island. The notion that they are merely the victims of necessity, concentrated in the only suitable localities, does not stand up to critical examination. Cases are known where new mixed colonies have grown from scratch, and where erosion or lighthouse-building has driven a colony away, only to settle as successfully as before on some previously neglected site.

Mixed colonies occur just as often among terns, boobies, cormorants and petrels; and equally again among marsh birds, especially herons, egrets, cormorants and darters, or different species of pelicans or flamingoes. The same is frequently true of sociable birds of prey, especially vultures, and it is very common in sociable weaver-birds, which could if need be nest in almost any suitable tree.

A full account of interspecific social groupings cannot of course be given here, because they are so numerous and diverse. But to round out a general impression, mention ought to be made of the notable multispecific associations at the breeding 'camps' of fruit-bats in Australia; of different species of seals, sea-lions and other pinnipedes in many parts of the world; and of ocean-going turtles on tropical shores and islands. Among insects the Hymenoptera once more give good examples, for instance the commonly found mixed breeding colonies of solitary bees such as the numerous species of *Halictus*.

Ornithologists have paid much attention to the subject of mixed feeding parties in birds without coming to any satisfactory conclusion regarding their function and adaptive value. The various other kinds of multispecific association, as phenomena similarly calling for explanation, have on the other hand received almost no critical consideration at all.

The dispersion of gregarious breeders during the breeding season is based on a pattern of colony positions which in the majority of cases are permanent and traditional. Once such a system has been pioneered by one species and a series of safe and well-provided sites established, it is perhaps to be expected that any related species with similar requirements, invading the region as a newcomer, will adopt what it finds ready made and, if allowed to do so, join in the existing system. This can be seen to happen, for instance in the now rapidly spreading little egret (*Egretta garzetta*) and cattle egret (*Bubulcus ibis*) in south Europe and the New World respectively, and with flying-foxes (*Pteropus*) in Australia. There are, therefore, other partial explanations for the origin of mixed colonies; and the same arguments would apply equally to mixed roosts. But if the underlying function of aggregating, when it concerns one species alone, is social and epideictic, there is at least a strong probability that a consortium of species with common interests is mutually epideictic; and that the population size of the colony as a whole can influence the homeostatic responses of all the constituent species.

The general principles of population homeostasis, with its necessary element of epideictic feedback, seem indeed to demand that such a mutual reaction should exist. When one looks for it, as we have done here, this is what is found. Beyond that the matter cannot be carried at this stage.

There is still another rather different point to be emphasized before leaving the subject of interspecific integration. It is that the members of a typical ecological community are to a profound extent interlocked and adapted for mutual advantage. The organic community is generally extremely complex, including green plants that trap energy from the sun, draw minerals from the substrate and carbon from the air, the animals that depend directly and indirectly on the plants, and the soil flora and fauna that decomposes their detritus and contributes to the humus. In complex communities all play their part in cycling the nutrient materials and no one section is completely independent of the others. There seems no doubt that natural selection can operate at the community level too, and that this has generally led to building up the efficiency of exploitation of habitats of many kinds and especially to their enrichment and conservation. There is usually a striking appearance of good order and high productivity in natural climax* communities, when the physical limitations of the site are taken into account. In addition to the development of mutually beneficial partnerships, the evidence points also, as we have seen, to the remarkable effect of the joint husbandry of commonly held resources on mutual toleration and non-aggression.

Genic and Traditional Inheritance

Very characteristically, the hereditary transmission of social behaviour patterns is not a wholly genic process, but involves an

---

* 'Climax' in the ecological sense refers to the relatively permanent and stable plant and animal communities that gradually become established in most habitats where environmental conditions are sufficiently constant; it is usually contrasted with the succession of transient stages that are typical of periods of pioneering and change.

important element of tradition. Social groups, primarily concerned with the regulation of their own numbers, are often tied to a definite home area. Within this they perpetuate themselves through succeeding generations, indefinitely retaining their continuous existence and identity. There is a mass of evidence relating to the high degree of *Ortstreue* or fidelity to their native and established homes in animals generally; and in some bird populations at least it might now be possible to guess at typical coefficients of inbreeding in a few cases, although not as yet to calculate them accurately. Gene flow with neighbouring groups occurs generally, through the sporadic acceptance of foreign immigrants, but it tends to be slow especially in stable habitats, and not strong enough to prevent the accumulation and retention of recognizable local differences in the gene pool.

These local stocks, sometimes referred to as demes, are the smallest interbreeding units possessed of potential immortality and thus able to undergo progressive modification through natural selection.

Much of the basic pattern of social conduct and conventional procedure can be genetically transmitted from one generation to the next. Highly complex responses, such as the computer-like ability to navigate automatically by the sun and stars, can be largely innate. In most forms of behaviour, whether social or not, the background physiology of the individual plays an important part, and much outward activity depends on the state of his food reserves, water balance, gonadal development, endocrine functions and so on. In a reciprocal relationship, his outward activity and sensory experiences exert powerful effects on his internal physiology: the two are inseparably integrated.

The topographic aspects of social behaviour involve tradition for their transmission from one generation to the next. For example, once an individual herring has arrived on the traditional spawning ground the whole complex sequence of his social and reproductive behaviour could conceivably be directed by innate information; but first he must reach the site, and for this purpose there is no alternative to learning its location from experienced individuals who already know where to find it. Genic inheritance cannot cope with this kind of unique and multifarious detail, nor with the rapidity with which traditional sites may, for accidental reasons, require to be changed.

The great advantage of traditional—as opposed to genetical—transmission is its speedy adaptability. In the higher vertebrates especially, where the capacity to learn without error or forgetfulness is so greatly developed, the use of tradition is no longer confined to the purely topographical aspects of social procedure. What we call customs, having the same character as traditions though they are generally less permanent and entrenched, are extensively involved in the day-to-day activities of the social group. Personal acquaintance with other members of the group, and knowledge of one's proper relationship and deportment towards them, are also acquired by experience, and not rarely by the imitation of one's elders.

The result is, in the mammals especially, that population homeostasis comes to depend on the most intricate mixtures of innate and traditional social behaviour. With increasing intellectual development there is probably a tendency for tradition to assume greater and greater importance, on account of its flexibility and spontaneous tendency to diversify. This would provide abundant raw material for group selection, and the correspondingly rapid perfecting of homeostatic performance. Certainly in primitive man the basis of population control had become almost completely traditional; and unfortunately the loss of these traditions has left his modern successors now faced with disaster, with—as yet—no detectable homeostatic protection.

Man's Plight

There is a wealth of evidence (cf. Carr-Saunders, 1922) to suggest that in palaeolithic tribes which survived into modern times, family limitation was universally practised, no matter whether they were eskimos, North or South American indians, bushmen, veddahs, or Australian aborigines. There were several standard methods, one or more of which appear to have been perpetuated by tradition in every tribe. These were, the deferment of marriage until difficult or dangerous conventions regarding eligibility had been satisfied; abstention from intercourse during lactation, which was often extremely prolonged; abortion and infanticide. Cannibalism, human sacrifice, witchcraft, head-hunting, or the rigours of ceremonial initiation, added to the toll of 'socially-induced mortality'. Surgery on the genitalia at initiation which may have impaired fertility was sometimes practised, for instance in Australia. There is no doubt that these primitive hunting peoples lived in balance with their food resources as animals do, similarly dispersed in small tribes each occupying its own exclusive communal territorial system; and that in the old-established parts of their geographical range, population density was effectively stationary for centuries and even millennia.

If this machinery of population homeostasis still existed in modern man, or had existed within popular memory, it would of course be familiar to everyone, and would long ago have been seen to obtain in the same way in the animal world. It is our own unfamiliarity with such mechanisms, and consequently incredulity that they ever existed, that provides the obstacle to present understanding.

In the broadest terms, what must have happened was that the traditional homeostatic conventions of the Old Stone Age fell into neglect with the neolithic rise of agriculture. Agricultural people, at least in the most favoured regions like Mesopotamia, concentrated upon the fertile soils, with crops and herds that allowed a density of settlement impossible to hunting tribesmen. They could produce surplus food and afford the services of craftsmen and traders. It was a world of expanding enterprise, able to support a rising population and consolidating its strength and specialized skills by its growing numbers and urbanization. The outdated 'heathen' practices of family limitation were no doubt simply discarded.

In the less progressive agricultural peoples, for example in Africa, living still in their small tribes and dependent on crops grown for a few seasons after a patch of forest had been burnt, some homeostatic practices were in fact retained right down to modern times. As with their paleolithic ancestors, they had no rational understanding of what they did nor why they did it. It was closely identified with spiritual instincts and beliefs and never exposed to question or rational analysis. Natural selection had shaped it and determined its obligatory transmission from parents and elders to their children.

When it comes to man, therefore, the social-homeostasis hypothesis suggests, even more forcibly than was realized already, how anomalous and extremely dangerous his position is. The small, largely isolated local group has now become a thing of the past; world population has a growing tendency to become a single super-group as far as natural selection is concerned, because we are able to trade and pool our food-resources and because we deliberately strive to minimize intergroup competition at the local, national and racial level.

Clearly it is necessary for us to assume deliberate control not only of world population but equally of our expanding economy, and to do it very fast, if we are to escape being caught by the ruthless agents of natural selection, before a man-made homeostatic system can be made to work.

References

Carr-Saunders, A. M. (1922). *The Population Problem, a Study in Human Evolution*, Clarendon Press, Oxford.

Carrick, R. (1963). 'Ecological significance of territory in the Australian magpie *Gymnorhina tibicen*.' *Proc. XIII Intern. Ornithol. Congr.*, 740–753.

Darwin, C. (1859). *The Origin of Species*, 6th ed. (1872), John Murray, London.

Gause, G. F. (1934). *The Struggle for Existence*, Williams & Wilkins, Baltimore, U.S.A.

Harris, V. A. (1964). *The Life of the Rainbow Lizard*, Hutchinson, London.

Jenkins, D., Watson, A. and Miller, G. R. (1963). 'Population studies on red grouse populations.' *J. Appl. Ecol.*, **1**, 183–195.

Kalmus, H. (1964). 'Navigation by animals.' *Ann. Rev. Physiol.*, **26**, 109–130.

Lack, D. (1954). *The Natural Regulation of Animal Numbers*, Clarendon Press, Oxford.

Lanyon, W. E. (1956). 'Ecological aspects of the sympatric distribution of meadowlarks in the north-central states.' *Ecology*, **37**, 98–108.

Leuckart, R. (1863). *Die menschlichen Parasiten und die von ihnen herrührenden Krankheiten*, C. F. Winter, Leipzig and Heidelberg.

Richards, O. W. (1953). *The Social Insects*, Macdonald, London.

Wynne-Edwards, V. C. (1962). *Animal Dispersion in Relation to Social Behaviour*, Oliver & Boyd, Edinburgh; Hafner Publishing Co., New York.

Wynne-Edwards, V. C. (1963). 'Intergroup selection in the evolution of social systems.' *Nature*, **200**, 623–626.

Schjelderup-Ebbe, Th. (1922). 'Beiträge zur Socialpsychologie des Haushuhns.' *Z. Psychol.*, **88**, 225–252.

V
CONCLUSION

16

Control Hierarchies

H. KALMUS

Department of Human Genetics and Biometry,
University College, London

The regulatory devices described in this book do not as a rule operate in isolation; mostly they form time sequences or interact in typical hierarchical arrangements. An important characteristic of these arrangements is that their structures, however complex they may turn out to be, can be built up from preliminary simpler components to large flow diagrams, without impairing the limited validity of the original partial schemes. In principle all control systems of organisms, and even of larger entities, can be presented in one flow diagram and such attempts are increasingly made. One, covering the cybernetic systems of the human body, has recently been published by Keidel (1963).

Alternatively, hierarchies of flow diagrams may be prepared. While the usefulness of some of the large diagrams may be dubious, flow diagrams giving summaries as well as detailed ones are useful in ecological and ethological and even some physiological studies.

Another striking feature of biological control hierarchies and indeed even of many simpler control systems is their 'messiness' from an engineering point of view. The higgledy-piggledy arrangements of control in an organism are often strikingly similar to assemblies of most diverse servomechanisms in manufacture, each controlling a partial function and yet combined for a single industrial purpose. The fuel supply to a boiler may for instance be controlled (optimized) by a combination of thermocouples, amplifiers, motors, flow meters, flues, mechanical hoppers and still other devices; similarly the respiratory system of a mammal is controlled by chemical, muscular, thermic and air flow regulating subsystems, which—together with yet others—tend jointly to ensure that respiration is

425

accomplished with the minimum consumption of energy (Priban and Fincham, 1965). A flow diagram illustrating the overall structure of such a respiratory system but omitting the detailed features of the subsystems is shown in Figure 1. Fuel control is of course only part of a production process involving many more technological, economic and social factors; and similarly respiration is only one function of an organism.

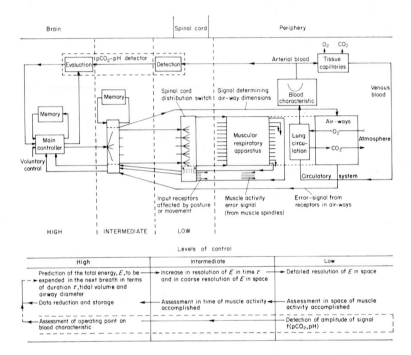

Figure 1. Overall control of a mammalian respiratory system. (From Priban and Fincham, 1965.)

The use of the same material structures as 'elements' in different regulatory systems, serving diverse ends, is perhaps more characteristic for organisms than for machines, with the possible exception of computers. One and the same muscle may at various times be implicated in maintaining posture or in a number of locomotor activities, such as walking, swimming or climbing, and thus become part of different control loops, serving these various activities.

The larynx is implicated in the control of swallowing, breathing and voice production (p. 147). On the molecular level an enzyme may be implicated in a number of metabolic cycles (see p. 70).

The multiplicity of means serving to control the same function and the use of one structure for different ends can both be explained by the opportunist character of evolution. A mammal's several cooling arrangements presumably arose independently of each other and became coordinated later on. Evolution may also explain the multiple functions of many anatomical entities, usually in strictly regulated ways. An example is the liver cell. A generation ago medical students were required to know about ten liver functions, of which the most important were the storage of glycogen, fats and proteins, iron metabolism, bile production and several detoxicating mechanisms. Today many more liver functions are known to exist, though their mutual interactions as well as their control by extrahepatic influences are still only partially understood. The persistence of so many functions in what we consider to be a specialized cell like a liver cell indicates that it has in fact retained a good proportion of the faculties of non-differentiated ancestral cells; the liver cell is not just an element in one chemical engineering process and the liver is not the site of one specialized process only. In some ways it rather resembles a 'cottage industry' in which diverse manufacturing processes, repair, maintenance as well as a good deal of control, administration and even sales are dispersed in large numbers of small more or less identical units.

Regulation hierarchies and time sequences in biological control systems have been described in several Chapters of this book, for instance for the bee colony (p. 362) or a neurohormonal system (p. 196). Three further examples will be used to describe some 'structural' and temporal aspects in the regulation of feeding, acclimatization, and exercise. *Feeding* serves three main purposes: (i) the growth and maintenance of the individual, (ii) the supply of sufficient energy (or as some would say negative entropy) for its functions, and (iii) the provision of material for gametes and embryos. All these requirements vary with time and situation and are more or less adequately controlled by various means. Regulation of feeding and some of its interactions with other activities will be briefly discussed following a behavioural study on a bird.

Some of the behavioural responses of wood pigeons (*Columba palumbus*) to seasonal changes in their food have been observed and recorded by Murton, Isaacson and Westwood (1963). In mid-winter these birds left their roosts in the early morning and returned at dusk. For much of the day they could be observed feeding on the fields. At other times of the year a varying amount of time was spent in the woods. On pastures in winter 95% of the feeding day (7·6 hours) was devoted to food searching. On clover leys, where more food could be obtained during a given period, only 69% of the day was spent feeding. On pastures in May the pigeons also showed less urgency in their food searching.

The number of pecks per minute varied throughout the day as did the area searched for food (estimated from the number of paces in time units). In February the pecking rate on pasture increased from 70 per minute in the early morning to 103 per minute in the evening; but the area searched decreased correspondingly.

In winter each pigeon collected an average of 34,900 food items (mainly clover leaves) per day from pastures, equivalent to a dry weight of 47 gm. Food was stored in the crop only in the afternoon, presumably in consequence of the increased feeding rate.

Less time was spent feeding when grain was available on sowings, stubbles or standing corn; the pigeons require about 50 gm of wheat or barley per day. The number of grains collected per minute increased while the area searched decreased with food density, though not proportionally. Food densities below 0·2 grains per square foot were too low for successful exploitation; on the other hand an increase above 15 grains per square foot did not result in greater amounts being collected.

The time required by wood pigeons to gather their food—depending on day length as well as the nature and variability of the food—limits the time available for other activities. Breeding would only have sufficient time between April and October; for other reasons actually only June–October are suitable.

Several of these findings have an important bearing on the control of these birds, which in England were estimated to cause about two million pounds' worth of damage in an average year, mainly to peas and brassicas. It is obvious that shooting or the reduction of food supplies, for instance by ploughing the stubbles, will in different months have quite different effects not only on the pigeon population at the time but also on the numbers in the following spring. By adapting these measures to local conditions an effective and economic control of wood pigeons might be possible in many places; the present method of shooting does not for any length of time reduce the number of wood pigeons, though of course it provides a permanent supply of birds and of sport being more suitable for 'game' than for 'vermin'.

Our more general considerations of the physiology of feeding control are based on a paper by De Ruiter (1963). His main point is that in the causation of feeding practically the entire body of an animal is involved. This we may add implies that at times most of an animal's brain and endocrine secretions can be employed in activities, which serve the acquisition, intake and digestion of food.

Figure 2 is a diagram of the main anatomical divisions in a verte-brate central nervous system implicated in the control of feeding responses. There is no trace of a particular overall feeding centre in this scheme. De Ruiter tries to unite the ethological and physiological

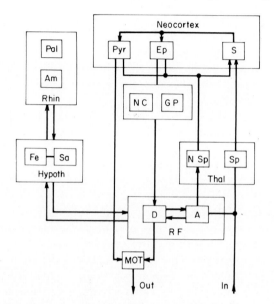

Figure 2. Interactions between parts of the central nervous system of a vertebrate, implicated in feeding and satiety responses. (From De Ruiter, 1963.)

A	=	ascending reticular system
Am	=	subcortical components of limbic system
D	=	descending reticular system
Ep	=	extrapyramidal system
Fe	=	lateral area of hypothalamus, induces feeding
RF	=	reticular formation
GP	=	globus pallidus
Hypoth	=	hypothalamus
MOT	=	lower motor mechanism
NC	=	nucleus candatus
N Sp	=	non-specific ascending system
Pal	=	cortical components of limbic system
Pyr	=	pyramidal system
Rhin	=	rhinencephalon
S	=	cortical sensory areas
Sa	=	ventromedial area of hypothalamus, induces satiety
Sp	=	specific ascending limb
Thal	=	thalamus

approaches to the phenomena of feeding, by considering the nervous system as a coherent network of reflex arcs acting upon each other and thus modifying the relationship between incoming information and outgoing command. This input and output may be described in electrical (discharges) or chemical (adrenergic, cholinergic) terms in respect to specified neural assemblies in definite brain regions or by the concomitant drives and effects, such as the induction of a specific feeding activity or of thirst, satiety etc. At present detailed descriptions of behaviour provide by far the best overall measure of the state of the brain. Monitoring the effects of local brain stimulation on the behavioural responses of the whole individual is not only a necessity in cortical surgery, but also a most potent method in the study of animal behaviour.

Among the variables which regulate feeding behaviour are the concentration of circulating nutrients, e.g. blood sugar, which in turn is under hormonal (insulin) control, the body reserve and weight, proprioception in the stomach and other parts of the digestive tract, the odour and taste (flavour) of food and its visual aspects and many other less obvious factors, e.g. conditioning.

The activities implicated in feeding range from mastication, swallowing and salivation at one end of the scale to the activities employed in the search for food, prey capture, begging or even sexual acts (birds, spiders) at the other. And it is these more rarely implicated behavioural modes which support the view mentioned above that most parts of an animal can serve its feeding urge. In a negative way of course all activities which exclude feeding, such as aggression or sexual activities, can also be said also to control feeding. Figure 3 is an attempt at synthesizing the network of some important control systems concerned with feeding. In spite of its extent and apparent complexity this scheme is still greatly simplified and thus incomplete: thus only one exteroreceptor, signalling the presence of food, is included, while obviously visual, tactile or chemical receptors may in fact be signalling; 'digestive system' or 'metabolic state' indicate very complex functional entities, which can—and in particular studies must—be replaced by detailed flow diagrams. Nevertheless such an overall scheme has its value in that it provides a definite model for one's general views on feeding behaviour and a framework for special investigation. It also provides some ideas on where to intervene in cases of feeding disturbances in animals or in men, e.g. in some cases of pathological aphagia. In theory one could in a patient who refused to eat stimulate the median hypothalamic area or inhibit the lateral one; in practice a lowering of the patient's

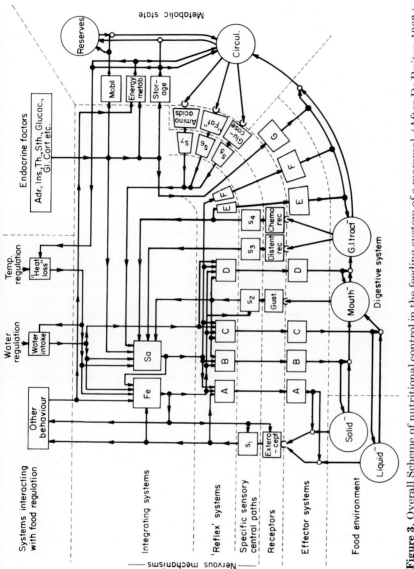

Figure 3. Overall Scheme of nutritional control in the feeding system of a mammal. (After De Ruiter, 1963.) A, B, C, D, represent motor patterns involved in the acquisition or ingestion of food (approach, mastication, sucking, and deglutition); E, F = peristalsis and digestion; G = intestinal absorption; Fe = hunger mechanism; Sa = satiety mechanism; S_1–S_7 = afferent pathways of various internal and external receptors. Circles indicate the sites, outside and inside the body, where the various nutrients occur.

blood sugar level by an insulin injection and inhibition of the neo-
cortex by some central depressant—thus removing some acquired
neurotic habit—may be more promising.

A consideration of the temporal constraints imposed on some processes involved in feeding control provides some insight into the
efficacy of the individual feedback loops at particular phases and
into the kind of sequence in which they must operate: for instance
the ultimate effects of ingested food on the metabolic state of the
body cannot be perceived by the receptors in the circulation or the
brain, before digestion and absorption is completed; yet feeding stops
as a rule long before this has been achieved; thus more rapid feed-
backs must operate at this time, e.g. stretch receptors in the intestinal tract (mainly the stomach) which indicate the volume of
ingested food and chemoreceptors in the mouth and possibly also in
the stomach, which indicate its composition. How these and other
control devices interact, and how the feeling of hunger or of satiety
is precisely produced, is at present quite unknown.

Hierarchical and temporal order of control mechanisms is not only
apparent in such universal 'physiological' activities as feeding or
breathing, but can also be observed in complex environmental circumstances requiring adaptation such as during acclimatization or
during training for extreme performance (exercise). Sometimes both
these adaptive changes occur simultaneously, as for instance in
preparing people for Himalayan expeditions. The entire, complex,
short and long term effects of lack of oxygen, carbon dioxide and
water vapour, cold, fatigue, and other factors on people climbing at
high altitudes cannot be disentangled here. Only some of the control
mechanisms and the order in which they are activated in the course
of acclimatization can be briefly described and the kinds of regulation
operating in people undergoing training listed.

Acclimatization may be illustrated by considering human thermal
adaptation in its wider context, including of course thermoregulation
proper as discussed in Chapter 5. Human communities survive in
localities showing summer extremes as great as $-40°$ and $+37°c$ and
winter extremes between $-51°$ and $+26°c$ (Harrison and coworkers,
1964). For short periods individuals can cope with even higher and
lower outside temperatures, while the deep body temperature of man
rarely rises by more than $4°c$ in severe sickness and by much less
normally.

As with many animals temperature loss or gain is to some extent
regulated by the temperature preferences of the people concerned,
which make them choose between outdoors and indoors or even

between warm and cold climates. Heating, clothing, and in a more indirect way diet and exercise provide the second line in regulating the temperature of the human body and only when most or all of these 'behavioural' means of regulation have been utilized do the mechanisms which constitute 'physiological temperature' regulation (see p. 87) come into play. A third line of regulatory defences against excessive temperature is provided by various adaptive semipermanent physiological changes in circulation and heat control, which result in a considerable improvement in working ability and 'acclimatization' in general.

Lastly, it is probable though not perhaps conclusively demonstrated that populations from cold and hot climates differ in several genetically determined adaptative characteristics, such as body size, proportion or composition. The great number of natural as well as of artificial methods for regulating our body temperature with which genetics, physiology and technology provides mankind has made our species one of the most eurythermous mammals on Earth—and is a prerequisite for our penetration into space and the interior of the Earth. In addition we have now learned to temporarily abolish temperature regulation and to use chilling for such purposes as sperm and tissue storage and heart surgery.

In considering human temperature regulation in its widest sense, we have to pay attention to complex regulatory systems in which man is but one component though the most important one. Such situations (p. 81) multiply with technological progress (Cherry, 1954; Shackel and Whitfield, 1963) and are only partially counteracted by automation (Krochmann, 1965). Somewhat opposed to these tendencies are areas of human endeavour where the capacities of the individual *per se* and thus his powers of control and regulation are put to the test. Foremost among these is possibly athletics as represented by certain motor activities. The complexity of regulatory processes implicated in extreme human motor activity may be illustrated by enumerating the most important components and bodily functions (Table 1) which it is necessary to consider when trying to improve athletic achievement (Chapman and Mitchell, 1965). No attempt will however be made to incorporate the totality of these components into an overall flow diagram, nor can the sequence in time of the various adaptative devices be discussed as this would in itself require a separate book.

When assessing the limitations to further improvements in athletic performance it is necessary to consider the individual constraints in the human control systems. There seems, for instance, to be no

Table 1. Components and functions implicated in athletic performance.
(Material from Chapman and Mitchell, 1965.)

Muscular system (mechanical output)
    Muscle cells (trophic reactions)
    Myoglobine (oxygen debt)

Respiratory system (ventilation)
    Haemoglobin (oxygen transport)
    Lungs (volume)

Circulatory system (metabolite transport)
    Heart (output, speed, depths)
    Heart muscle (adapt. contraction)
    Peripheral circulation (blood distribution)
    Muscle arterioli (blood supply)

Nervous system (coordination)
    Muscle coordination
    Cardiac regulation
    Respiratory centres
    Blood distribution (see above)

Hormonal system (change of condition)
    Adrenals (anticipation, stress)

Sense organs (behavioural coordination)
    Visual + acoustic (anticipation of needs, assessment of competitions etc.)

Haemapoetic system (long term regulations)
    Climatic adaptation (temperature, altitude)

Genetical system (selection)

difficulty in the exchange of oxygen and carbon dioxide through the
alveoli of the lungs, except in certain rare diseases, but many of
the other components may—according to circumstances—become
limiting factors usually in combination. Such breakdowns have considerable similarities with some manifestations of decompensation
in certain diseases, e.g. cardiac failure, suffocation or metabolic
poisoning; they may show a complex and variable picture of 'regulatory retreat' to a second or third line of defence and ultimately end
in death; or they may produce oscillatory states.

The uncertainty concerning the role of the individual constraints
in limiting adaptation to athletic performance makes predictions of
future athletic achievements rather hazardous. In the past most of
these predictions have been considerably surpassed. The same cir-
cumstance also renders the interpretation of the improvements
difficult. It is in fact possible that the greater proportion of improvements in athletic records stems not from any particular advances in
physiological knowledge, training or equipment, but simply from a
sort of numerical regulation: athletes are selected from an ever

widening circle of people, that is from larger numbers, and among larger samples more extreme variants are likely to occur.

The last pages of this chapter are concerned with some aspects of the extension of man's control over his environment and over himself. This quest is going to continue and may be regarded as part of life's perpetual struggle against constraint and death. The purpose in the planning of this book was to show what forms biological control and regulation can take and the role they play in life.

References

Chapman, C. B. and Mitchell, T. H. (1965). 'The physiology of exercise.' *Sci. Am.*, **212**, 88–96.

Cherry, C. E. (1954). Die Beziehung zwischen Mensch und Maschine. *Veröff. Arbgem. Forsch. Nordrhein-Westfahlen*, **38**, 1–95.

De Ruiter, L. (1963). 'The physiology of vertebrate feeding behaviour: towards a synthesis of ethological and physiological approaches to problems of behaviour.' *Z. Tierpsychol.*, **20**, 498–516.

Holst, E. v. and Saint Paul, Ursula. (1963). 'On the functional organisation of drives.' *Anim. Behav.*, **11**, 1–21.

Harrison, G. A., Weiner, J. S., Tanner, J. M. and Barnicot, N. A. (1964). *Human Biology*, Clarendon Press, Oxford.

Keidel, W. D. (1963). 'Kybernetische Systeme des menschlichen Organismus.' *Veröff. Arbgem. Forsch. Nordrhein-Westfahlen*, **118**, 99.

Krochmann, E. (1965). 'Der Einsatz von Prozenrechnein.' *Umschau*, 657–664.

Murton, R. K., Isaacson, A. J. and Westwood, N. J. (1963). 'The feeding ecology of the woodpigeon.' *Brit. Birds*, **56**, 345–375.

Priban, I. P. and Fincham, W. F. (1965). 'Self-adaptive control and the respiratory system.' *Nature*, **208**, 339–343.

Shackel, B. and Whitfield, D. (1963). 'Instruments and people. Ergonomics for Industry, 2.' *DSIR Information Div.*, London.

Further Reading*

Some Books, Symposia, Papers and Journals dealing with Biological Control

Books

Arbib, M. A. (1964). *Brains, Machines and Mathematics*, McGraw-Hill, London.

Ashby, W. Ross (1964). *An Introduction to Cybernetics*, Methuen University Paperback, London.

Bayliss, L. E. (1966). *Living Control Systems*, English Universities Press, London.

Bernard, C. (1859). *Leçons sur la Physiologie et des Alterations Pathologiques des Liquides de L'Organisme*, Ballière, Paris.

Bernard, C. (1876). *Leçons sur la Chaleur Animale*, Ballière, Paris.

Bonnes, D. M. (1961). *Control Mechanisms in Cellular Processes*, Ronald Press, New York.

Cannon, W. B. (1939). *The Wisdom of the Body* (revised ed.), Norton, New York.

Delius, L. (1962). *Zentrarnervöse Regulation*, Springer, Berlin.

Hassenstein, B. (1965). *Kybernetik in der Biologie*, Quelle und Meyer, Heidelberg.

Lerner, M. (1964). *Genetic Homeostasis*, Oliver & Boyd, Edinburgh.

Linnaeus, C. (1749). *Oeconomia Naturae etc.*, Ammoenitates Academic, Uppsala.

Stanley-Jones, D. and K. (1960). *The Kybernetics of Natural Systems*, Pergamon Press, Oxford.

Steinbuch, K. (1963). *Automat und Mensch*, Springer, Berlin.

Wiener, N. (1961). *Cybernetics: or Control and Communication in the Human and the Machine*, 2nd ed., Cambridge, Massachusetts.

Symposia

Grant, J. K. (1963). 'The control of lipid metabolism.' *Biochem. Soc. Symp.*, No. 24, Academic Press, London.

'Homeostasis.' *Symp. Soc. Exp. Biol.*, No. 18 (1964), Cambridge University Press, Cambridge.

'Integrative Mechanisms.' *Symp. Soc. Exp. Biol.*, No. 20 (1966), Cambridge University Press, Cambridge.

* See also p. 10.

Papers and Journals

Ingram, V. M. (1964). 'Control mechanisms in hemoglobin synthesis.' *Medicine*, **43**, 6.

Kapp, R. O. (1954). 'Living and lifeless machines.' *Brit. J. Philos. Sci.*, **5**, No. 18.

Söderberg, U. (1964). 'Neurophysiological aspects of homeostasis.' *Ann. Rev. Physiol.*, **26**, 271–288.

Kybernetik, Springer-Verlag, Berlin, Göttingen, Heidelberg.

Progr. Biocybernetics, Elsevier Publ. Co., Barking, Essex.

Author Index

Page numbers in ordinary type are those of citations in the text. Page numbers in *italics* are those where bibliographical references to the stated authors' works are listed at the ends of chapters.

439

Subject Index